Nanotechnology

Volume 6: Nanoprobes
Edited by Harald Fuchs

Related Titles

Nanotechnologies for the Life Sciences

Challa S. S. R. Kumar (ed.)

Volume 1: Biofunctionalization of Nanomaterials
2005
978-3-527-31381-5

Volume 2: Biological and Pharmaceutical Nanomaterials
2005
978-3-427-31382

Volume 3: Nanosystem Characterization Tools in the Life Sciences
2005
978-3-527-31383-9

Volume 4: Nanodevices for the Life Sciences
2006
978-3-527-31384-6

Volume 5: Nanomaterials - Toxicity, Health and Environmental Issues
2006
978-3-527-31385-3

Volume 6: Nanomaterials for Cancer Therapy
2006
978-3-527-31386-0

Volume 7: Nanomaterials for Cancer Diagnosis
2006
978-3-527-31387-7

Volume 8: Nanomaterials for Biosensors
2006
978-3-527-31388-4

Volume 9: Tissue, Cell and Organ Engineering
2006
978-3-527-31389-1

Volume 10: Nanomaterials for Medical Diagnosis and Therapy
2007
978-3-527-31390-7

Nanotechnology

Günter Schmid (ed.)

Volume 1: Principles and Fundamentals
2008
978-3-527-31732-5

Harald Krug (ed.)

Volume 2: Environmental Aspects
2008
978-3-527-31735-6

Rainer Waser (ed.)

Volume 3: Information Technology I
2008
978-3-527-31738-7

Rainer Waser (ed.)

Volume 4: Information Technology II
2008
978-3-527-31737-0

Viola Vogel (ed.)

Volume 5: Nanomedicine and Nanobiotechnology
2009
978-3-527-31736-3

Harald Fuchs (ed.)

Volume 6: Nanoprobes
2009
978-3-527-31733-2

Michael Grätzel, Kuppuswamy Kalyanasundaram (eds.)

Volume 7: Light and Energy
2009
978-3-527-31734-9

Lifeng Chi (ed.)

Volume 8: Nanostructured Surfaces
2009
978-3-527-31739-4

www.wiley.com/go/nanotechnology

G. Schmid, H. Krug, R. Waser, V. Vogel, H. Fuchs,
M. Grätzel, K. Kalyanasundaram, L. Chi (Eds.)

Nanotechnology

Volume 6: Nanoprobes

Edited by Harald Fuchs

WILEY-VCH Verlag GmbH & Co. KGaA

The Editor

Prof. Dr. Harald Fuchs
University of Münster
Institute of Physics
Wilhelm-Klemm-Str. 10
48149 Münster
Germany

Cover: Nanocar reproduced with kind permission of Y. Shirai/Rice University

All books published by **Wiley-VCH** are carefully produced. Nevertheless, authors, editors, and publisher do not warrant the information contained in these books, including this book, to be free of errors. Readers are advised to keep in mind that statements, data, illustrations, procedural details or other items may inadvertently be inaccurate.

Library of Congress Card No.: applied for

British Library Cataloguing-in-Publication Data
A catalogue record for this book is available from the British Library.

Bibliographic information published by the Deutsche Nationalbibliothek
The Deutsche Nationalbibliothek lists this publication in the Deutsche Nationalbibliografie; detailed bibliographic data are available in the Internet at http://dnb.d-nb.de.

© 2009 WILEY-VCH Verlag GmbH & Co. KGaA, Weinheim

All rights reserved (including those of translation into other languages). No part of this book may be reproduced in any form – by photoprinting, microfilm, or any other means – nor transmitted or translated into a machine language without written permission from the publishers. Registered names, trademarks, etc. used in this book, even when not specifically marked as such, are not to be considered unprotected by law.

Typesetting Thomson Digital, Noida, India
Printing Strauss GmbH, Mörlenbach
Binding Litges & Dopf Buchbinderei GmbH, Heppenheim
Printed in the Federal Republic of Germany
Printed on acid-free paper

ISBN: 978-3-527-31733-2

Contents

Preface *XIII*
List of Contributors *XV*

1 Spin-Polarized Scanning Tunneling Microscopy *1*
Mathias Getzlaff
1.1 Introduction and Historical Background *1*
1.2 Spin-Polarized Electron Tunneling: Considerations Regarding Planar Junctions *2*
1.3 Spin-Polarized Electron Tunneling in Scanning Tunneling Microscopy (STM): Experimental Aspects *4*
1.3.1 Probe Tips for Spin-Polarized Electron Tunneling *5*
1.3.1.1 Ferromagnetic Probe Tips *5*
1.3.1.2 Antiferromagnetic Probe Tips *8*
1.3.1.3 Optically Pumped GaAs Probe Tips *10*
1.3.1.4 Nonmagnetic Probe Tips *12*
1.3.2 Modes of Operation *13*
1.3.2.1 Constant Current Mode *13*
1.3.2.2 Spectroscopy of the Differential Conductance *14*
1.3.2.3 Local Tunneling Magnetoresistance *14*
1.4 Magnetic Arrangement of Ferromagnets *15*
1.4.1 Rare-Earth Metals: Gd/W(110) *15*
1.4.2 Transition Metals: Co(0001) *17*
1.5 Spin Structures of Antiferromagnets *18*
1.5.1 Topological Anti-Ferromagnetism of Cr(001) *19*
1.5.2 Magnetic Spin Structure of Mn with Atomic Resolution *23*
1.6 Magnetic Properties of Nanoscaled Wires *26*
1.7 Nanoscale Elements with Magnetic Vortex Structures *29*
1.8 Individual Atoms on Magnetic Surfaces *31*
1.9 Domain Walls *35*
1.10 Chiral Magnetic Order *39*
References *42*

Nanotechnology. Volume 6: Nanoprobes. Edited by Harald Fuchs
Copyright © 2009 WILEY-VCH Verlag GmbH & Co. KGaA, Weinheim
ISBN: 978-3-527-31733-2

2	**Nanoscale Imaging and Force Analysis with Atomic Force Microscopy** *49*
	Hendrik Hölscher, André Schirmeisen, and Harald Fuchs
2.1	Principles of Atomic Force Microscopy *49*
2.1.1	Basic Concept *49*
2.1.2	Current Experimental Set-Ups *50*
2.1.2.1	Sensors *50*
2.1.2.2	Detection Methods *52*
2.1.2.3	Scanning and Feedback System *53*
2.1.3	Tip–Sample Forces in Atomic Force Microscopy *53*
2.1.3.1	Van der Waals Forces *55*
2.1.3.2	Capillary Forces *55*
2.1.3.3	Pauli or Ionic Repulsion *55*
2.1.3.4	Elastic Forces *56*
2.1.3.5	Frictional Forces *57*
2.1.3.6	Chemical Binding Forces *57*
2.1.3.7	Magnetic and Electrostatic Forces *57*
2.2	Modes of Operation *58*
2.2.1	Static or Contact Mode *58*
2.2.1.1	Force versus Distance Curves *59*
2.2.2	Dynamic Modes *60*
2.3	Amplitude Modulation (Tapping Mode) *61*
2.3.1	Experimental Set-Up of AM-Atomic Force Microscopy *61*
2.3.1.1	Theory of AM-AFM *63*
2.3.1.2	Reconstruction of the Tip–Sample Interaction *69*
2.3.2	Frequency-Modulation or Noncontact Mode in Vacuum *70*
2.3.2.1	Set-Up of FM-AFM *72*
2.3.2.2	Origin of the Frequency Shift *73*
2.3.2.3	Theory of FM-AFM *75*
2.3.2.4	Applications of FM-AFM *77*
2.3.2.5	Dynamic Force Spectroscopy *78*
2.4	Summary *81*
	References *82*
3	**Probing Hydrodynamic Fluctuations with a Brownian Particle** *89*
	Sylvia Jeney, Branimir Lukic, Camilo Guzman, and László Fórró
3.1	Introduction *89*
3.2	Theoretical Model of Brownian Motion in an Optical Trap *90*
3.2.1	The General Langevin Equation for a Brownian Sphere in an Incompressible Fluid *90*
3.2.1.1	The Random Thermal Force $F_{th}(t)$ *91*
3.2.1.2	The Friction Force $F_{fr}(t)$ *91*
3.2.1.3	The External Force $F_{ex}(t)$ *94*
3.2.2	Solutions to the Different Langevin Equations for Cases Observable by OTI *94*

3.2.2.1	Free Brownian Motion	94
3.2.2.2	Optically Confined Brownian Motion	97
3.2.3	Time Scales of Brownian Motion	99
3.3	Experimental Aspects of Optical Trapping Interferometry	100
3.3.1	Experimental Set-Up	100
3.3.1.1	Optical Trapping Interferometry and Microscopy Light Path	101
3.3.1.2	Sample Preparation	103
3.3.2	Position Signal Detection and Acquisition	103
3.3.3	Position Signal Processing	105
3.3.4	Temporal and Spatial Resolution of the Instrument	105
3.4	High-Resolution Analysis of Brownian Motion	108
3.4.1	Calibration of the Instrument	108
3.4.2	Influence of Different Parameters on Brownian Motion	109
3.4.2.1	Changing the Trap Stiffness	110
3.4.2.2	Changing the Fluid	111
3.4.2.3	Changing the Particle Density	113
3.4.3	Implications of the Existence of Long-time Tails in Nanoscale Experiments	113
3.4.3.1	Single Particle Tracking by OTI	114
3.4.3.2	Diffusion in OTI	114
3.4.3.3	Thermal Noise Statistics	116
3.5	Summary and Outlook	116
	References	118
4	**Nanoscale Thermal and Mechanical Interactions Studies using Heatable Probes**	**121**
	Bernd Gotsmann, Mark A. Lantz, Armin Knoll, and Urs Dürig	
4.1	Introduction	121
4.2	Heated Probes	122
4.3	Scanning Thermal Microscopy (SThM)	126
4.4	Heat-Transfer Mechanisms	129
4.4.1	Heat Transport Through the Cantilever Legs and Air	129
4.4.2	Heat Transfer Through Radiation	131
4.4.3	Thermal Resistance of a Water Meniscus	132
4.4.4	Heat Transfer Through a Silicon Tip	132
4.4.5	Thermal Spreading Resistance	138
4.4.6	Interface Thermal Resistance	139
4.4.7	Combined Heat Transport Through Tip, Interface and Sample	140
4.4.8	Heat-Transport Experiments Through a Tip–Surface Point Contact	141
4.5	Thermomechanical Nanoindentation	144
4.6	Application in Data Storage: The 'Millipede' Project	155
4.6.1	Writing	155
4.6.2	Reading	156
4.6.3	Erasing	156

4.6.4	Medium Endurance	158
4.6.5	Bit Retention	159
4.6.6	Tip Endurance	159
4.6.7	Data Rate	159
4.7	Nanotribology and Nanolithography Applications	161
4.7.1	Nanowear Testing	161
4.7.2	Nanolithography Applications	163
	References	166

5 Materials Integration by Dip-Pen Nanolithography 171

Steven Lenhert, Harald Fuchs, and Chad A. Mirkin

5.1	Introduction	171
5.2	Ink Transport	172
5.2.1	Theoretical Models for Ink Transport	173
5.2.2	Experimental Parameters Affecting Ink Transport	176
5.2.2.1	Driving Forces	176
5.2.2.2	Covalent Reaction with the Substrate	176
5.2.2.3	Noncovalent Driving Forces	177
5.2.2.4	Tip Geometry and Substrate Roughness	177
5.2.2.5	Humidity and Meniscus Formation	178
5.2.2.6	External Driving Forces	179
5.2.2.7	Thermal DPN	179
5.2.2.8	Electrochemical DPN	180
5.3	Parallel DPN	181
5.3.1	Passive Arrays	181
5.3.2	Active Arrays	181
5.4	Tip Coating	182
5.4.1	Methods for Inking Multiple Tips with the Same Ink	182
5.4.2	Ink Wells	183
5.4.3	Fountain Pens	184
5.4.4	Nanopipettes	184
5.5	Characterization	185
5.6	Applications Based on Materials Integration by DPN	187
5.6.1	Selective Deposition	187
5.6.2	Combinatorial Chemistry	189
5.6.3	Biological Arrays	190
5.7	Conclusions	193
	References	193

6 Scanning Ion Conductance Microscopy of Cellular and Artificial Membranes 197

Matthias Böcker, Harald Fuchs, and Tilman E. Schäffer

6.1	Introduction	197
6.1.1	Scanning Ion Conductance Microscopy	198
6.2	Methods	199

6.2.1	The Basic Set-Up	199
6.2.2	Nanopipettes	200
6.3	Description of Current–Distance Behavior	201
6.4	Imaging with SICM	202
6.4.1	Modulated Scan Technique	202
6.4.2	Cellular Membranes	203
6.4.3	Artificial Membranes	204
6.4.4	SICM with Shear Force Distance Control	206
6.5	Outlook	207
	References	209
7	**Nanoanalysis by Atom Probe Tomography**	**213**
	Guido Schmitz	
7.1	Introduction	213
7.2	Historical Development	215
7.3	The Physical Principles of the Method	216
7.3.1	Field Ionization and Evaporation	216
7.3.2	Ion Trajectories and Image Magnification	220
7.3.3	Tomographic Reconstruction	223
7.3.4	Accuracy of the Tomographic Reconstruction	226
7.4	Experimental Realization of Measurements	230
7.4.1	Position-Sensitive Ion Detector Systems	230
7.4.2	Instrumental Design of 3-D Atom Probes	233
7.4.3	Specimen Preparation	236
7.5	Exemplary Studies Using Atom Probe Tomography	238
7.5.1	Nucleation of the First Product Phase	239
7.5.2	Thermal Stability of Giant Magnetoresistance Sensor Layers	242
7.5.3	Influence of Grain Boundaries and Curved Interfaces	245
7.6	Approaching Nonconductive Materials: Pulsed Laser Atom Probe Tomography	248
7.6.1	The Limitations of High-Voltage Pulsing	248
7.6.2	The Mechanism of Pulsed Laser Evaporation	249
7.6.3	Application to Microelectronic Devices	252
	References	255
8	**Cryoelectron Tomography: Visualizing the Molecular Architecture of Cells**	**259**
	Dennis R. Thomas and Wolfgang Baumeister	
8.1	Introduction	259
8.2	Basic Principles and Challenges of Electron Tomography	260
8.3	Automated Cryoelectron Tomography	262
8.4	Resolution, Signal-to-Noise Ratio and Visualization of Tomograms	263
8.5	Merging High Resolution with Low: The Molecular Interpretation of Cryotomograms	265

8.5.1	Specific Labeling	266
8.5.2	Pattern Recognition	268
8.6	Creating Template Libraries	269
8.7	Outlook	270
	References	270

9 Time-Resolved Two-Photon Photoemission on Surfaces and Nanoparticles 273
Martin Aeschlimann and Helmut Zacharias

9.1	Introduction	273
9.2	Theoretical Background	274
9.2.1	Electron–Electron Interaction	275
9.2.2	Plasmonic Processes	276
9.2.3	Two-Temperature Model	277
9.2.4	Electron–Phonon Coupling	278
9.3	Experimental	280
9.4	Relaxation of Excited Carriers	285
9.5	Volume Excitation in Metallic Nanostructures Investigated by TR-PEEM	289
9.6	Long-Lived Resonances in Adsorbate/Substrate Systems	293
9.7	Outlook: Spatial and Temporal Control of Nano-Optical Fields	298
	References	300

10 Nanoplasmonics 307
Gerald Steiner

10.1	Introduction	307
10.2	Single Clusters	308
10.3	Nanoshells	312
10.4	Layer of Clusters	312
10.5	Surface-Enhanced Spectroscopy	316
10.5.1	Surface-Enhanced Raman Scattering	316
10.5.2	Surface-Enhanced Fluorescence	319
10.5.3	Surface-Enhanced Infrared Absorption Spectroscopy	320
10.6	Biosensing	321
	References	323

11 Impedance Analysis of Cell Junctions 325
Joachim Wegener

11.1	A Short Introduction to Cell Junctions of Animal Cells	325
11.1.1	Cell Junctions for Mechanical Stability of the Tissue	326
11.1.1.1	Adherens Junctions	327
11.1.1.2	Desmosomes	327
11.1.1.3	Focal Contacts	327
11.1.1.4	Hemi-Desmosomes	328
11.1.1.5	Less-Prominent Types of Mechanical Junctions	328

11.1.2	Cell Junctions Sealing Extracellular Pathways: Tight Junctions *328*	
11.1.3	Communicating Junctions: Gap Junctions and Synapses *329*	
11.1.3.1	Chemical Synapses *329*	
11.1.3.2	Gap Junctions *329*	
11.2	Established Physical Techniques to Study Cell Junctions *330*	
11.2.1	Cell–Matrix Junctions *330*	
11.2.1.1	Scanning Probe Techniques *330*	
11.2.1.2	Nonscanning Microscopic Techniques *331*	
11.2.1.3	Fluorescence Interference Contrast Microscopy *331*	
11.2.1.4	Total Internal Reflection (Aqueous) Fluorescence Microscopy *331*	
11.2.1.5	Quartz Crystal Microbalance *332*	
11.2.1.6	Other Techniques *332*	
11.2.2	Cell–Cell Junctions *333*	
11.2.2.1	Tight Junctions *333*	
11.2.2.2	Gap Junctions *334*	
11.3	Impedance Spectroscopy *335*	
11.3.1	Fundamental Relationships in Impedance Analysis *336*	
11.3.2	Data Representation and Analysis *337*	
11.4	Impedance Analysis of Cell Junctions *340*	
11.4.1	General Remarks about Experimental Issues *340*	
11.4.1.1	Two-Probe versus Four-Probe Measurement *340*	
11.4.1.2	Introducing Electrodes for Impedance Readings into an Animal Cell Culture *340*	
11.4.1.3	Experimental Set-Up *343*	
11.4.2	Time-Resolved Impedance Measurements at Designated Frequencies *344*	
11.4.2.1	*De novo* Formation of Cell–Matrix and Cell–Cell Junctions *347*	
11.4.2.2	Modulation of Established Cell Junctions *350*	
11.4.3	Modeling the Complex Impedance of Cell-Covered Electrodes *352*	
11.4.4	Spectroscopic Characterization of Cell–Cell and Cell–Matrix Junctions *354*	
	References *356*	

Index *359*

Preface

Volume 6 of the Nanotechnology Series, which focuses on nanoprobes and imaging techniques for local quantitative analysis and surface modification, describes the recent progress that has been made in a variety of areas of high-resolution surface and interface characterization, and functional surface modification for novel storage concepts and sensor applications.

With the invention of the scanning tunneling microscope by G. Binnig and H. Rohrer in 1981, the base was laid for a rapidly growing family of different tip/probe microscopy techniques which are now regarded as the 'eyes and tools of nanotechnology.' In fact, these techniques have also led to a significant acceleration in the development of many other fields of nanoscience and technology, when it became possible to investigate surfaces under a variety of ambient conditions at atomic and molecular resolution. Yet, beyond imaging, a number of unprecedented local spectroscopy techniques also became possible that complemented any previously existing surface-analytical tools. Far beyond conventional surface physics, this group of methods – which included scanning tunneling microscopy, atomic force microscopy and scanning near-field optical microscopy – have provided many new insights into materials science, surface chemistry and biology. It is not surprising, therefore, that over half of the chapters in this volume are dedicated to tip/probe techniques. Initially, however, the main aim of the volume was not to focus on these techniques exclusively, but rather to present the state of the art of the complementary methods available for high-resolution surface and interface analysis. A second aim was to provide examples of local surface modification techniques since, in many cases, it is only by applying such different techniques that we can provide a complete picture of matter at, and even below, the nanometer scale.

Within this volume, details of the scanning tunneling microscope, in spectroscopic mode, are provided in Chapter 1, while the atomic force microscope, in its various static and dynamic modes, is described in Chapter 2. In Chapter 3, the details of three-dimensional atomic force microscopy technology, as represented by the photonic force microscopy which utilizes focused light as a very soft 'spring' that allows the probe to move in three dimensions, are described. In Chapter 4, the application of massively parallel atomic force microscopy-based technology is outlined, including the 'millipede' device for storage application, while the technique of dip-pen lithography, which is used to write complex functional patterns with differ-

Nanotechnology. Volume 6: Nanoprobes. Edited by Harald Fuchs
Copyright © 2009 WILEY-VCH Verlag GmbH & Co. KGaA, Weinheim
ISBN: 978-3-527-31733-2

ing molecules, is described in Chapter 5. Investigations into local ionic conductivity, both on biological samples and synthetic membrane systems, can be carried out using scanning ion conductance microscopy, and these are outlined in Chapter 6.

In contrast to scanning tip/probe techniques, sharp tips may also be used in atom probe tomography (Chapter 7), which employs field ion microscopy to visualize the tip structure and thus provide a chemical analysis of the surface layers, atom by atom. Yet, the surface volume and interface analysis would not be complete without electron microscopy (Chapter 8), a technique which has undergone vast improvement over the past 20 years with the introduction of aberration-corrected lenses and energy filter imaging. In addition, the dynamics of chemical reactions on the surfaces can be monitored using laser spectroscopy, the technique of which has been developed relatively recently (Chapter 9). Likewise, interface analysis using plasmon excitation represents an additional valuable method that complements local imaging methods by using electrons or freely propagating photons (Chapter 10). Finally, the complex behavioral properties of biological cell–cell contacts under external chemical or nanoparticle loading can be studied with improved technologies based on impedance spectroscopy (Chapter 11).

The collection of complementary nanoscopic methods described in this volume represents the current state of the art to inspect the mechanical, electrical, optical and dynamic properties of subnanometer (and in some cases atomic-scale) structures, with high precision and accuracy. This in turn allows the quantitative retrieval of information for each of these properties, at predefined locations on a surface or at an interface. In the future, a strong collaboration between materials scientists, chemists, engineers and physicists working in the nanosciences will undoubtedly lead to further improvements in these technologies, due to the co-development of both instrumentation and nanomaterials.

Finally, it is with great pleasure that I thank all of my colleagues who readily accepted the considerable additional workload in preparing their contributions for this volume.

Münster, August 2008   H. Fuchs

List of Contributors

Martin Aeschlimann
University of Kaiserslautern
Department of Physics
Erwin-Schrödinger-Str. 46
67663 Kaiserslautern
Germany

Wolfgang Baumeister
Max-Planck-Institute of Biochemistry
Department of Structural Biology
Am Klopferspitz 18
82152 Martinsried
Germany

Matthias Böcker
University of Erlangen-Nürnberg
Institute of Applied Physics
Staudtstr. 7, Bld. A3
91058 Erlangen
Germany

Urs Dürig
IBM Research GmbH
IBM Zurich Research Laboratory
Saeumerstrasse 4
8803 Rueschlikon
Switzerland

László Fórró
Ecole Polytechnique Fédérale de Lausanne
EPFL SB-IPMC
Station 3
1015 Lausanne
Switzerland

Harald Fuchs
University of Münster
Institute of Physics
Wilhelm-Klemm-Str.10
48149 Münster
Germany

and

University of Münster
Institute of Physics & Center for Nanotechnology (CeNTech)
Heisenbergstr. 11
48149 Münster
Germany

Mathias Getzlaff
University of Düsseldorf
Institute of Applied Physics / Nanotechnology
Universitäts Str. 1
40225 Düsseldorf
Germany

List of Contributors

Bernd Gotsmann
IBM Research GmbH
IBM Zurich Research Laboratory
Saeumerstrasse 4
8803 Rueschlikon
Switzerland

Camilo Guzman
Ecole Polytechnique Fédérale de Lausanne
EPFL SB-IPMC
Station 3
1015 Lausanne
Switzerland

Hendrik Hölscher
Forschungszentrum Karlsruhe GmbH
Institut für Mikrostrukturtechnik (IMT)
P.O. Box 36 40
76021 Karlsruhe
Germany

Sylvia Jeney
Ecole Polytechnique Fédérale de Lausanne
EPFL SB-IPMC
Station 3
1015 Lausanne
Switzerland

Armin Knoll
IBM Research GmbH
IBM Zurich Research Laboratory
Saeumerstrasse 4
8803 Rueschlikon
Switzerland

Mark A. Lantz
IBM Research GmbH
IBM Zurich Research Laboratory
Saeumerstrasse 4
8803 Rueschlikon
Switzerland

Steven Lenhert
Forschungszentrum Karlsruhe GmbH
Institut für NanoTechnologie
Hermann-von-Helmholtz Platz 1
76344 Eggenstein – Leopoldshafen
Germany

and

University of Münster
Institute of Physics & Center for Nanotechnology (CeNTech)
Wilhelm-Klemm-Str.10
48149 Münster
Germany

Branimir Lukic
Ecole Polytechnique Fédérale de Lausanne
EPFL SB-IPMC
Station 3
1015 Lausanne
Switzerland

Chad A. Mirkin
Northwestern University
Department of Chemistry and International Institute for Nanotechnology
2145 Sheridan Road
Evanston
IL 60208-3113
USA

André Schirmeisen
University of Münster
Institute of Physics & Center for Nanotechnology (CeNTech)
Wilhelm-Klemm-Str.10
48149 Münster
Germany

Tilman E. Schäffer
University of Erlangen-Nürnberg
Institute of Applied Physics
Staudtstr. 7, Bld. A3
91058 Erlangen
Germany

Guido Schmitz
University of Münster
Institute of Material Physics
Wilhelm-Klemm-Str. 10
48149 Münster
Germany

Gerald Steiner
Dresden University of Technology
Faculty of Medicine 'Carl Gustav Carus'
Clinical Sensoring and Monitoring
Fetscher Str. 74
01307 Dresden
Germany

Dennis R. Thomas
Max-Planck-Institute of Biochemistry
Department of Structural Biology
Am Klopferspitz 18
82152 Martinsried
Germany

Joachim Wegener
University of Regensburg
Institute of Analytical Chemistry,
Chemo- and Biosensors
Universitätsstr. 31
93053 Regensburg
Germany

Helmut Zacharias
University of Münster
Institute of Physics
Wilhelm-Klemm-Str. 10
48149 Münster
Germany

and

University of Münster
Center for Nanotechnology (CeNtech)
Heisenbergstr. 11
48149 Münster
Germany

1
Spin-Polarized Scanning Tunneling Microscopy
Mathias Getzlaff

1.1
Introduction and Historical Background

Until the 1980s an idealized and rather unrealistic view was found in surface physics for a lack of techniques which allowed real-space imaging. During this time, surfaces were often assumed to be perfect – that is, imperfections such as step edges, dislocations or adsorbed atoms were neglected. Most of the important information was gained rather indirectly by spatially averaging methods or experimental techniques with insufficient resolution.

However, in 1982, with the invention of the scanning tunneling microscope by G. Binnig and H. Rohrer [1, 2], the situation changed dramatically. This instrument allowed, for the first time, the topography of surfaces to be imaged in real space with both lateral and vertical atomic resolution.

Subsequently, a number of different spectroscopic modes were introduced which provided additional access to electronic behavior, thus allowing the correlation of topographic and electronic properties down to the atomic scale.

In 1988, Pierce considered the possibility of making the scanning tunneling microscope sensitive towards the spin of tunneling electrons by using spin-sensitive tip materials as a further development [3], and this was also predicted – theoretically – by Minakov *et al.* [4]. As a step towards this aim, Allenspach *et al.* [5] replaced the electron gun of a scanning electron microscope with a scanning tunneling microscope tip. Thus, in field emission mode the electrons impinged on a magnetic surface, and the spin polarization of the emitted electrons was subsequently monitored; this, at least in principle, would allow magnetic imaging with nanometer resolution.

However, it was the first 'direct' realization by Wiesendanger *et al.* [6] that opened the possibility of imaging the magnetic properties at atomic resolution. Moreover, the importance of this proposal was not restricted only to basic studies but was also applicable to research investigations. Meanwhile, a rapidly increasing interest emerged from an industrial point of view, a concept which became even more important when considering the need for dramatic increases in the storage density of devices such as computer hard drives. Clearly, further developments in this area will require tools that allow high spatial resolution magnetic imaging for an improved understanding of nanoscaled objects such as magnetic domains and domain wall structures.

In this chapter, we will describe the successful development and implementation of spin-polarized scanning tunneling microscopy (SP-STM), and will also show – by means of selected examples – how our understanding of surface magnetic behavior has vastly increased in recent years.

1.2
Spin-Polarized Electron Tunneling: Considerations Regarding Planar Junctions

First, let us assume that two ferromagnetically or antiferromagnetically coupling layers are separated by an insulator (Figure 1.1, left part). The following discussion can also be extended to ferromagnetic nanoparticles located within an insulating matrix (Figure 1.1, right part).

In order for an electric current to occur, there is the prerequisite that the thickness of the barrier should be small enough so as to allow quantum mechanical tunneling. It is also essential for this discussion that this process is assumed to conserve the spin orientation.

The dependence of the tunneling current on the relative magnetization is shown in Figure 1.2, assuming two ferromagnetic thin layers. The total resistivity for the parallel alignment is less than for the antiparallel orientation. This effect, which is known as tunneling magnetoresistance (TMR), represents a band structure effect that relies on the spin resolved density of states (DOS) at the Fermi level. In comparison, giant magnetoresistance (GMR) is caused by a spin-dependent scattering at the interfaces (further information is available in Ref. [7]).

In order to discuss the behavior of two ferromagnetic electrodes separated by an insulating barrier, the model of Jullière [8] is used; this employs the assumptions that the tunneling process is spin-conserving, and that the tunneling current is proportional to the density of states of the corresponding spin orientation in each electrode.

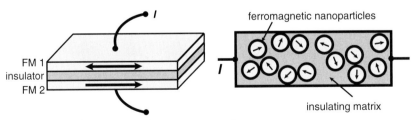

Figure 1.1 Tunneling magnetoresistance can occur when ferromagnetic thin films are separated by an insulating layer (left), and when ferromagnetic nanoparticles are embedded in an insulating matrix (right).

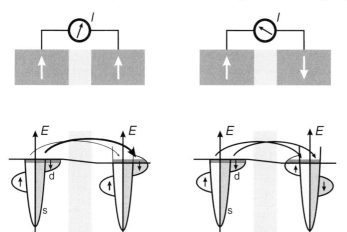

Figure 1.2 Dependence of the tunneling current on the relative magnetization of two ferromagnetic layers. For a parallel orientation a large quantity of spin down electrons at the Fermi energy can tunnel into empty down states; this results in a high tunneling current. In contrast, for an antiparallel orientation the quantity of empty down states is significantly lower, leading to a reduced tunneling current.

In this situation, the tunneling current for a parallel magnetization is given by:

$$I^{\uparrow\uparrow} \propto n_1^\uparrow n_2^\uparrow + n_1^\downarrow n_2^\downarrow \tag{1.1}$$

with n_i being the electron density of electrode i at the Fermi level E_F. For the antiparallel orientation, the tunneling current amounts to:

$$I^{\uparrow\downarrow} \propto n_1^\uparrow n_2^\downarrow + n_1^\downarrow n_2^\uparrow \tag{1.2}$$

With $a_i = n_i^\uparrow/(n_i^\uparrow + n_i^\downarrow)$ being the part of majority electrons of electrode i, and $1 - a_i = n_i^\downarrow/(n_i^\uparrow + n_i^\downarrow)$ that of the minority electrons, the spin polarization P_i of electrode i is given by:

$$P_i = \frac{n_i^\uparrow - n_i^\downarrow}{n_i^\uparrow + n_i^\downarrow} = 2a_i - 1 \tag{1.3}$$

This allows the differential conductance for a parallel orientation to be expressed as:

$$G^{\uparrow\uparrow} = G_p \propto a_1 a_2 + (1-a_1)(1-a_2) = \frac{1}{2}(1 + P_1 P_2) \tag{1.4}$$

and for an antiparallel orientation as:

$$G^{\uparrow\downarrow} = G_{ap} \propto a_1(1-a_2) + (1-a_1)a_2 = \frac{1}{2}(1 - P_1 P_2) \tag{1.5}$$

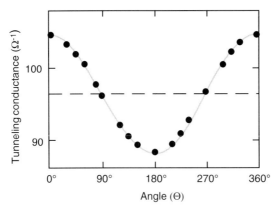

Figure 1.3 Dependence of the tunneling conductance (inverse resistance) of a planar Fe-Al$_2$O$_3$-Fe junction on the angle Θ between the magnetization vectors of both electrodes. (Data taken from Ref. [9]).

The magnitude of the TMR effect is given by:

$$\text{TMR} = \frac{G^{\uparrow\uparrow} - G^{\uparrow\downarrow}}{G^{\uparrow\downarrow}} = \frac{G_p - G_{ap}}{G_{ap}} = \frac{R_{ap} - R_p}{R_p} \tag{1.6}$$

where R_p (R_{ap}) is the resistance for the (anti)parallel orientation, respectively. By using the spin polarization, the TMR effect can be written as:

$$\text{TMR} = \frac{\Delta R}{R_p} = \frac{2 P_1 P_2}{1 - P_1 P_2} \tag{1.7}$$

In the above considerations, it has been assumed that the magnetizations in both ferromagnetic electrodes are oriented parallel or antiparallel. However, the differential conductance also depends on the angle Θ between both directions of magnetization. This behavior is shown in Figure 1.3 for an Fe–Al$_2$O$_3$–Fe junction. Thus, until now the situation has been discussed for $\Theta = 0°$ and $\Theta = 180°$, respectively. For an arbitrary angle Θ, the differential conductance can be expressed as:

$$G = G_0 \cdot (1 + P_1 P_2 \cos\Theta) \tag{1.8}$$

with $G_0 = (G_p + G_{ap})/2$ being the spin-averaged conductance.

1.3
Spin-Polarized Electron Tunneling in Scanning Tunneling Microscopy (STM): Experimental Aspects

The substitution of a ferromagnetic electrode (as discussed above) with a ferromagnetic probe tip represents the situation in SP-STM. The insulating barrier is realized by the vacuum between the sample and the tip, which are separated by a distance of several Ångstroms, thus allowing the laterally resolved determination of magnetic properties. As a consequence, the zero bias anomaly – that is, the decrease in the TMR with increasing bias voltage in planar junctions – is not present for SP-STM investigations because the anomaly can be attributed to scattering of electrons at defects in amorphous barriers [10].

The following sections describe the two fundamental experimental aspects concerning spin-polarized electron tunneling in an STM experiment. Here, different probe tips and modes of operation are employed in order to obtain magnetic information from a sample.

1.3.1
Probe Tips for Spin-Polarized Electron Tunneling

In order to realize SP-STM, the probe tip should fulfill most of the following conditions:

- The higher the spin polarization of the apex atom, the more pronounced the magnetic information (cf. Equations 1.4–1.7) in comparison to electronic and topographic information. Due to the typical reduction of this spin polarization by adsorbates from the residual gas, even under ultra-high vacuum (UHV) conditions, a clean environment or an inert tip material is certainly advantageous [11].

- The sensitivity can also be improved by periodically reversing the magnetization direction, thus *directly* probing the local tunneling magnetoresistance.

- The interaction between tip and sample should be as low as possible because the stray field of a ferromagnetic tip may modify or destroy the sample's domain structure.

- Controlling the orientation of the magnetization axis of the tip parallel or perpendicular to the sample surface allows the domain structure of any sample to be imaged, independent of whether its easy axis is in-plane or out-of-plane.

1.3.1.1 Ferromagnetic Probe Tips

With regards to stray field minimization, bulk tips made from ferromagnetic $3d$ transition metals (see Refs. [12, 13]), with their high content of magnetic material and high saturation magnetization are an unfavorable choice. This mostly restricts their application to ferri- and antiferromagnetic samples [14], which are practically insensitive to external fields. Nevertheless, the stray field can be reduced either by using a material which exhibits a low saturation magnetization, or by using thin-film tips with a film thickness comparable to (or less than) the tip–sample separation [15]. The spin dependence of image potential states can also be used as a sensitive probe of surface magnetism [16], allowing high-resolution magnetic imaging at tip–sample distances larger than in normal tunneling experiments, and thereby reducing the stray field of the ferromagnetic tip.

The first path was realized by Wulfhekel and Kirschner [17–23], who periodically switched the magnetization direction of an amorphous CoFeNiSiB tip with a small coil wound around the tip (see Figure 1.4). Such a tip must exhibit a low coercivity, as this allows minimization of the coil size and thus the coil's stray field at the sample

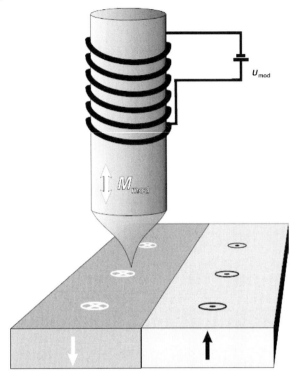

Figure 1.4 Schematic representation of the SP-STM as operated by Wulfhekel and Kirschner [17]. A soft magnetic tip is periodically magnetized in opposite directions along the tip axis by a small coil wound around the tip thus being sensitive to an out-of-plane magnetization. The resultant variation in tunneling current is measured using a lock-in technique. (From Ref. [24], with permission of IOP Publishing Ltd.)

position, as well as a low saturation magnetization. The most important precondition is an extremely low magnetostriction in order to suppress any modulation of the tip length due to the periodic remagnetization process. Consequently, the tips will be sensitive to the perpendicular component of the sample magnetization [21]. In order to realize an in-plane sensitivity, the same type of technique is applicable; however, the 'tip' would now consist of a ferromagnetic ring, the magnetization of which is also periodically switched by using a coil [25, 26].

In order to realize the second situation (see Ref. [27]), an *in situ* preparation of magnetic thin films is necessary. Typically, a polycrystalline tungsten (W) wire is electrochemically etched (this is important for tip stability). The W tip is heated to at least 2200 K upon introduction into the UHV chamber; otherwise, the magnetic coating material is frequently lost during the approach. This high-temperature flash removes oxides and other contaminants, thereby enhancing the binding between the

1.3 Spin-Polarized Electron Tunneling in Scanning Tunneling Microscopy (STM): Experimental Aspects

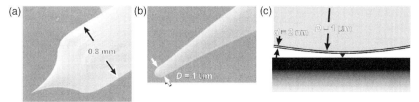

Figure 1.5 Scanning electron microscopy (SEM) images of an electrochemically etched, polycrystalline W tip after a high-temperature flash at $T > 2200$ K. (a) The overview shows the shaft of the tip, which exhibits a diameter of 0.8 mm; (b) High-resolution image of the very end of the tip. The tip apex has an angle of about 15° and the tip diameter amounts to approximately 1 μm; (c) Schematic representation of the tip apex (in scale). The magnetic film is very thin compared to the curvature of the tip. Most likely, a small magnetic cluster protrudes from the tip, and this is responsible for the lateral resolution of the SP-STM. (From Ref. [24], with permission of IOP Publishing Ltd.)

tip surface and the magnetic overlayer. While the overall shape of the tip [as shown in the scanning electron microscopy (SEM) overview image of Figure 1.5a] remains almost unaffected by the high-temperature treatment, the high-resolution SEM image shown in Figure 1.5b reveals that the tip diameter is increased to 1 μm, most likely due to melting of the tip apex. Following this high-temperature treatment, the tips are magnetically coated with a magnetic film exhibiting a thickness of several monolayers (MLs). In contrast to bulk tips, the magnetization direction is governed by the shape anisotropy. The anisotropy of thin film tips can thus be adjusted by selecting an appropriate film material. For example, while 3–10 ML Fe [27] and <50 ML Cr [28] are almost always sensitive to the in-plane component of the magnetization, 7–9 ML Gd [29], 10–15 ML $Gd_{90}Fe_{10}$ [30] and 25–45 ML Cr [30] are usually perpendicularly magnetized at low temperature. The well-known spin reorientation transition of Co films on Au (see Refs [31, 32]) which occur with increasing thickness of the magnetic material allows tuning of the magnetically sensitive direction of the tip with the *same* set of coating materials. For thin Co coverages (<8 ML) on a Au-coated W tip, an out-of-plane magnetic sensitivity is achieved, whereas for thicker Co films the in-plane component of the sample magnetization can be probed [33].

At least qualitatively, this observation can easily be understood. Two anisotropy terms are relevant: (i) the shape anisotropy which arises due to the pointed shape of the tip; and (ii) the surface and interface anisotropy of the film. The first term will always try to force the magnetization along the tip axis – that is, perpendicular to the sample surface. In contrast, the second term is material-dependent. Due to the rather large curvature of the tip as compared to the thickness of the coating film (see Figure 1.5c), the effective surface and interface anisotropy of a thin film can be deduced from an equivalent film on a flat W(110) substrate. For instance, in the case of 10 ML Fe the two anisotropy terms favor different directions. While the shape anisotropy still tries to force the magnetization along the tip axis, the ferromagnetic film on W(110) exhibits a strong in-plane anisotropy [35] which obviously overcomes the shape anisotropy. An external field of 2 T is required to force the tip magnetization out of the easy (in-plane) into the hard (out-of-plane) direction [34]; this is consistent with results of Elmers and Gradmann [35] concerning thin film systems.

Even at room temperature magnetic thin film tips can be used for several days without losing their spin sensitivity. Initially, this is surprising as any surface is continuously exposed to the residual gas in the vacuum chamber which, depending on the reactivity of the sample under investigation, leads to a more or less rapid – but continuous – degradation of the surface spin polarization [36]. However, the geometry of the tunnel junction must be taken into account as it differs from that of an open surface. While residual gas particles may impinge onto a flat, uncovered surface from the entire half-space, the tip apex is almost completely shadowed by the sample, as shown schematically in Figure 1.5c. Thereby, gas transport onto the tip apex is dramatically reduced. Of course, the same argument can be applied to the sample surface which is, however, only valid for the particular location of the sample surface that is right under the tip apex. As this location varies when scanning the tip across the sample, the shadowing is only temporarily effective for any particular site of the sample surface, whereas the tip is shadowed at all times.

1.3.1.2 Antiferromagnetic Probe Tips

Despite the fact that the magnetic dipole interaction between the sample and the tip is considerably reduced for ferromagnetic ultrathin film coatings on a nonmagnetic tip, in comparison to thicker coatings or even bulk ferromagnetic tips, an additional influence cannot be ruled out. One straightforward and experimentally feasible solution, however, is to use an antiferromagnetically coated (see Ref. [30]) or a bulk antiferromagnetic tip consisting of, for example, a MnNi alloy [37–40]. The tip should exhibit no significant stray field, since opposite contributions compensate on an atomic scale, thus allowing the nondestructive imaging and investigation of spin structures even for magnetically soft samples. The spin sensitivity is determined solely by the orientation of the magnetic moment of the atom that forms the very end of the tip apex; the orientation of all other magnetic moments plays no role. Furthermore, the tip is insensitive to external fields, which allows direct access to intrinsic sample properties in field-dependent studies.

In order to demonstrate this insensitivity, we can refer to an investigation conducted by Kubetzka *et al.* [30]. Here, the response of an identically prepared system to an applied perpendicular magnetic field is shown using a ferromagnetic tip on the one hand, and an antiferromagnetic tip on the other hand. Figure 1.6a shows a series of dI/dU maps – that is, maps of the differential conductance, of 1.95 ML Fe on W(110) recorded with a ferromagnetic GdFe tip. Figure 1.6a(i) shows an overview of the initial state, while Figure 1.6a(ii) is taken at higher resolution, as indicated by the frame in Figure 1.6a(i). Because the coverage is slightly below 2 ML, narrow ML areas can be seen with a bright appearance at the chosen bias voltage. These ML areas efficiently decouple double layer (DL) regions on adjacent terraces. At 350 mT [see Figure 1.6a(iii)] the domain distribution is asymmetric; the bright domains have

1.3 Spin-Polarized Electron Tunneling in Scanning Tunneling Microscopy (STM): Experimental Aspects

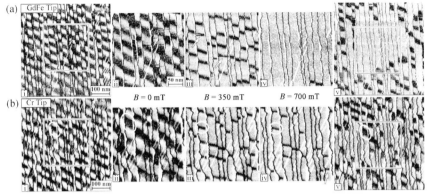

Figure 1.6 (a) dI/dU maps of 1.95 ML Fe/W (110) recorded with a GdFe tip (out-of-plane contrast): (i) 500 × 500 nm² overview of magnetic initial state; (ii) 250 × 250 nm² zoom-in; (iii) Asymmetry at $B = 350$ mT: dark domains are compressed and form 360° walls; (iv) saturation is observed within the field of view; (v) the influence of the tip's stray field becomes obvious in the overview recorded at $B = 0$ mT; (b) Analogous series of an identically prepared sample, recorded with a Cr-coated tip: (i, ii) magnetic initial state; (iii) asymmetry at $B = 350$ mT; (iv) a large fraction of the walls has survived at 700 mT, in contrast to (a); (v) the scanned area exhibits no significant difference in comparison to its surrounding. (Reprinted with permission from Ref. [30]; copyright (2002) American Physical Society.)

grown and the dark domains have shrunk. In some places the magnetic contrast changes abruptly from one horizontal scan line to the next (see arrows), this being the result of a rearrangement of the sample's magnetic state during the imaging process. At 700 mT [see Figure 1.6a(iv)] the sample has almost reached saturation within the field of view. However, it becomes obvious in the overview image, subsequently recorded at the same location in zero applied field [see Figure 1.6a(v)], that this field value does not reflect intrinsic sample properties. A large fraction of the dark domains has survived outside the region which was scanned previously at 700 mT. Thus, superpositioning of the applied field and the additional field emerging from the magnetic coating of the tip is much more efficient than the applied field alone.

Figure 1.6b shows an analogous series of images of a sample which was identically prepared but imaged with an antiferromagnetic Cr-covered tip. This exhibits an out-of-plane sensitivity, like the GdFe tips [30]. A dark domain is marked as an example to be recognized in all five images. The domain structure in Figure 1.6b(i)–(iii) displays no significant difference to the corresponding structures in Figure 1.6a. Since a rearrangement of the domain structure during imaging is not observed throughout this series, the occurrence of such events in Figure 1.6a can be attributed to the GdFe tip's stray field. As in Figure 1.6a(iii), the dark domains are compressed at 350 mT, which proceeds at 700 mT. At this field value, and in contrast to Figure 1.6a, a large fraction of the dark domains has survived. In the overview image of Figure 1.6b(v), which was taken again subsequently in a zero-applied field, the previously scanned area exhibits no significant difference in domain distribution in

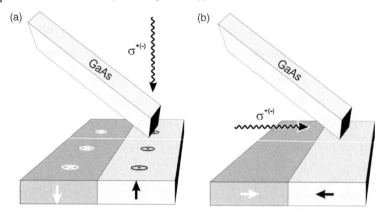

Figure 1.7 Schematic experimental set-up to realize spin-polarized electron tunneling using optically pumped GaAs tips, as proposed in Ref. [3]. (a) Spins aligned perpendicular to the sample surface can be detected by a GaAs tip excited by helical light incident along the surface normal; (b) When the incident light is along the sample surface the experiment is sensitive to electron spins parallel to the surface plane. (Reprinted from Ref. [24], with permission of IOP Publishing Ltd.)

comparison to its surrounding. This result directly demonstrates the advantage of a stray field free tip.

1.3.1.3 Optically Pumped GaAs Probe Tips

The concept of spin-polarized tunneling between a magnetic surface and an optically pumped GaAs tip (as discussed below) was first proposed by Pierce [3]. This experimental approach allows both the sign and the polarization direction of photoexcited electrons to be modified, simply by choosing an appropriate laser light helicity and experimental geometry. Varying the spin polarization of the tunneling electrons with a simultaneously constant intensity of the incident light enables the magnetic effects to be separated from the topographic and electronic effects. The corresponding schematic arrangement (see Figure 1.7) proves that the experiment can be made sensitive to either the out-of-plane or the in-plane magnetization direction by changing the direction of the incident light to be parallel and perpendicular to the sample surface, respectively.

Optically pumped p-type GaAs is widely used as a source to produce spin-polarized electrons close to the Fermi level. The physical principle is based on two properties of this material: (i) It is a direct band-gap semiconductor; and (ii) the degeneracy of the p-like valence band is lifted by spin–orbit interaction into a fourfold degenerate $p_{3/2}$ level (Γ_8 band edge) and a twofold degenerate $p_{1/2}$ level (Γ_7 band edge). The spin–orbit splitting amounts to $\Delta_{so} = 0.34$ eV. If circularly polarized light (σ^+ or σ^-) with an energy slightly above the energy gap of $E_{gap} = 1.52$ eV of p-GaAs is irradiated onto the sample, the electronic transition in GaAs must fulfill the optical selection rule $\Delta m_j = m_f - m_i = \pm 1$, where $m_{f,i}$ is the angular momentum of the final and initial states, respectively (see Figure 1.8). When using σ^+ light, the relative transmission probability into $m_j = -1/2$ states is threefold higher than that into $m_j = -3/2$ states

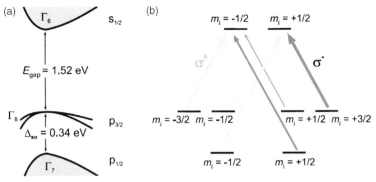

Figure 1.8 (a) Schematic band structure of GaAs in the vicinity of the Γ-point of the Brillouin zone. The width of the band gap between the conduction (Γ_6) and the fourfold degenerate $p_{3/2}$ valence band edge (Γ_8) amounts to 1.52 eV. Another 0.34 eV lower there is the twofold degenerate $p_{1/2}$ level (Γ_7); (b) Allowed transitions between different m_j sublevels for circularly polarized light of opposite helicity (σ^+ and σ^-). The transition probability is represented by the thickness of the arrows. (Reprinted from Ref. [24], with permission of IOP Publishing Ltd.)

(and vice versa for σ^- light). As a result, while the theoretical limit of the electron spin polarization in photoemission is ±50%, values as high as 43% have been achieved experimentally [41].

The preliminary experiments on GaAs–insulator–ferromagnet tunnel junctions were reported in Ref. [42]. The specimen was prepared by cleaving a GaAs crystal in air along the (1 1 0) plane. A 20–40 Å layer of Al was then evaporated onto the GaAs (1 1 0) surface and subsequently oxidized. Onto this insulating barrier was then deposited a 150 Å thick Co film, which was itself protected by a 50 Å Au cap layer. The chosen experimental set-up required that the light would traverse the ferromagnetic layer and the insulator before reaching the semiconductor. After magnetization of the Co film perpendicular to the plane, a dependence of the tunneling current on the helicity of the light was measured, which suggested the existence of spin-polarized transport. However, an even stronger signal was detected when there was no tunneling barrier between the semiconductor and the ferromagnet' this was explained by "…an intensity modulation of the circularly polarized light upon transmission through the magnetically ordered layer, determined by the polar magneto-optic coefficients" [42]. The reduction of the helical asymmetry when using a tunnel barrier, compared to a situation without any barrier, was explained in a later analysis [43, 44] by a negative tunneling conductance. As this method to create spin-polarized electrons involves neither a magnetic material nor magnetic fields, it offers excellent conditions for application in SP-STM.

In spite of this favorable situation, GaAs tips have not yet been used successfully for the imaging of magnetic surfaces. Similar to the experiments performed with planar junctions, this may be caused by difficulties in separating spin-polarized tunneling from magneto-optical effects [45–49].

Compared to the proposal of Pierce [3] (as shown in Figure 1.7), the first successful observation of spin-polarized electron tunneling with the scanning tunneling microscope, using a GaAs electrode [50], was obtained by exchanging the role of tip and sample – that is, by using a Ni tip and a GaAs(110) surface. Moreover, instead of optically pumping the GaAs sample and thereby producing spin-polarized charge carriers in the GaAs, the reverse process was used, with spin-polarized electrons being injected from the Ni tip into the conduction band of GaAs. Upon transition of the injected electrons from this metastable state in the conduction band into the final state in the valence band recombination, luminescence was seen to occur and the circular polarization of the emitted photons was analyzed.

Evidence for a second explanation for the failure of magnetic imaging with GaAs electrodes – namely an insufficient lifetime of the spin carriers at the tip apex – comes from analogous STM-excited luminescence experiments performed with single crystalline Ni(110) tips and a stepped GaAs(110) sample [51]. With the tip positioned above flat terraces, a high-spin injection efficiency was measured. However, the intensity of the recombination luminescence on the upper terrace was found to have decreased by a factor of 1000. Simultaneously, the polarization decreased by a factor of 6. This observation was explained by a reduction of either the spin injection efficiency or of the spin relaxation lifetime, and attributed to the metallic nature of the step edge caused by midgap states of the (111) surface. As a sharp tip must possess numerous step edges around the apex atom, it is a straightforward conclusion that the spin relaxation lifetime may be drastically reduced at the very end of the tip.

1.3.1.4 Nonmagnetic Probe Tips

Surprisingly, even nonmagnetic tips can be used to image certain magnetic sample properties, as demonstrated by Bode *et al.* [52, 53] and by Pietzsch *et al.* [54]. The images shown in Figure 1.9a and b [54] were taken for slightly less than 2 ML Fe on

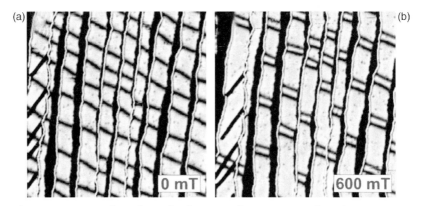

Figure 1.9 Domain walls as observed with a nonmagnetic W tip. (a) No external field applied; (b) Taken at 600 mT. The external field enforces pair formation. Sensitivity to the spin orientation inside the walls is lost, and all walls are imaged as dark lines. Thus, the image provides information on the magnetization lying along an easy or a hard direction. Note the five lines in the left stripe running in a direction bottom-left to top-right; these are not domain walls but are dislocation lines. (Reprinted from Ref. [54], with permission of Springer. Copyright (2004).)

1.3 Spin-Polarized Electron Tunneling in Scanning Tunneling Microscopy (STM): Experimental Aspects

W(110), which allows a comparison with the results shown in Figure 1.6 (which were obtained using a magnetic tip). However, a tungsten tip without any magnetic coating was now used, whereupon the dark lines revealed the presence of domain walls. The main difference between the two measurements was that the nonmagnetic tip did not provide sensitivity to the spin orientation inside the walls. Instead, both walls of a pair were imaged equally, in contrast to the observation made with the ferromagnetic tip [55] (cf. Figure 1.26). This is a consequence of the fact that the measurement made with the W tip does not involve spin-polarized tunneling; rather, it is the spin-averaged electronic structure of the sample that gives rise to the signal. The electronic structure of the DL stripes is locally modified due to the presence of a domain wall. In other words, the electronic structure is sensitive to whether the magnetization is in an easy or a hard direction. First-principle calculations have shown [52, 53] that the spin–orbit-induced mixing of different d-states depends on the magnetization direction, and changes the local density of those states that are detectable by non-spin-polarized STS.

As an important implication of this effect, the magnetic nanostructure of surfaces can be investigated with conventional nonmagnetic tips. This has the clear advantage that there is definitely no dipolar magnetic stray field from the tip that could interfere with the sample. In addition, the preparation of a magnetic tip is omitted.

1.3.2 Modes of Operation

In the following sections, different modes of operation enabling to achieve a magnetic contrast using magnetic probe tips will briefly be discussed from a theoretical point of view according to Ref. [56].

1.3.2.1 Constant Current Mode

In the situation when the tip and sample are magnetic, the tunneling current can be described as a sum of a spin-averaged I_0 and a spin-dependent term I_{sp} [56] (cf. Equation 1.8):

$$I(\vec{r}_t, U, \theta) = I_0(\vec{r}_t, U) + I_{sp}(\vec{r}_t, U, \theta) \tag{1.9}$$

$$= const. \cdot (n_t \tilde{n}_s(\vec{r}_t, U) + \vec{m}_t \tilde{\vec{m}}_s(\vec{r}_t, U)) \tag{1.10}$$

with n_t being the non-spin-polarized local density of states (LDOS) at the tip apex, $\tilde{n}_s$ the energy-integrated LDOS of the sample, and $\vec{m}_t$ and $\tilde{\vec{m}}_s$ the vectors of the (energy-integrated) spin-polarized LDOS with:

$$\tilde{\vec{m}}_s(\vec{r}_t, U) = \int \vec{m}_s(\vec{r}_t, E) dE \tag{1.11}$$

and θ the angle between the magnetization direction of tip and sample. For a non-spin-polarized STM experiment – that is, using either a nonmagnetic tip or a nonmagnetic sample – the second term, I_{sp}, vanishes.

The constant current mode is restricted to some limited cases, which is partly due to the integral in Equation 1.11 and taken over all energies between the Fermi energy E_F and eU, with U being the applied bias voltage because I_{sp} becomes reduced if the spin polarization changes sign between E_F and eU. Furthermore, the magnetically induced corrugation is small compared to the topographic and electronic corrugation; this is due to the exponential dependence of the tunneling current on the distance between tip and sample. Nevertheless, it is still possible to obtain information of complex atomic-scale spin structures at ultimate magnetic resolution (as shown in Ref. [56]), although it is necessary to understand the influence of the tip [57, 58].

1.3.2.2 Spectroscopy of the Differential Conductance

The difficulties of separating the topographic, electronic and magnetic information can be overcome by measuring the differential conductance, dI/dU, with a magnetic tip [56]:

$$dI/dU(\vec{r}_t, U) \propto n_t \tilde{n}_s(\vec{r}_t, E_F + eU) + \vec{m}_t \vec{\tilde{m}}_s(\vec{r}_t, E_F + eU) \tag{1.12}$$

In this situation, the measured quantity no longer depends on the energy-integrated spin polarization, but rather on the spin polarization in a narrow energy window ΔE around $E_F + eU$.

The differential conductance is determined experimentally by adding a small ac-voltage to the dc-bias voltage at a frequency which is significantly above the cut-off frequency of the feedback loop that ensures a constant current. The amplitude of the ac-voltage is responsible for the width ΔE. The current modulation is amplified by means of a lock-in technique.

The electronically homogeneous surfaces maps of differential conductance reflect the magnetic behavior, since any variation of the signal must be due to the second spin-dependent term, I_{sp}. The situation becomes more complicated for electronically heterogeneous surfaces; nevertheless, a careful comparison between spin-averaged and spin-resolved measurements often allows a distinction to be made between topographic and electronic contrast compared to the magnetically induced information. However, this set of experiment requires measurements to be made with both nonmagnetic and magnetic probe tips.

The determination of differential conductance also provides access to the spin polarization of a surface which is *locally resolved* [59–61].

The recording of inelastic tunneling spectra (i.e. the second derivative of the conductance d^2I/dU^2) with a magnetic tip in an external magnetic field, it becomes possible to study directly – that is, without a separating insulating layer – magnon excitations in magnetic nanostructures [62].

1.3.2.3 Local Tunneling Magnetoresistance

As an alternative method, the local tunneling magnetoresistance $dI/d\vec{m}_t$ between the magnetic tip and magnetic sample can be determined by modulation of the tip magnetization direction and determining the variation of the tunneling current using a lock-in technique. This type of measurement was first proposed by Johnson and Clarke [63] and later accomplished by Wulfhekel and Kirschner [17]. By taking the derivative of Equation 1.10 one obtains:

$$dI/d\vec{m}_t \propto \tilde{\vec{m}}_s \qquad (1.13)$$

Thus, the signal is proportional to the energy-integrated spin-polarized LDOS. One significant advantage of this technique relates to the detailed knowledge of the magnetization of the probe tip. A nonvanishing signal is also obtained only if a local magnetization is present. Furthermore, this method allows the investigation of samples in a single domain state; this situation differs from the spectroscopy of differential conductance, which demands that different magnetic domains are simultaneously visible in a single image. Consequently, due to the direct detection of magnetic information, knowledge of the electronic properties is no longer required.

It must be borne in mind, however, that the interpretation of chemically heterogeneous surfaces (e.g. of alloys) remains difficult. Both, the sign and magnitude, of the element-specific spin polarization may vary, thereby avoiding any direct identification of the domain structure. Nonetheless, the chemical contrast plays an additional role.

1.4
Magnetic Arrangement of Ferromagnets

In this section, we will demonstrate the magnetic arrangement of ferromagnetic systems for systems which exhibit localized magnetic moments, and also which represent an itinerant or band magnet. A typical representative of the former group is the rare-earth metal gadolinium, while the transition metal cobalt is typical of the latter group.

1.4.1
Rare-Earth Metals: Gd/W(110)

In analogy to the low-temperature experiments performed with ferromagnet–insulator–superconductor planar tunneling junctions [64, 65], where the quasiparticle density of states of superconducting aluminum is split by a magnetic field into spin up and spin down parts, two spin-polarized electronic states with opposite polarization can be used to probe the magnetic orientation of the sample relative to the tip, thus enabling spin-polarized scanning tunneling spectroscopy (SP-STS) [60].

The principle of SP-STS is shown schematically in Figure 1.10, using a sample which exhibits an exchange split-surface state with a relatively small exchange splitting, Δ_{ex}. This situation is, for example, realized for the Gd(0001) [66–70], Tb (0001) [71] and Dy(0001) surfaces [72]. If Δ_{ex} is too large, then one spin component would be too far from the Fermi level and not accessible by STS, as for example in the case of Fe(001), where Δ_{ex} amounts to 2.1 eV and only the minority band appears as a

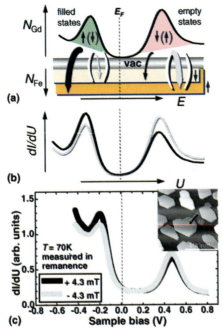

Figure 1.10 (a) The principle of SP-STS using a sample with an exchange split surface state, for example, Gd(0001), and a magnetic Fe tip with a constant spin polarization close to E_F. Due to the spin valve effect the tunneling current of the surface state spin component being parallel to the tip is enhanced at the expense of its spin counterpart; (b) This should lead to a reversal in the dI/dU signal at the surface state peak position upon switching the sample magnetically; (c) Exactly this behavior could be observed in the tunneling spectra measured with the tip positioned above an isolated Gd island (see arrow in the inset). (Reprinted with permission from Ref. [27]; copyright (1998), American Physical Society.)

peak in the dI/dU spectra just above the Fermi level [73]. In contrast, the majority (minority) part of the Gd(0001) surface state has a binding energy of -220 meV (500 meV) at 20 K [71]; that is, the exchange splitting only amounts to 700 meV, far below the Curie temperature of 293 K.

In the following section we consider vacuum tunneling between a Gd(0001) surface and a tip material; for simplicity, we assume a constant spin polarization (see Figure 1.10a, lower part). If the magnetization direction of the tip remains constant, then two possible magnetic orientational relationships occur between the tip and sample – parallel or antiparallel – under the assumption that the magnetization of the tip and sample is aligned. Since, however, both the majority and the minority component of the Gd(0001) surface state appear in the tunneling spectra, the spins of one component of the surface state will in any case be *parallel* with the tip, while the spins of the other component will be *antiparallel*. Therefore, the spin valve effect will act differently on the two spin components; due to the strong spin dependence of the density of states, the spin component of the surface state parallel to the tip magnetization is enhanced at the expense of its counterpart being antiparallel.

Consequently, by comparing tunneling dI/dU spectra measured above domains with opposite magnetization, one expects a reversal in contrast at the majority and minority peak positions (see Figure 1.10b).

Tunneling spectra measured in an external magnetic field with an in-plane sensitive ferromagnetic probe tip positioned above an isolated Gd(0001) island show exactly the expected behavior (see Figure 1.10c). The sample was magnetized in a magnetic field of $+4.3$ mT applied parallel to the sample surface, and the spectra were subsequently measured in remanence with the tip positioned above the Gd island (this is marked by an arrow in the inset of Figure 1.10c). The direction of the magnetic field was then reversed (to -4.3 mT) and further dI/dU spectra were monitored at the same location. Figure 1.10c shows the tunneling spectra measured in remanence after the application of a positive or negative field. A comparison of the spectra reveals that for a positive field, the differential conductance dI/dU measured at a sample bias which corresponds to the binding energy of the occupied (majority) part of the surface state, is higher than for negative field. The opposite is true for the empty (minority) part. Freestanding Gd islands on W(110) were chosen for this experiment, since it is known from Kerr effect measurements [74] that the coercivity is only 1.5 mT – that is, much lower than the applied field. Thus, it can be safely concluded that the magnetization of the sample was switched by the external field while the tip magnetization remained unchanged.

Further information relating to magnetic imaging of the Gd(0001) surface can be found in Refs [36, 68, 70], while data concerning the surface of another rare earth metal, Dy(0001), are available in Ref. [72].

1.4.2
Transition Metals: Co(0001)

The domain structure of the surface of a Co(0001) single crystal has been studied by Wulfhekel *et al.* [17–23]. The uniaxial magnetocrystalline anisotropy of hcp-Co points along the *c*-axis – that is, perpendicular to the (0001) surface. However, the total energy of the sample is minimized by the formation of surface closure domains where the magnetization locally tilts towards the surface plane, thereby reducing the dipolar energy. As the magnetocrystalline anisotropy energy and dipolar energy are similar in size, and the in-plane components of the magnetocrystalline anisotropy energy are almost degenerate, a complicated dendritic pattern is formed at the surface. Figure 1.11a shows the typical dendritic-like perpendicular domain pattern of Co(0001) as measured by Wulfhekel *et al.* [23]. Due to the fact that the tip magnetization is intentionally modulated by a small coil, the bright and dark locations in Figure 1.11a can be assigned to specific magnetic orientations, namely the magnetization vector points out of or into the surface, respectively. A sharp contrast can be recognized in Figure 1.11b, which is completely absent in the topographic image [21]. The absence of any correlation confirms that this contrast is not caused by different local structural or electronic properties; rather, it represents a domain wall separating two regions of different magnetization directions. This

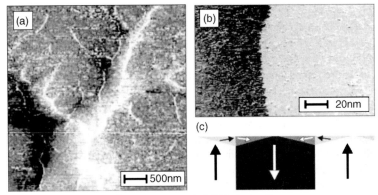

Figure 1.11 (a) Magnetic domain image on Co(0001) obtained by SP-STM; (b) A sharp domain wall at the end of a branch, at high magnification; (c) Schematic cross-section of the closure domain pattern of Co. (Reprinted with permission from Ref. [23]; copyright (2003), EDP Sciences.)

domain wall is found to correspond to the domain wall across two canted surface domains (see Figure 1.11c).

Further information concerning the ferromagnetic transition metal Fe on W(110) and Mo(110) is provided below. The spin-resolved electronic properties of dislocation lines that occur during thin film growth of Fe films on W(110) are described in Ref. [75], while details of the complex magnetic structure of Fe on W(001) are reported in Refs [76, 77]. The easy magnetization axis was shown to be layer-dependent, whereas the second and third Fe layers were magnetized along ⟨110⟩ or equivalent directions; the easy axis of the fourth layer was rotated by 45°.

1.5
Spin Structures of Antiferromagnets

The lateral averaging of magnetically sensitive techniques often fails in the imaging of antiferromagnetic surfaces because the overall detected magnetization is equal to zero. Here, we will show how SP-STM can be used to overcome this difficulty. The first example is the topological anti-ferromagnetism of Cr(001); a second example, namely the atomic resolution of magnetic behavior, will be demonstrated using the antiferromagnetic Mn monolayer on W(110) and Mn_3N_2 films on MgO(001).

The antiferromagnetic nature was additionally reported for the first monolayer of Fe on W(001) [77]. The antiferromagnetic coupling between a whole atomic layer and a ferromagnetic substrate was investigated for Mn on Fe(001) [25, 26, 78]; a Co layer on a Cu-capped Co substrate exhibits a ferromagnetic or antiferromagnetic coupling, depending on the interlayer thickness [79]. Magnetite Fe_3O_4 represents a ferrimagnet with a high spin polarization near the Fermi level. SP-STM was used to investigate the corresponding (001) [39, 40, 80–83] and (111) surfaces [40, 84].

1.5.1
Topological Anti-Ferromagnetism of Cr(001)

The Cr(001) surface for which the topological step structure is directly linked to the magnetic structure represents a topological anti-ferromagnet; that is, each terrace exhibits a ferromagnetic alignment of the magnetic moments, although between two adjacent terraces the magnetization possesses an antiparallel orientation that was predicted, on a theoretical basis, by Blügel et al. [85].

By using the scanning possibilities, the antiferromagnetic coupling between neighbored terraces of a Cr(001) surface can be imaged directly [70, 86–94]. The topography (see Figure 1.12a) presents a regular step structure with terrace widths of about 100 nm [88]. The line section in the bottom of Figure 1.12a shows that all step edges in the field of view are of single atomic height – that is, 1.4 Å. This topography should lead to a surface magnetization that periodically alternates between adjacent terraces and, indeed, this was observed experimentally (see Figure 1.12b). The line section of the differential conductance drawn along the same path as in Figure 1.12a indicates two discrete levels with sharp transitions at the positions of the step edges.

The typical domain wall width, as measured on a stepped Cr(001) surface, amounts to 120–170 nm [86]. In analogy to ferromagnetic domain walls (these are discussed in detail in Chapter 9), this value is determined by intrinsic material parameters – that is, the strength of the exchange coupling and the magnetocrystalline anisotropy. Clearly, the domain wall width cannot remain unchanged very close to a screw

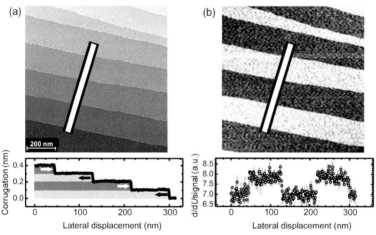

Figure 1.12 (a) Topography and (b) spin resolved map of the dI/dU signal of a clean and defect free Cr(0 0 1) surface as measured with a ferromagnetic Fe-coated tip. The bottom panels show averaged sections drawn along the line. Adjacent terraces are separated by steps of monatomic height. (Reprinted with permission from [88]; copyright (2003) by the American Physical Society.).

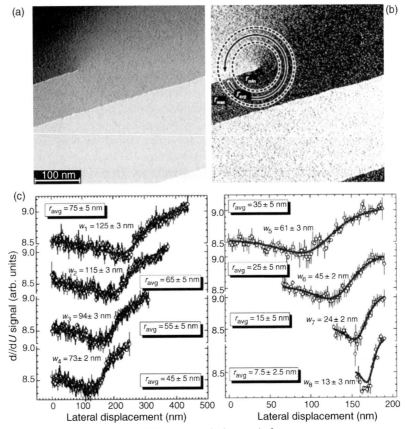

Figure 1.13 (a) Topography and (b) magnetic dI/dU signal of a Cr(001) surface with a single screw dislocation. The magnetic frustration leads to the formation of a domain wall between the dislocation; (c) Circular sections drawn at different radii around the center of the screw dislocation. (Reprinted with permission from Ref. [88]; copyright (2003), American Physical Society.)

dislocation where the circumference becomes comparable with or smaller than the intrinsic domain wall width.

The dependence of domain wall width on the distance from the screw dislocation of the Cr(001) surface is shown in Figure 1.13a [88]. Here, approximately 100 nm from the next step edge, a single screw dislocation can be recognized in the upper left corner of the image. The magnetic dI/dU map of Figure 1.13b reveals that this screw dislocation is the starting point of a domain wall which propagates towards the upper side of the image. Starting at the tail of the arrow (zero lateral displacement), eight circular line sections are drawn counterclockwise around the screw dislocation at different radii r_{avg}, from 75 nm down to 7.5 nm; these data are plotted in Figure 1.13c. The domain walls were fitted using the model provided in Chapter 9. The results are shown as gray lines in Figure 1.13c; except for the smallest average

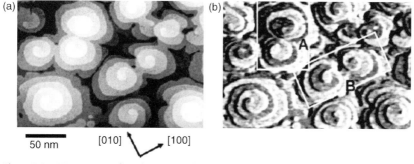

Figure 1.14 STM images of (a) the topography and (b) dI/dU magnetic signal obtained simultaneously from the same area of a 9 nm-thick Cr(001) film. (Reprinted with permission from Ref. [93]; copyright (2007), Elsevier.)

radius (r_{avg} = 7.5 nm), an excellent agreement with the experimental data was found. At an average radius r_{avg} = 75 nm, the domain wall width amounted to 125 nm, being in close agreement with the intrinsic domain wall width of Cr(001) as determined far away from screw dislocations. This was not surprising, as the circumference amounted to about 500 nm – much larger than the intrinsic domain wall width. However, as soon as r_{avg} was reduced below 60 nm a significant reduction in domain wall width was observed, although the circumference still exceeded the intrinsic domain wall width. The results showed clearly that the domain wall width was always considerably narrower than the circumference of the cross-section.

We can now discuss a more complex structure, namely the influence of the distance and chirality between two adjacent spiral terraces on magnetic structures on Cr(001) films [90, 93]. Figure 1.14a shows a topographic STM image of a 9 nm-thick Cr(001) film [93] where the feature of the surface morphology is that the Cr layers form high-density, self-organized spiral terraces. Each terrace is displaced by a monatomic step height, and a screw dislocation is clearly visible in the center of each spiral pattern. The typical diameter of these spiral terraces is 50 nm. A complex spin frustration and characteristic magnetic ordering is present, being restricted by the topological asymmetry of the spiral terraces. Figure 1.14b shows the dI/dU magnetic image obtained simultaneously in the same area of Figure 1.14a, exhibiting a magnetic contrast. A comparison of the two images of Figure 1.14a and b reveals that most parts of the observed magnetic contrasts are consistent with a topological antiferromagnetic structure. The maximum magnetic contrast corresponds to the topological antiferromagnetic order, and a deviation from the maximum magnetic contrast can be recognized as the spin frustration which appear in the region near the screw dislocations and between two spirals.

The magnetic structure can be deduced from the observed dI/dU magnetic signal intensity by assuming the orientation of the tip magnetization parallel to the bcc [100] direction. For example, the derived magnetic structures of two adjacent spirals (the regions A and B indicated in Figure 1.14b) are shown in Figure 1.15 by arrows. Although the two adjacent spirals have the same chirality, the sign is opposite

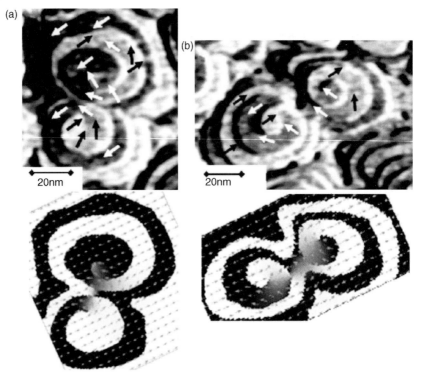

Figure 1.15 Observed (upper) and simulated (lower) magnetic structures of two adjacent spirals: (a) region A and (b) region B indicated in Figure 1.14b. The directions of the magnetization are represented by the arrows. The distance between the two screw dislocations is 20 nm (region A) and 32 nm (region B). (Reprinted with permission from Ref. [93]; copyright (2007), Elsevier.)

between regions A and B; the distance between the two screw dislocations d_s is 20 and 32 nm for the regions A and B, respectively. Although the magnetization rotates gradually around the center of the spiral in the case of $d_s = 75$ nm [90], this does not occur for $d_s = 20$ nm and 32 nm, which suggests that the spin-frustrated region decreases with decreasing d_s. There seems to be little difference in the sign of chirality of spirals.

In order to understand its origin, we can calculate the magnetic structure of these spiral terraces [93]; the result is shown in the two lower panels in Figure 1.15. The white (black) contrast represents the magnetization to be parallel (antiparallel) to the [100] direction. The topological antiferromagnetic order appears in a series of adjacent terraces, as well as in the most part of spiral terraces; the frustrated regions (the gray regions in the simulated figures) are evident between the center of adjacent spirals. It should be noted that the observed and calculated magnetic structures are clearly asymmetric with respect to the straight line connecting two screw dislocations, in spite of the different d_s-values. The simulated spin alignments are in good qualitative agreement with the observed results.

1.5.2
Magnetic Spin Structure of Mn with Atomic Resolution

The deposition of Mn on W(110) in the submonolayer regime results in a pseudomorphic growth; that is, Mn mimics the bcc symmetry as well as the lattice constant of the underlying substrate [95]. By using a clean W tip, atomic resolution could be achieved on the Mn islands, as demonstrated by Heinze et al. [96] (see Figure 1.16a). Additional information is provided in Refs [70, 97]. The diamond-shaped unit cell of the (1×1)-grown Mn ML is clearly visible. The line section drawn along the dense-packed [1 1̄ 1] direction exhibits a periodicity of 0.27 nm, which almost perfectly fits the expected nearest-neighbor distance of 0.274 nm. The measured corrugation amplitude amounts to 15 pm. A calculated STM image for a conventional tip without spin polarization is given for comparison (see inset of Figure 1.16a). Clearly, the qualitative agreement between theory and experiment is excellent.

In a second set of experiments [96], different ferromagnetic tips were used. Since it is known from first-principles computations that the easy magnetization axis of the Mn ML on W(110) is in-plane [96], the experiment required a magnetic tip with a magnetization axis in the plane of the surface in order to maximize the effects. This condition is fulfilled by Fe-coated probe tips [27]. Figure 1.16b shows an STM image taken with such a tip, where the periodic parallel stripes along the [001] direction of the surface can be recognized. The periodicity along the [1 1̄ 0] direction amounts to 4.5 Å, which corresponds to the size of the magnetic c(2 × 2) unit cell. The inset in Figure 1.16b shows the calculated STM image for the magnetic ground state, that is, the c(2 × 2)-antiferromagnetic configuration. Thus, theory and experiment give a consistent picture. Even the predicted faint constrictions of the stripes along the [001] direction related to the pair of second-smallest reciprocal lattice vectors are visible in the measurement. Again, experimental and theoretical data can be compared more quantitatively by drawing line sections along the dense-packed [1 1̄ 1] direction (see Figure 1.16b). The result, which is plotted in Figure 1.16c, reveals that the periodicity, when measured with a Fe-coated probe, is twice the nearest-neighbor distance (i.e. 0.548 nm).

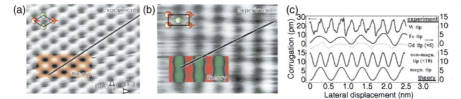

Figure 1.16 Comparison of experimental and theoretical STM images of a Mn ML on W(110) with (a) a nonmagnetic W tip and (b) a magnetic Fe tip; (c) Experimental and theoretical line sections for the images in (a) and (b). The unit cell of the calculated magnetic ground-state configuration is shown in (a) and (b) for comparison. The image size is 2.7 × 2.2 nm. (Reprinted with permission from Ref. [96]; AAAS.)

The pronounced dependence of the effect on the magnetization direction of the tip can be exploited to gain further information on the magnetization direction of the sample. This is done by using a tip that exhibits an easy magnetization axis that is almost perpendicular to the one of the sample surface. This condition is fulfilled by a W tip coated with about 7 ML Gd [29]. The gray line in Figure 1.16c represents a typical line section as measured with such a Gd-coated probe tip. Indeed, the corrugation amplitude was always much smaller than that for Fe-coated tips and never exceeded 1 pm, thus supporting the theoretical results that the easy axis of the Mn atoms is in-plane.

In the following section it will be shown, by reference to the studies of Yang et al. [98] and Smith et al. [99], that both magnetic and nonmagnetic atomic-scale information can be obtained simultaneously in the constant current mode for another Mn-based system which consists of Mn_3N_2 films grown on MgO(001) with the c axis parallel to the growth surface, which is (010). The surface geometrical unit cell, containing six Mn atoms and four N atoms (3 : 2 ratio), can be denoted as c(1 × 1), whereas the surface magnetic unit cell is just (1 × 1).

The bulk structure of Mn_3N_2 exhibits a face-centered tetragonal (fct) rock salt-type structure. The bulk magnetic moments of the Mn atoms are ferromagnetic within (001) planes, lie along the [100] direction, and are layerwise antiferromagnetic along [001]. Besides the magnetic superstructure, every third (001) layer along the c direction has all N sites vacant, which results in a bulk unit cell exhibiting $c = 12.13$ Å (six atomic layers).

Figure 1.17a presents a SP-STM image [98] of the surface acquired using a Mn-coated W tip thus being sensitive to the in-plane direction. Although the row structure with period $c/2$ is observed, a modulation with period c of the height of the rows is additionally obvious. The modulation shown in Figure 1.17b is evident for both domains D1 and D2 by the area-averaged line profiles taken from inside the boxed regions on either side of the domain boundary. For the domain D1 (red line), the modulation amplitude is about a factor of 2 larger than for the domain D2 (blue line). As the height modulation is proportional to $m_t m_s \cos\theta$, with m_t and m_s being the

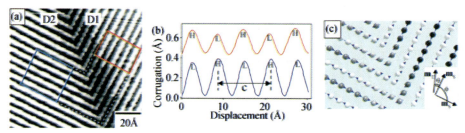

Figure 1.17 (a) SP-STM image acquired using a Mn-coated W tip; (b) Two area-averaged line profiles (red and blue) corresponding to the regions inside the red and blue rectangles in (a); (c) Simulated SP-STM map: contrast: white ↔ black ⇒ θ: 0 ↔ π. The inset shows the moments of tip ($\vec{m}_t$) and the sample ($\vec{m}_s$) for the two different domains and the angles between them. Each ball represents a magnetic atom. (Reprinted with permission from Ref. [98]; copyright (2002), American Physical Society.)

moment of the tip and sample, respectively, and θ the angle between them (cf. Equation 1.8), it is simple to show that $\theta = \arctan(\Delta z_2/\Delta z_1)$, where z_1 and z_2 are the height modulation in domains D1 and D2, respectively. In the case shown here, with $\Delta z_1 = 0.04$ Å and $\Delta z_2 = 0.02$ Å, θ amounts to ≈27°.

A high peak (H) on one side of the domain boundary converts to a low peak (L) on the other side. This inversion is simulated in Figure 1.17c by a simple antiferromagnetic model configuration of spin moments and a tip spin at the angle $\theta = 27°$. The gray scale for each magnetic atom is proportional to $m_t m_s \cos\theta$ (white: $\theta = 0$; black: $\theta = \pi$). Clearly, the inversion occurs when the difference $\phi - \theta = \pi/2$, where θ and φ are the two different angles between tip and sample moments in domains D1 and D2, respectively.

The data can now be used to separate the magnetic and nonmagnetic components. Beginning with the SP-STM image shown in Figure 1.18a [98], the average height profile $z(x)$ where x is along [001] (Figure 1.18b, dark blue line) and also $z(x + c/2)$ (Figure 1.18b, light blue line) are plotted. Clearly, by taking the difference and sum of these two functions, the magnetic component with periodicity c and the nonmagnetic component with period $c/2$ can be extracted: $m_t m_s \cos[\theta(x)] \sim [z(x) - z(x + c/2)]/2$. This is further justified if it is assumed that the bulk magnetic symmetry is maintained at the surface. When using this procedure, the resulting magnetic profile for the data of Figure 1.18 has a period of c and a trapezoidal wave shape, as shown in Figure 1.18c (violet line). The nonmagnetic profile is also shown in

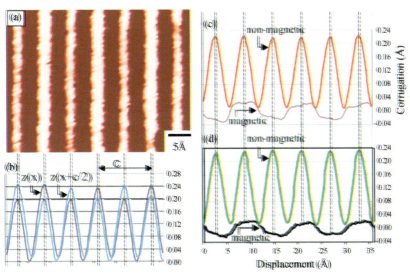

Figure 1.18 (a) SP-STM image acquired using a Mn-coated W tip; (b) Area-averaged line profile $z(x)$ of the whole image of (a) (dark blue), and $z(x + c/2)$ (light blue); (c) The resulting nonmagnetic component (red) and magnetic component (violet) for the Mn-coated tip; (d) Nonmagnetic (green) and magnetic (black) components for the Fe-coated tip on a similar sample region. (Reprinted with permission from Ref. [98]; copyright (2002), American Physical Society.)

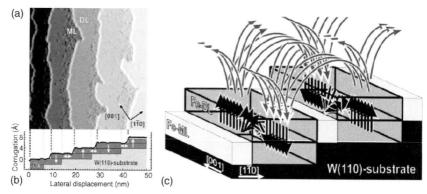

Figure 1.19 (a) Topographic STM image (scan range: 50 nm × 50 nm) of 1.6 ML Fe/W(110) after annealing to 450 K; (b) Line section measured at the bottom edge of the STM image. The local coverage alternates between one and two atomic layers. White arrows symbolize the easy magnetization directions of the mono- and the double layer; that is, in-plane and perpendicular to the surface, respectively; (c) Adjacent perpendicularly magnetized DL stripes exhibit an antiparallel dipolar coupling. Within domain walls the Fe DL on W(110) locally exhibits an in-plane magnetization. (Reprinted with permission from Ref. [104]; Elsevier.)

Figure 1.18c (red line) exhibiting a period of $c/2$ and a sinusoidal shape, the same as for the average line profile acquired with a nonmagnetic tip. The magnetic component amplitude is about 20% of the nonmagnetic component amplitude.

1.6
Magnetic Properties of Nanoscaled Wires

The behavior of perpendicularly magnetized Fe double layer nanowires epitaxially grown on a stepped W(110) single crystal [29, 55, 69, 100–103] with an average terrace width of about 9 nm is presented as an example of magnetic wires exhibiting a width in the nanometer range (see Figure 1.19a) [104]. This study was carried out at low temperatures, below about 10 K. At higher temperatures a reorientation to an in-plane easy axis occurs with the spin reorientation temperature being coverage-dependent for samples with a coverage between 1.5 and 2.2 atomic layers [105]. The sample was prepared by the deposition of 1.6 ML Fe on the W(110) substrate held at elevated temperature of 450 K. This preparation procedure leads to step-flow growth of the second Fe ML on top of the closed first Fe layer, thereby creating a system of nanowires with alternating Fe ML and DL coverage elongated along the step edges of the substrate (this situation is shown schematically in Figure 1.19b, which also contains the line section corresponding to the line shown in Figure 1.19a). The coverage range between 1.4 and 1.8 ML Fe/W(110) is characterized by magnetic saturation at relatively low external perpendicular fields combined with the absence of a hysteresis – that is, zero remanence. As shown schematically in Figure 1.19c, this antiparallel order is a consequence of the dipolar coupling which reduces the

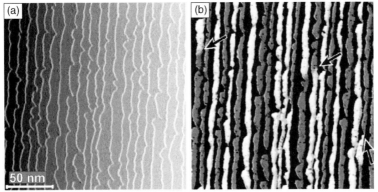

Figure 1.20 (a) STM topograph and (b) magnetic dI/dU image of Fe nanowires on W(110). Both images were measured simultaneously. The sample exhibits a demagnetized antiferromagnetic ground state which is energetically favorable due to flux closure between adjacent perpendicularly magnetized Fe nanowires [106]. (Reprinted with permission from Ref. [104]; Elsevier.)

magnetic stray field of the perpendicularly magnetized Fe DL. At domain walls the magnetization vector may locally be oriented along the hard magnetic axis – that is, in-plane.

Tunneling spectroscopy was used to image the corresponding magnetic domain structure. Since it is known from full dI/dU spectroscopy curves how the contrast must be interpreted (see Section 1.3.2), it is no longer necessary to measure the entire spectra at every pixel of the image as this is very time-consuming (about 10–20 h per image for the investigation discussed here [104]). Instead, the dI/dU signal at a fixed sample bias already gives a good contrast. Figure 1.20 shows the simultaneously recorded topography (panel a) and the dI/dU signal (panel b) of 1.5 ML Fe/W(110). The measurement time for this image was about 30 min. Due to its different electronic properties, the Fe ML appears dark, but this is not related to the magnetic properties. Instead, the ML is known to exhibit an in-plane magnetization [107] which cannot be detected using Gd-tips which are sensitive only to out-of-plane magnetization [29]. Clearly, the magnetic domain structure is dominated by DL nanowires which are magnetized alternately up and down, although exceptions from this model can easily be recognized. Several domain walls within single Fe nanowires are visible; some of these are marked with arrows in Figure 1.20b. There are also numerous adjacent nanowires which couple ferromagnetically rather than antiferromagnetically. It is likely that these DL nanowires approach very close to each other – or may even touch – so that the exchange coupling dominates.

Imaging by SP-STM can also be used to deduce macroscopic magnetic properties, a situation demonstrated by Pietzsch *et al.* [104, 108] for a system of Fe nanowires as just discussed above. These authors showed that spin-resolved dI/dU maps in a varying external field could be used to obtain the magnetic hysteresis curve of the

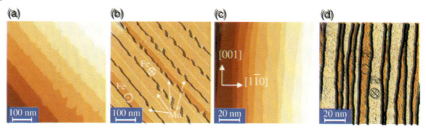

Figure 1.21 (a, c) Topographic STM images and (b, d) simultaneously measured differential conductance d*I*/d*U* maps of 0.5 ps-Fe (a, b) and of 1.0 ps-ML Fe (c, d) grown on Mo(110) at 700 K, respectively. Images (c, d) have been measured on the vicinal surface of the Mo substrate. Black and white lines in (b, d) located at step edges are artifacts due to scanning too quickly over the step edges. The black dots observed in (d) are due to adsorbates [24, 107]. Images and conductance maps were measured using a W/10 ML Au/4 ML Co magnetic tip. The Fe nanostructures reveal an out-of-plane magnetic contrast. (Reprinted with permission from Ref. [109]; copyright (2005), American Physical Society.)

surface area under investigation; that is, SP-STM enables the observation of magnetic hysteresis down to the nanometer scale.

Replacing W(110) with Mo(110) provides the unique possibility of observing the modification of magnetic properties of the Fe nanostructures, but leaving the structure and morphology almost unaffected [33, 109]. The magnetic easy axis is directed along the [001] direction for Fe/Mo(110) [110], while the easy axis is [1 1̄ 0] for Fe/W(110) films [111]. The pseudomorphic ML (ps-ML) Fe/Mo(110) nanostructures are perpendicularly magnetized at low temperatures [112], whereas the ps-ML Fe/W (110) is magnetized in-plane along the [1 1̄ 0] direction [113].

Figure 1.21a shows the topography and Figure 1.21b the simultaneously recorded d*I*/d*U* map of 0.5 ps-ML Fe deposited onto the Mo(110) single crystal at 700 K [109]. Monatomic Mo terraces decorated with the regular narrow Fe nanostripes grown by step-flow growth at the step edges are visible. The location of the Fe atoms on the Mo (110) surface is better visible on the d*I*/d*U* map (see Figure 1.21b) due to the element specific contrast resulting from the differences of the spin-averaged d*I*/d*U* signal which are connected with the local electronic surface properties that are different for Fe and Mo [112]. Uncovered Mo surfaces are indicated in Figure 1.21b by white arrows. The Fe nanowires show two different colors, representing two different values of the local d*I*/d*U* signal for equivalent surface regions (ML Fe/Mo(110)) for which the spin-averaged conductance signals should be the same [24]. All STM images and conductance maps shown in Figure 1.21 were measured using W tips covered by 10 ML Au and subsequently by 4 ML Co. It is known [114] that 4 ML Co prepared on W(110)/Au reveal an out-of-plane magnetic easy axis. Therefore, it may be expected that the magnetization of the tip would show perpendicular to the front plane of the tip – that is, along the tip axis – leading to an out-of-plane magnetic sensitivity of the tip. The contrast observed for the Fe nanostructures results from the perpendicularly magnetized Fe nanostripes, in agreement with Ref. [112].

The perpendicularly magnetized ML Fe nanostripes shown in Figure 1.21b are not antiferromagnetically ordered; that is, only two of the stripes are magnetized 'up', whereas the orientation of the magnetization for the remaining stripes shows in the opposite direction ('down'). This means that the dipolar coupling between adjacent ML Fe nanowires is weak. The strength of the dipolar coupling between adjacent stripes increases with the stripe width and decreases with the distance between adjacent stripes [113]. The distance between adjacent ML Fe nanowires can be diminished down to a minimum by an increase of Fe coverage up to 1 ML. The topography of the 1 ML Fe deposited onto the vicinal surface of the Mo(110) crystal is presented in Figure 1.21c. Narrow ML Fe nanowires obtained on the vicinal surface are antiferromagnetically ordered, as demonstrated on the conductivity map (see Figure 1.21d).

1.7
Nanoscale Elements with Magnetic Vortex Structures

The domain structure of nanoscale magnetic elements has attracted considerable interest. The dependence of the domain structure on shape [115, 116], size [117] and edge structure [118] has been explored in many experiments. Further information on nanoislands gained by SP-STM are provided concerning Fe islands on W (110) [102, 119, 120], Fe islands on Mo(110) [112, 121], Co islands on Cu(111) [122, 123], Co islands on Pt(111) [124–126], Co islands on Au(11) [14], and FeAu alloy islands on Mo(110) [127, 128].

Due to their small dimensions, these elements are often superparamagnetic. This issue is addressed in Refs [112, 121], making use of time-dependent SP-STM studies. In a further development, Krause *et al.* showed [129] that superparamagnetic Fe nanoislands with typical sizes of 100 atoms could be addressed and locally switched using a magnetic probe tip. SP-STM thus provides an improved understanding of the underlying mechanism due to the feasibility to separate and quantify three fundamental contributions involved in magnetization switching, namely the current-induced spin torque, heating the island by the tunneling current, and Oerstedt field effects.

However, the influence of the thickness in conjunction with the magnetocrystalline anisotropy concerning nanoscale elements has rarely been studied [130]. Micromagnetic calculations have shown [131] that the lowest-energy domain configuration of permalloy rectangles depends critically on the thickness. With increasing thickness, a transition from the so-called C-state via the Landau-type or vortex configuration into a diamond state (double-vortex) was found. This behavior is caused by the thickness-dependent contribution of the magnetostatic energy, which must be paid wherever the magnetization is perpendicular to the element's rim. At a certain critical thickness it is energetically favorable to avoid the stray field by magnetizing the element along the edges throughout the entire particle, leading to a so-called *flux closure arrangement*.

This thickness-dependent behavior was corroborated experimentally by Bode *et al.* [120] using Fe islands on W(110). The left column of Figure 1.22 [120] shows the topography for different heights of the Fe islands, with the mean island height h

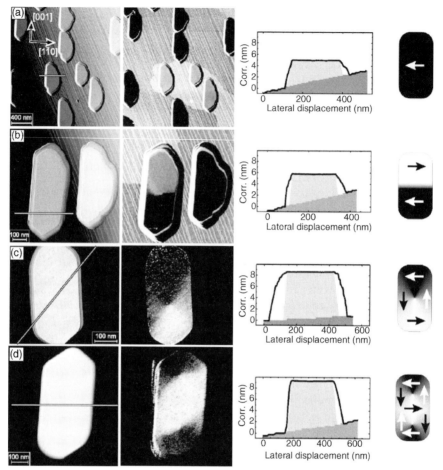

Figure 1.22 Topographic images (first column), spin-resolved dI/dU maps (second column) and topographic line sections (third column) of Fe islands on W(110) with different mean island heights h: (a) 53.5 nm, (b) 54.5 nm, (c) 57.5 nm, and (d) 58.5 nm. The data were obtained at $T = 14$ K. The resulting island domain configurations are schematically represented in the fourth column. (Reprinted with permission from Ref. [120]; copyright (2004), American Institute of Physics.)

varying between 3.5 and 8.5 nm. The lateral dimensions of the islands, irrespective of their height, are almost equal. The islands shown in Figure 1.22a exhibit an average height of approximately 3.5 nm (see line section). In the right-hand panel of Figure 1.22a the different magnetization states of islands can be distinguished by means of different dI/dU intensities. This variation results from spin-polarized tunneling between the magnetic STM tip (due to a coating with more than 100 ML Cr the tip is sensitive to the in-plane direction [28]) and the magnetic sample, and therefore represents different relative in-plane orientations of the magnetization in tip and sample. As no significant variation was found on top of the atomically flat island surface, it would appear that these Fe islands are single domain, as shown schematically in the right-hand panel of Figure 1.22a. Evidently, there exists a close correlation between the magnetization direction of individual Fe islands and the surrounding Fe ML; dark (bright) Fe islands are always surrounded by a dark (bright) ML. With increasing thickness the magnetic pattern of the Fe islands becomes more and more complex such that, at $h = 54.5$ nm (see Figure 1.22b) a two-domain state is present. The island in Figure 1.22c exhibits a height h of 57.5 nm, while the corresponding spin-resolved dI/dU map shows the typical pattern of a single vortex state. A diamond state is found on the even higher island shown in Figure 1.22d ($h = 58.5$ nm).

Thus, the magnetic ground state is expected to be a vortex, as can be understood by the following consideration. If the dimensions of the particles are too small they do not form a single domain state, as this would require a relatively high stray field (or dipolar) energy. On the other hand, if the dimensions were too large, such domains would be formed such as those found in macroscopic pieces of magnetic material, because the additional cost of domain wall energy cannot be compensated by the reduction in stray field energy. By exhibiting a vortex configuration, the magnetization curls continuously around the particle center, drastically reducing the stray field energy and avoiding domain wall energy. Yet, the question arises as to the diameter of this core. An earlier investigation conducted by Shinjo et al. [132], using magnetic force microscopy, suggested an upper limit of about 50 nm caused by the intrinsic lateral resolution that was due to detection of the stray field. In the following section, it will be seen that an enhanced lateral resolution can be obtained using the technique of SP-STM, as shown by Wachowiak et al. [28, 102].

In order to gain a detailed insight into the magnetic behavior of the vortex core, a zoom into the central region was carried out where the rotation of the magnetization into the surface normal is expected. Maps of the dI/dU signal measured with different Cr-coated tips that are sensitive to the in-plane and out-of-plane component of the local sample magnetization are shown in Figure 1.23a and b, respectively [28]. The dI/dU signal as measured along a circular path around the vortex core (the circle in Figure 1.23a) exhibits a cosine-like modulation, indicating that the in-plane component of the local sample magnetization curls continuously around the vortex core. Figure 1.23b, which was measured with an out-of-plane sensitive tip on an identically prepared sample, exhibits a small bright area approximately in the center of the island. Therefore, the dI/dU map of Figure 1.23b confirms that the local magnetization in the vortex core is tilted normal to the surface (Figure 1.23c). The line section across the vortex core enables the width to be determined at about 9 nm, which is not restricted due to lateral resolution.

1.8
Individual Atoms on Magnetic Surfaces

The measurement of spin polarization states of individual atoms and understanding how atomic spins behave in a condensed matter environment, are essential steps

Figure 1.23 Magnetic dI/dU maps as measured with an (a) in-plane and an (b) out-of-plane sensitive Cr tip. The curling in-plane magnetization around the vortex core is recognizable in (a) and the perpendicular magnetization of the vortex core is visible as a bright area in (b); (c) Schematic arrangement of a magnetic vortex core. Far from the vortex core the magnetization curls continuously around the center with the orientation in the surface plane. In the center of the core the magnetization is perpendicular to the plane (highlighted). (Reprinted with permission from Ref. [28]; AAAS.)

towards the creation of devices, the functionality of which can be engineered at the level of individual atomic spins.

The direct observation of a spin polarization state of isolated adatoms remains challenging because isolated atoms have a low magnetic anisotropy energy that causes their spin to fluctuate over time due to environmental interactions. In the following section, measurements made by Yayon *et al.* [133] concerning the spin polarization state of *individual* Fe and Cr adatoms on a metal surface, are described. In order to fix the adatom spin in time, the adatoms were deposited onto ferromagnetic Co nanoislands, thereby coupling the adatom spin to the island magnetization through the direct exchange interaction.

Cobalt islands were chosen as a calibrated substrate where different magnetization states ('up' ↑ and 'down' ↓ with respect to the surface plane) are easily accessed [122]. The left-hand part of Figure 1.24 shows a representative topograph of Fe adatoms adsorbed onto triangular Co islands on the Cu(111) surface. The spatial oscillations seen on the Cu(111) surface are due to interference of surface state electrons scattered from the adatoms and Co islands [134].

In the right-hand part of Figure 1.24, panel (a) shows a color-scaled spin-polarized dI/dU map, together with topographic contour lines (measured simultaneously) for Fe and Cr atoms codeposited on two Co islands with opposite magnetization. The Fe and Cr atoms can easily be distinguished by their topographic signatures (Cr atoms

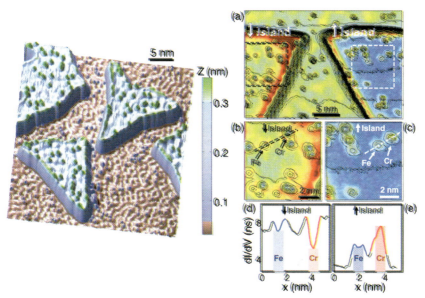

Figure 1.24 Left: Topograph of Fe adatoms adsorbed onto triangular Co islands on Cu(111) at $T = 4.8$ K. Fe adatoms are seen as green protrusions on the Co islands and blue protrusions on the bare Cu(111) surface. Right: (a) Spin-polarized dI/dU map of Fe and Cr adatoms on ↓ and ↑ Co islands on Cu(111); (b, c) Zoom-ins of areas marked by dashed lines on ↓ and ↑ islands in (a); (d, e) Line scans through the centers of Fe and Cr adatoms on ↓ and ↑ islands, respectively (marked by dashed lines in b and c). (Reprinted with permission from Ref. [133]; copyright (2007), American Physical Society.)

protrude 0.07 nm from the island surface, while Fe atoms protrude 0.04 nm). Spin-contrast between adatoms sitting on the two islands is seen as line cuts through Fe and Cr atoms (see Figure 1.24b–e). Fe atoms sitting on the ↓ island exhibit a larger dI/dU signal than those on the ↑ island, while Cr atoms on the ↓ island show a smaller dI/dU signal than those on the ↑ island. This confirms the parallel nature of the Fe/Co island spin coupling and the antiparallel nature of the Cr/Co island spin coupling over this energy range. SP-STS thus clearly reveals single adatom spin contrast: Each type of adatom yields a distinct spectrum, and over the energy range of the Co island surface-state Fe and Cr adatoms show opposite spin polarization. However, this measurement does not unambiguously determine the direction of the total spin of the adatom because the total spin is an integral over all filled states, whereas the spectra shown here were recorded over a finite energy range.

For a better understanding of the magnetic coupling between adatoms and islands, a density functional theory (DFT) calculation was also carried out [133]. The adsorption energies of Fe and Cr atoms on a ferromagnetic 2 ML film of Co on Cu(1 1 1) were calculated with the adatom moment fixed parallel and antiparallel to the magnetization of the Co film. The resulting values (see Figure 1.25) showed that Fe adatoms preferred a ferromagnetic alignment to the Co film, while Cr adatoms preferred an antiferromagnetic alignment. Comparison with the spin-polarized measurements implied that the Fe and Cr adatoms exhibit a negative spin polarization over the energy range of the Co island surface state.

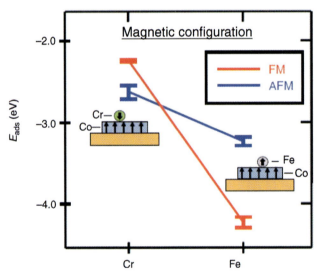

Figure 1.25 Calculated binding energies of ferromagnetic and antiferromagnetic configurations for Fe and Cr adatoms on a 2 ML high Co film on Cu(111). Error bars indicate the energy difference between hcp and fcc adatom adsorption sites. Cartoons depict the lowest-energy magnetic coupling configuration for Fe and Cr adatoms on the Co film. (Reprinted with permission from Ref. [133]; copyright (2007), American Physical Society.)

As discussed in Section 1.5.1, the Cr(001) surface exhibits a topological antiferromagnetic order. By increasing the number of adatoms, however, a small proportion of the Fe atoms on this surface will also exhibit an antiferromagnetic coupling to the underlying Cr(001) terraces [135, 136].

It is known from spin-polarized photoemission experiments that even nonmagnetic atoms such as oxygen (see Ref. [137] for O/Fe(110) and Ref. [138] for O/Co (0001)), sulfur [139] and iodine [140] become polarized upon chemisorption onto ferromagnetic surfaces. For each of these systems, SP-STM allows a deeper insight on the basis of its atomic resolution. For example, it was found that individual oxygen atoms on an Fe DL would induce highly anisotropic scattering states which were of minority spin character only [141]. This spin-dependent electron scattering at the single impurity level opens the possibility of understanding the origin of magnetoresistance phenomena on the atomic scale.

In the case of Fe islands, it has been reported that magnetic domains can be observed even after the deposition of a sulfur layer [142], and can act as a passivation species. These findings can be understood on the basis of the above discussion, also taking into account the fact that spin-polarized electrons from the interface with binding energies near the Fermi level are not fully damped but rather exhibit an attenuation length of at least several monolayers. Additionally, this mean free path is spin-dependent [143], such that an appropriate adsorbate layer may allow to extend the SP-STM to operate even under ambient conditions.

1.9
Domain Walls

The motion of domain walls is often hindered by lattice defects, leading to Barkhausen jumps in magnetic hysteresis curves. By using a high lateral resolution, the effective pinning of domain walls by screw-and-edge dislocations was first presented by Krause *et al.* [144] for Dy films on W(110).

Here, we will describe the details of two further aspects concerning domain walls which require an effective lateral resolution of SP-STM. First, we report on the behavior of domain walls in external magnetic fields, as investigated by Kubetzka *et al.* [103]; second, we will outline details of the widths of domain walls in nanoscale systems, referring to the studies conducted by Pratzer *et al.* [107].

The formation and stability of 360° domain walls plays a crucial role in the remagnetization processes of thin ferromagnetic films, with possible implications for the performance and development of magnetoresistive and magnetic random access memory (MRAM) devices. These are formed in external fields applied along the easy direction of the magnetic material when pairs of 180° walls with the same sense of rotation are forced together. Their stability against remagnetization into the uniform state is a manifestation of a hard axis anisotropy perpendicular to the rotational plane of the wall. This anisotropy may be of crystalline origin or – in films with an in-plane magnetic easy direction – due to the shape anisotropy.

Within Fe DL wires on W(110) that are separated by narrow regions with ML coverage (see Figure 1.26a [103]), two types of 180° walls can be distinguished using a ferromagnetic probe tip prepared by coating a W tip with several ML of Fe and therefore being sensitive to the in-plane magnetization component [27]. The wall width amounts to $w = 7$ nm. The intermediate dI/dU signal (gray) corresponds to a perpendicular magnetization oriented either up or down. These two cases cannot be distinguished with a tip exhibiting a pure in-plane sensitivity unless the symmetry is broken by an external field. Figure 1.26 shows dI/dU maps in an increasing perpendicular magnetic field of up to 800 mT. Areas magnetized parallel to the field direction grow at the expense of antiparallel ones, and pairs of 180° walls are forced together, which is equivalent to the formation of 360° walls. As expected, their lateral extension decreases with increasing field value. A closer inspection of these field-dependent measurements reveals that: (i) the magnetization rotates along every single nanowire with a defined chirality; and that (ii) the rotational sense is the same in each of the 12 wires within the imaged area. However, as the azimuthal angle of the tip magnetization is unknown, the absolute sense of rotation cannot be determined. For the same reason, it cannot be decided on the basis of these data alone whether the walls are of Bloch or Néel type, although the facts that the closed DL film is magnetized in-plane along [1 1̄ 0] at elevated temperatures and the domain walls are oriented along the same direction, are indicative of their Bloch-type character. The first of the above observations is to be expected for stability reasons, as neighboring walls of opposite chirality (unwinding or untwisted walls [108]) will attract each other and can easily annihilate, in contrast to winding walls. As a consequence, the cooling process of the sample from above the Curie temperature to the measurement temperature of 14 K will result in a defined chirality within every individual wire, since such a structure is more stable against thermal fluctuations. The observed average distance between neighboring walls does not therefore necessarily reflect the magnetic ground state at low temperatures, as it might be a relic from the cooling process which is effectively frozen in a metastable state. The second of the observations will be discussed in Section 1.10.

With increasing external magnetic field, the tip's magnetization is successively rotated from in-plane towards the perpendicular direction. Its in-plane direction is also reversed during data acquisition at 400 mT (see Figure 1.26, black arrow), and this causes an inverted contrast for the remaining upper part of the image. At this field value a group of five 360° walls has formed a row (see ellipse), a correlation that might arise from their in-plane stray field. At 800 mT most of the 360° walls within the scanned area have been remagnetized by a rotation via the hard [001] in-plane direction.

Attention is now drawn to the Fe ML located in-between the Fe DL. Figure 1.27 [107] shows the topography (panel a) and the magnetic dI/dU signal (panel b) of 1.25 ML Fe/W(110) grown at $T = 500$ K. Several domain walls separating dark and bright domains of the Fe ML can clearly be recognized in the overview of Figure 1.27b. Because of their different electronic properties, the DL stripes appear dark at this particular sample bias. At approximately the center of the white box in Figure 1.27b can be seen a bright spot; this is caused by a domain wall in this particular DL. The

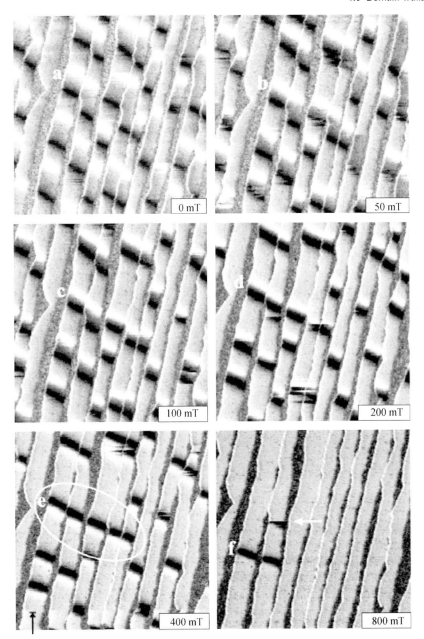

Figure 1.26 d*I*/d*U* maps of the surface area exhibiting Fe DL wires on W(110) imaged in an increasing perpendicular magnetic field. Pairs of 180° domain walls are gradually forced together, which is equivalent to the formation and compression of 360° walls. At 800 mT, most of these have vanished; that is, the Fe film is in magnetic saturation. With increasing external magnetic field the tip's magnetization is gradually forced from the in-plane towards the perpendicular direction. (Reprinted with permission from Ref. [103]; copyright (2003), American Physical Society.)

1 Spin-Polarized Scanning Tunneling Microscopy

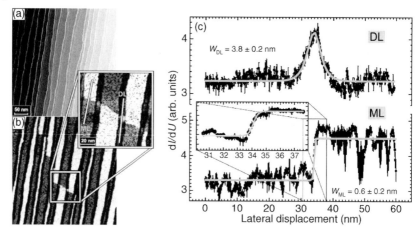

Figure 1.27 (a) Topographic and (b) spin-resolved dI/dU image showing the in-plane magnetic domain structure of 1.25 ML Fe/W (110). Several ML and DL domain walls can be recognized in the higher magnified inset; (c) Line sections showing domain wall profiles of the ML (bottom) and the DL (top). The inset shows that the ML domain wall width is on the atomic scale, with $w_{ML} = 6 \pm 2$ Å. (Reprinted with permission from Ref. [107]; copyright (2001), American Physical Society.)

inset of Figure 1.27b presents this location at higher magnification. Averaged line sections drawn along the white lines across domain walls in the ML and the DL are plotted in Figure 1.27c lower and upper, respectively, where clearly the ML domain wall is much narrower than the DL wall. The inset of Figure 1.27c shows the data in the vicinity of the ML domain wall in greater detail, and reveals a domain wall width $w < 1$ nm. In order to allow a more quantitative discussion the measured data have been fitted with a theoretical tanh function of a 180° wall profile [145]. This can be extended to an arbitrary angle between the magnetization axis of tip and sample ϕ [107, 146] by

$$y(x) = y_0 + y_{sp} \cos\left\{\arccos\left[\tanh\left(\frac{x-x_0}{w/2}\right)\right] + \phi\right\} \tag{1.14}$$

where $y(x)$ is the dI/dU signal measured at position x, x_0 is the position of the domain wall, w is the wall width, and y_0 and y_{sp} are the spin-averaged and spin-polarized dI/dU signals, respectively. Due to the Fe-coated tip which exhibits in-plane sensitivity, it is known that $\phi_{DL} = \pi/2$ and $\phi_{ML} = 0$. The best fit to the wall profile of the DL is achieved with $w_{DL} = 3.8 \pm 0.2$ nm. The profile of the ML domain wall is much narrower. If the fit procedure is performed over the full length of the line section $w_{ML} = 0.50 \pm 0.26$ nm, whereas $w_{ML} = 0.66 \pm 0.18$ nm is found if the fit is applied to the data in the inset of Figure 1.27c; this confirms the result of the analysis of the magnetization curves – that is, an almost atomically sharp domain wall.

A domain wall width of only six to eight atomic rows was also observed for an antiferromagnetic Fe monolayer on W(001) [147]. Such a narrow domain wall width can, in theory, be understood to arise from band structure effects, also taking into account intra-atomic noncollinear magnetism [148].

With regards to noncollinear effects it is important to distinguish between inter-atomic and intra-atomic magnetism. The first type is well known, and has been observed experimentally for small magnetic clusters [149] and in magnetic layers [150]; it has also been described on a theoretical basis [56, 151–157]. Inter-atomic magnetism can be understood within the Heisenberg model, taking into account *atomic* magnetic moments which are nonparallel for different atoms. Intra-atomic noncollinear effects arise from the tunneling current which flows through orbitals of the *same* atom, whilst that directly at the tip apex possesses a spin density orientation that is *noncollinear* [55].

1.10
Chiral Magnetic Order

Due to the inversion symmetry of bulk crystals, homochiral spin structures are unable to exist. However, as low-dimensional systems lack structural inversion symmetry, these single-handed spin arrangements may occur due to the Dzya-loshinskii–Moriya interaction [158, 159] arising from the spin-orbit scattering of electrons in an inversion asymmetric crystal field. This observation is now discussed with reference to the studies of Bode *et al.* [160], which were carried out in the same system Mn/W(110) for which atomic resolution of the magnetic properties had been demonstrated [96] (see Section 1.5.2).

In metallic itinerant magnets, spin-polarized electrons of the valence band hop across the lattice and exert the Heisenberg exchange interaction between magnetic spin moments $\vec{S}$ located on atomic sites i and j:

$$E_{exch} = \sum_{i,j} J_{ij}\, \vec{S}_i \cdot \vec{S}_j \tag{1.15}$$

As a consequence of a coulombic interaction, the exchange interaction is isotropic. Owing to the presence of a spin–orbit interaction, which connects the lattice with the spin symmetry, the broken parity of the lattice at an interface or surface gives rise to an additional interaction that breaks the inversion invariance of the Heisenberg Hamiltonian in Equation 1.15. This Dzyaloshinskii–Moriya interaction [158, 159]

$$E_{DM} = \sum_{i,j} \vec{D}_{ij} \cdot (\vec{S}_i \times \vec{S}_j) \tag{1.16}$$

arises from the spin–orbit scattering of hopping electrons in an inversion asymmetric crystal field (where $\vec{D}_{ij}$ is the Dzyaloshinskii vector). In such an environment the scattering sequence of spin-polarized electrons, for example $i \rightarrow j \rightarrow i$ versus $j \rightarrow i \rightarrow j$, is noncommutative. The presence of this interaction has far-reaching consequences. Depending on the sign, the symmetry properties and the magnitude of $\vec{D}_{ij}$, uniaxial ferromagnetic or antiferromagnetic structures fail to exist and are instead replaced by a directional noncollinear magnetic structure of one specific chirality $\vec{C} = \vec{S}_i \times \vec{S}_{i+1}$, being either right-handed ($C > 0$) or left-handed

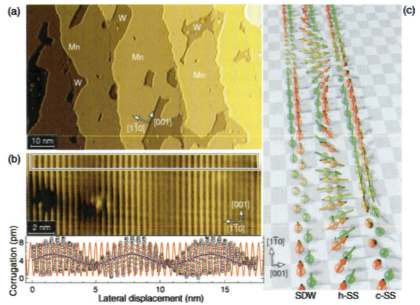

Figure 1.28 SP-STM of the Mn monolayer on W(110) and potential spin structures. (a) Topography of 0.77 atomic layers of Mn on W(110); (b) High-resolution constant-current image (upper panel) of the Mn monolayer taken with a Cr-coated tip. The stripes along the [001] direction are caused by spin-polarized tunneling between the magnetic tip and the sample. The averaged line section (lower panel) reveals a magnetic corrugation with a nominal periodicity of 0.448 nm and a long wavelength modulation. Comparison with a sine wave (red), expected for perfect antiferromagnetic order, reveals a phase shift of π between adjacent antinodes. In addition, there is an offset modulation (blue line) which is attributed to a varying electronic structure owing to spin–orbit coupling; (c) Artist's impression of the considered spin structures: a spin density wave (SDW), a helical spin spiral (h-SS) and a left-handed cycloidal spin spiral (c-SS). (Reprinted with permission from Ref. [160]; copyright (2007), MacMillan Publishers Ltd.)

($C<0$). In fact, J, D and also the anisotropy constants K, create a parameter space containing magnetic structures of unprecedented complexity [161], including 2-D and 3-D cycloidal, helicoidal or toroidal spin structures, or even vortices.

Figure 1.28a shows the topography of 0.77 atomic layers of Mn grown on a W(110) substrate [160]. The magnetic structure can be directly imaged with SP-STM using magnetically coated W tips. Figure 1.28b shows a high spatial resolution constant-current image measured on the atomically flat Mn layer using a Cr-coated probe tip which is sensitive to the in-plane magnetization [28]. The SP-STM data reveal periodic stripes running along the [001] direction, with an inter-stripe distance matching the surface lattice constant along the [1 $\bar{1}$ 0] direction (this was discussed earlier in Section 1.5.2). The additional important observation is, however, that the line section in the lower panel of Figure 1.28b, representing the magnetic amplitude, is not *constant* but is rather *modulated*, with a period of about 6 nm. Further, the magnetic corrugation is not simply a symmetric modulation superimposed on a constant offset I_0. Instead, an additional long-wave modulation of I_0 (blue line) is present which is ascribed to spin–orbit coupling-induced variations of the spin-averaged electronic structure. When using in-plane-sensitive tips, the minima of the magnetic corrugation are found to coincide with the minima of the long-wave modulation of the spin-averaged local density of states. Within the field of view (see Figure 1.28b), three antinodes of the magnetic corrugation are visible. Comparing the experimental data with a sine function (red), representing perfect antiferromagnetic order, reveals a phase shift of π between adjacent antinodes.

The long wavelength modulation of the magnetic amplitude observed in Figure 1.28b may be explained by two fundamentally different spin structures: (i) a spin density wave (SDW) as it occurs, for example, in bulk Cr; or (ii) a spin spiral. Whereas, a SDW is characterized by a sinusoidal modulation of the size of the magnetic moments and the absence of spin rotation, the spin spiral consists of magnetic moments of approximately constant magnitude but whose directions rotate continuously. Spin spirals that are confined to a plane perpendicular or parallel to the propagation direction are denoted as either helical spirals (h-SS) or cycloidal spirals (c-SS). Figure 1.28c shows an artist's impression of a SDW, a h-SS and a c-SS. The magnetic contrast vanishes in either case twice over one magnetic period because: (i) the sample magnetic moments themselves vanish periodically; or (ii) the magnetic moments beneath the tip apex are orthogonal with respect to the tip magnetization m_t. The two cases can, however, be distinguished by addressing different components of the sample magnetization. Whereas, in case (i) a maximum spin contrast is always achieved at lateral positions where the magnetic moments are largest, and independent of m_t, in case (ii) a rotation of m_t can shift the position of maximum spin contrast.

Such a rotation of m_t can be achieved by subjecting an in-plane-sensitive Fe-coated tip to an appropriate external magnetic field (see sketches in Figure 1.29 [160]). For samples without a net magnetic moment, it is expected that the sample magnetization would remain unaffected. The SP-STM images and line sections of Figure 1.29 show data taken at a perpendicular field of 0 T (Figure 1.29a), 1 T (Figure 1.29b) and 2 T (Figure 1.29c). By using the encircled adsorbate as a marker, a maximum

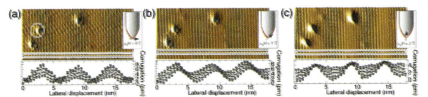

Figure 1.29 Field-dependent SP-STM measurements. Magnetically sensitive constant-current images of the Mn monolayer on W(110) (top panels) and corresponding line sections (bottom panels) taken with a ferromagnetic Fe-coated tip at external fields of (a) 0 T, (b) 1 T and (c) 2 T. As sketched in the insets, the external field rotates the tip magnetization from in-plane (a) to out-of-plane (c), shifting the position of maximum spin contrast. This proves that the Mn layer does not exhibit a spin density wave but rather a spin spiral rotating in a plane orthogonal to the surface. (Reprinted with permission from Ref. [160]; copyright (2007), MacMillan Publishers Ltd.)

magnetic contrast at this lateral position in zero field is observed, indicating large in-plane components of the sample magnetization here. This is also corroborated by the line section, which – in agreement with the in-plane-sensitive measurements of Figure 1.28b – shows a high magnetic corrugation at the maximum of the spin-averaged long-wave modulation. With an increasing external field the position of high magnetic corrugation shifts to the left (see Figure 1.29b) until a node reaches the adsorbate at 2 T (Figure 1.29c). The line sections reveal that the magnetic field shifts the position of high magnetic corrugation, but leaves the long-wave spin-averaged modulation unaffected. At 2 T – that is, with an almost perfectly out-of-plane magnetized tip – a maximum magnetic contrast is achieved and the spin-averaged signal exhibits a minimum (see line section of Figure 1.29c). Although this observation rules out a SDW, it provides clear proof of a spin spiral with magnetic moments rotating from in-plane (imaged in Figure 1.29a) to out-of-plane (imaged in Figure 1.29c).

The islands exhibit a spin spiral of only one chirality, as would be expected for a Dzyaloshinskii–Moriya interaction-driven magnetic configuration. The azimuthal orientation of the tip magnetization, however, cannot be reliably controlled, and consequently it is not possible to test experimentally whether the observed spin spiral is helical or cycloidal.

References

1 Binnig, G. and Rohrer, H. (1982) *Helvetica Physica Acta*, **55**, 726.
2 Binnig, G. and Rohrer, H. (1987) *Reviews of Modern Physics*, **59**, 615.
3 Pierce, D.T. (1988) *Physica Scripta*, **38**, 291.
4 Minakov, A.A. (1990) *Surface Science*, **236**, L377.
5 Allenspach, R. and Bischof, A. (1988) *Applied Physics Letters*, **54**, 587.
6 Wiesendanger, R., Güntherodt, H.-J., Güntherodt, G., Gambino, R.J. and Ruf, R. (1990) *Physical Review Letters*, **65**, 247.
7 Getzlaff, M. (2007) *Fundamentals of Magnetism*, Springer, Berlin.
8 Julliére, M. (1975) *Physics Letters A*, **54**, 225.
9 Miyazaki, T. and Tezuka, N. (1995) *Journal of Magnetism and Magnetic Materials*, **139**, L231.
10 Ding, H.F., Wulfhekel, W., Henk, J., Bruno, P. and Kirschner, J. (2003) *Physical Review Letters*, **90**, 116603.
11 Wiesendanger, R., Bürgler, D., Tarrach, G., Schaub, T., Hartmann, U., Güntherodt, H.-J., Shvets, I.V. and Coey, J.M.D. (1991) *Applied Physics A: Materials Science and Processing*, **53**, 349.
12 Wiesendanger, R., Bürgler, D., Tarrach, G., Wadas, A., Brodbeck, D., Güntherodt, H.-J., Güntherodt, G., Gambino, R.J. and Ruf, R. (1991) *Journal of Vacuum Science and Technology B*, **9**, 519.
13 Wiesendanger, R., Bode, M., Kleiber, M., Löhndorf, M., Pascal, R., Wadas, A. and Weiss, D. (1997) *Journal of Vacuum Science and Technology B*, **15**, 1330.
14 Rastei, M.V. and Bucher, J.P. (2006) *Journal of Physics: Condensed Matter*, **18**, L619.
15 Wadas, A. and Hug, H.J. (1992) *Journal of Applied Physics*, **72**, 203.
16 Kubetzka, A., Bode, M. and Wiesendanger, R. (2007) *Applied Physics Letters*, **91**, 012508.
17 Wulfhekel, W. and Kirschner, J. (1999) *Applied Physics Letters*, **75**, 1944.

18 Wulfhekel, W., Ding, H.F. and Kirschner, J. (2000) *Journal of Applied Physics*, **87**, 6475.
19 Wulfhekel, W., Ding, H.F., Lutzke, W., Steierl, G., Vázquez, M., Marin, P., Hernando, A. and Kirschner, J. (2001) *Applied Physics A: Materials Science and Processing*, **72**, 463.
20 Ding, H.F., Wulfhekel, W., Chen, C., Barthel, J. and Kirschner, J. (2001) *Materials Science and Engineering B*, **84**, 96.
21 Ding, H.F., Wulfhekel, W. and Kirschner, J. (2002) *Europhysics Letters*, **57**, 100.
22 Wulfhekel, W., Hertel, R., Ding, H.F., Steierl, G. and Kirschner, J. (2002) *Journal of Magnetism and Magnetic Materials*, **249**, 368.
23 Ding, H.F., Wulfhekel, W., Schlickum, U. and Kirschner, J. (2003) *Europhysics Letters*, **63**, 419.
24 Bode, M. (2003) *Reports on Progress in Physics*, **66**, 523.
25 Schlickum, U., Wulfhekel, W. and Kirschner, J. (2003) *Applied Physics Letters*, **83**, 2016.
26 Schlickum, U., Janke–Gilman, N., Wulfhekel, W. and Kirschner, J. (2004) *Physical Review Letters*, **92**, 107203.
27 Bode, M., Getzlaff, M. and Wiesendanger, R. (1998) *Physical Review Letters*, **81**, 4256.
28 Wachowiak, A., Wiebe, J., Bode, M., Pietzsch, O., Morgenstern, M. and Wiesendanger, R. (2002) *Science*, **298**, 577.
29 Pietzsch, O., Kubetzka, A., Bode, M. and Wiesendanger, R. (2000) *Physical Review Letters*, **84**, 5212.
30 Kubetzka, A., Bode, M., Pietzsch, O. and Wiesendanger, R. (2002) *Physical Review Letters*, **88**, 057201.
31 Allenspach, R., Stampanoni, M. and Bischof, A. (1990) *Physical Review Letters*, **65**, 3344.
32 Pütter, S., Ding, H.F., Millev, Y.T., Oepen, H.P. and Kirschner, J. (2001) *Physical Review B - Condensed Matter*, **64**, 092409.
33 Prokop, J., Kukunin, A. and Elmers, H.J. (2006) *Physical Review B - Condensed Matter*, **73**, 014428.
34 Kubetzka, A., Ferriani, P., Bode, M., Heinze, S., Bihlmayer, G., von Bergmann, K., Pietzsch, O., Blügel, S. and Wiesendanger, R. (2005) *Physical Review Letters*, **94**, 087204.
35 Elmers, H.J. and Gradmann, U. (1990) *Applied Physics A: Materials Science and Processing*, **51**, 255.
36 Bode, M., Getzlaff, M. and Wiesendanger, R. (1999) *Journal of Vacuum Science and Technology A - Vacuum Surfaces and Films*, **17**, 2228.
37 Murphy, S., Osing, J. and Shvets, I.V. (2003) *Surface Science*, **547**, 139.
38 Murphy, S., Osing, J. and Shvets, I.V. (1999) *Applied Surface Science*, **144–145**, 497.
39 Ceballos, S.F., Mariotto, G., Murphy, S. and Shvets, I.V. (2003) *Surface Science*, **523**, 131.
40 Murphy, S., Ceballos, S.F., Mariotto, G., Berdunov, N., Jordan, K., Shvets, I.V. and Mukovskii, Y.M. (2005) *Microscopy Research and Technique*, **66**, 85.
41 Pierce, D.T., Celotta, R.J., Wang, G.C., Unertl, W.N., Galejs, A., Kuyatt, C.E. and Mielczarek, S. (1980) *Review of Scientific Instruments*, **51**, 478.
42 Prins, M.W.J. and Abraham, D.L. (1993) H. van Kempen. *Journal of Magnetism and Magnetic Materials*, **121**, 152.
43 Prins, M.W.J., van Kempen, H., van Leuken, H., de Groot, R.A., van Roy, W. and de Boeck, J. (1995) *Journal of Physics - Condensed Matter*, **7**, 9447.
44 Jansen, R., Prins, M.W.J. and van Kempen, H. (1998) *Physical Review B - Condensed Matter*, **57**, 4033.
45 Mukasa, K., Sueoka, K., Hasegawa, H., Tazuke, Y. and Hayakawa, K. (1995) *Materials Science and Engineering B*, **31**, 69.
46 Prins, M.W.J., Jansen, R. and van Kempen, H. (1996) *Physical Review B - Condensed Matter*, **53**, 8105.
47 Suzuki, Y., Nabhan, W. and Tanaka, K. (1997) *Applied Physics Letters*, **71**, 3153.

48 Nabhan, W., Suzuki, Y., Shinohara, R., Yamaguchi, K. and Tamura, E. (1999) *Applied Surface Science*, **144–145**, 570.

49 Jansen, R., Schad, R. and van Kempen, H. (1999) *Journal of Magnetism and Magnetic Materials*, **198–199**, 668.

50 Alvarado, S.F. and Renaud, P. (1992) *Physical Review Letters*, **68**, 1387.

51 LaBella, V.P., Bullock, D.W., Ding, Z., Emery, C., Venkatesan, A., Oliver, W.F., Salamo, G.J., Thibado, P.M. and Mortazavi, M. (2001) *Science*, **292**, 1518.

52 Bode, M., Heinze, S., Kubetzka, A., Pietzsch, O., Nie, X., Bihlmayer, G., Blügel, S. and Wiesendanger, R. (2002) *Physical Review Letters*, **89**, 237205.

53 Bode, M., Kubetzka, A., Heinze, S., Pietzsch, O., Wiesendanger, R., Heide, M., Nie, X., Bihlmayer, G. and Blügel, S. (2003) *Journal of Physics: Condensed Matter*, **15**, S679.

54 Pietzsch, O., Kubetzka, A., Bode, M. and Wiesendanger, R. (2004) *Applied Physics A: Materials Science and Processing*, **78**, 781.

55 Bode, M., Pietzsch, O., Kubetzka, A., Heinze, S. and Wiesendanger, R. (2001) *Physical Review Letters*, **86**, 2142.

56 Wortmann, D., Heinze, S., Kurz, P., Bihlmayer, G. and Blügel, S. (2001) *Physical Review Letters*, **86**, 4132.

57 Hofer, W.A. and Fisher, A.J. (2002) *Surface Science*, **498**, L65.

58 Hofer, W.A. and Fisher, A.J. (2003) *Journal of Magnetism and Magnetic Materials*, **267**, 139.

59 Kubetzka, A., Pietzsch, O., Bode, M. and Wiesendanger, R. (2003) *Applied Physics A: Materials Science and Processing*, **76**, 873.

60 Wiesendanger, R., Bode, M. and Getzlaff, M. (1999) *Applied Physics Letters*, **75**, 124.

61 Yamada, T.K., Vázquez de Parga, A.L., Bischoff, M.M.J., Mizoguchi, T. and van Kempen, H. (2005) *Microscopy Research and Technique*, **66**, 93.

62 Balashov, T., Takács, A.F., Wulfhekel, W. and Kirschner, J. (2006) *Physical Review Letters*, **97**, 187201.

63 Johnson, M. and Clarke, J. (1990) *Journal of Applied Physics*, **67**, 6141.

64 Tedrow, P.M. and Meservey, R. (1973) *Physical Review B - Condensed Matter*, **7**, 318.

65 Meservey, R. and Tedrow, P.M. (1994) *Physics Reports - Review Section of Physics Letters*, **238**, 173.

66 Bode, M., Getzlaff, M., Heinze, S., Pascal, R. and Wiesendanger, R. (1998) *Applied Physics A: Materials Science and Processing*, **66**, S121.

67 Getzlaff, M., Bode, M., Heinze, S., Pascal, R. and Wiesendanger, R. (1998) *Journal of Magnetism and Magnetic Materials*, **184**, 155.

68 Getzlaff, M., Bode, M. and Wiesendanger, R. (1999) *Surface Review and Letters*, **6**, 591.

69 Bode, M., Pietzsch, O., Kubetzka, A. and Wiesendanger, R. (1055) *Journal of Electron Spectroscopy and Related Phenomena*, **2001**, 114–16.

70 Wiesendanger, R. and Bode, M. (2001) *Solid State Communications*, **119**, 341.

71 Bode, M., Getzlaff, M., Kubetzka, A., Pascal, R., Pietzsch, O. and Wiesendanger, R. (1999) *Physical Review Letters*, **83**, 3017.

72 Berbil-Bautista, L., Krause, S., Bode, M. and Wiesendanger, R. (2007) *Physical Review B - Condensed Matter*, **76**, 064411.

73 Stroscio, J.A., Pierce, D.T., Davies, A., Celotta, R.J. and Weinert, M. (1995) *Physical Review Letters*, **75**, 2960.

74 Farle, M. and Lewis, W.A. (1994) *Journal of Applied Physics*, **75**, 5604.

75 Bode, M., von Bergmann, K., Pietzsch, O., Kubetzka, A. and Wiesendanger, R. (2006) *Journal of Magnetism and Magnetic Materials*, **304**, 1.

76 von Bergmann, K., Bode, M. and Wiesendanger, R. (2004) *Physical Review B - Condensed Matter*, **70**, 174455.

77 von Bergmann, K., Bode, M., Kubetzka, A., Pietzsch, O. and Wiesendanger, R. (2005) *Microscopy Research and Technique*, **66**, 61.

78 Yamada, T.K., Bischoff, M.M.J., Heijnen, G.M.M., Mizoguchi, T. and

van Kempen, H. (2003) *Physical Review Letters*, **90**, 056803.
79 Okuno, S.N., Kishi, T. and Tanaka, K. (2002) *Physical Review Letters*, **88**, 066803.
80 Wiesendanger, R., Shvets, I.V., Bürgler, D., Tarrach, G., Güntherodt, H.-J., Coey, J.M.D. and Gräser, S. (1992) *Science*, **255**, 583.
81 Wiesendanger, R., Shvets, I.V., Bürgler, D., Tarrach, G., Güntherodt, H.-J. and Coey, J.M.D. (1992) *Zeitschrift für Physik D*, **86**, 1.
82 Wiesendanger, R., Shvets, I.V., Bürgler, D., Tarrach, G., Güntherodt, H.-J. and Coey, J.M.D. (1992) *Europhysics Letters*, **19**, 141.
83 Shvets, I.V., Wiesendanger, R., Bürgler, D., Tarrach, G., Güntherodt, H.-J. and Coey, J.M.D. (1992) *Journal of Applied Physics*, **71**, 5489.
84 Berdunov, N., Murphy, S., Mariotto, G. and Shvets, I.V. (2004) *Physical Review Letters*, **93**, 057201.
85 Blügel, S., Pescia, D. and Dederichs, P.H. (1989) *Physical Review B - Condensed Matter*, **39**, 1392.
86 Kleiber, M., Bode, M., Ravlić, R. and Wiesendanger, R. (2000) *Physical Review Letters*, **85**, 4606.
87 Kleiber, M., Bode, M., Ravlić, R., Tezuka, N. and Wiesendanger, R. (2002) *Journal of Magnetism and Magnetic Materials*, **240**, 64.
88 Ravlić, R., Bode, M., Kubetzka, A. and Wiesendanger, R. (2003) *Physical Review B - Condensed Matter*, **67**, 174411.
89 Kawagoe, T., Suzuki, Y., Bode, M. and Koike, K. (2003) *Journal of Applied Physics*, **93**, 6575.
90 Kawagoe, T., Iguchi, Y., Miyamachi, T., Yamasaki, A. and Suga, S. (2005) *Physical Review Letters*, **95**, 207205.
91 Hänke, T., Krause, S., Berbil-Bautista, L., Bode, M., Wiesendanger, R., Wagner, V., Lott, D. and Schreyer, A. (2005) *Physical Review B - Condensed Matter*, **71**, 184407.
92 Kawagoe, T., Iguchi, Y., Yamasaki, A., Suzuki, Y., Koike, K. and Suga, S. (2005) *Physical Review B - Condensed Matter*, **71**, 014427.
93 Kawagoe, T., Iguchi, Y. and Suga, S. (2007) *Journal of Magnetism and Magnetic Materials*, **310**, 2201.
94 Dreyer, M., Lee, J., Krafft, C. and Gomez, R. (2005) *Journal of Applied Physics*, **97**, 10E703.
95 Bode, M., Hennefarth, M., Haude, D., Getzlaff, M. and Wiesendanger, R. (1999) *Surface Science*, **432**, 8.
96 Heinze, S., Bode, M., Kubetzka, A., Pietzsch, O., Nie, X., Blügel, S. and Wiesendanger, R. (2000) *Science*, **288**, 1805.
97 Bode, M., Heinze, S., Kubetzka, A., Pietzsch, O., Hennefarth, M., Getzlaff, M., Wiesendanger, R., Nie, X., Bihlmayer, G. and Blügel, S. (2002) *Physical Review B - Condensed Matter*, **66**, 014425.
98 Yang, H., Smith, A.R., Prikhodko, M. and Lambrecht, W.R.L. (2002) *Physical Review Letters*, **89**, 226101.
99 Smith, A.R., Yang, R., Yang, H., Dick, A., Neugebauer, J. and Lambrecht, W.R.L. (2005) *Microscopy Research and Technique*, **66**, 72.
100 Bode, M., Kubetzka, A., Pietzsch, O. and Wiesendanger, R. (2001) *Applied Physics A: Materials Science and Processing*, **72**, S149.
101 Vedmedenko, E.Y., Kubetzka, A., von Bergmann, K., Pietzsch, O., Bode, M., Kirschner, J., Oepen, H.P. and Wiesendanger, R. (2004) *Physical Review Letters*, **92**, 077207.
102 Wiesendanger, R., Bode, M., Kubetzka, A., Pietzsch, O., Morgenstern, M., Wachowiak, A. and Wiebe, J. (2004) *Journal of Magnetism and Magnetic Materials*, **272–276**, 2115.
103 Kubetzka, A., Pietzsch, O., Bode, M. and Wiesendanger, R. (2003) *Physical Review B - Condensed Matter*, **67**, 020401.
104 Bode, M., Kubetzka, A., Pietzsch, O. and Wiesendanger, R. (2002) *Surface Science*, **514**, 135.

105 von Bergmann, K., Bode, M. and Wiesendanger, R. (2006) *Journal of Magnetism and Magnetic Materials*, **305**, 279.
106 Hauschild, J., Gradmann, U. and Elmers, H.J. (1998) *Applied Physics Letters*, **72**, 3211.
107 Pratzer, M., Elmers, H.J., Bode, M., Pietzsch, O., Kubetzka, A. and Wiesendanger, R. (2001) *Physical Review Letters*, **87**, 127201.
108 Pietzsch, O., Kubetzka, A., Bode, M. and Wiesendanger, R. (2001) *Science*, **292**, 2053.
109 Prokop, J., Kukunin, A. and Elmers, H.J. (2005) *Physical Review Letters*, **95**, 187202.
110 Usov, V., Murphy, S. and Shvets, I.V. (2004) *Journal of Magnetism and Magnetic Materials*, **283**, 357.
111 Gradmann, U., Korecki, J. and Waller, G. (1986) *Applied Physics A: Materials Science and Processing*, **39**, 101.
112 Bode, M., Pietzsch, O., Kubetzka, A. and Wiesendanger, R. (2004) *Physical Review Letters*, **92**, 067201.
113 Elmers, H.J., Hauschild, J. and Gradmann, U. (1999) *Physical Review B - Condensed Matter*, **59**, 3688.
114 Allenspach, R., Stampanoni, M. and Bischof, A. (1990) *Physical Review Letters*, **65**, 3344.
115 Yi, G., Aitchison, P.R., Doyle, W.D., Chapman, J.N. and Wilkinson, C.D.W. (2002) *Journal of Applied Physics*, **92**, 6087.
116 Kirk, K.J., McVitie, S., Chapman, J.N. and Wilkinson, C.D.W. (2001) *Journal of Applied Physics*, **89**, 7174.
117 Gomez, R.D., Luu, T.V., Pak, A.O., Kirk, K.J. and Chapman, J.N. (1999) *Journal of Applied Physics*, **85**, 6163.
118 Herrmann, M., McVitie, S. and Chapman, J.N. (2000) *Journal of Applied Physics*, **87**, 2994.
119 Kubetzka, A., Pietzsch, O., Bode, M. and Wiesendanger, R. (2001) *Physical Review B - Condensed Matter*, **63**, 140407.
120 Bode, M., Wachowiak, A., Wiebe, J., Kubetzka, A., Morgenstern, M. and Wiesendanger, R. (2004) *Applied Physics Letters*, **84**, 948.
121 Bode, M., Kubetzka, A., von Bergmann, K., Pietzsch, O. and Wiesendanger, R. (2005) *Microscopy Research and Technique*, **66**, 117.
122 Pietzsch, O., Kubetzka, A., Bode, M. and Wiesendanger, R. (2004) *Physical Review Letters*, **92**, 057202.
123 Pietzsch, O., Okatov, S., Kubetzka, A., Bode, M., Heinze, S., Lichtenstein, A. and Wiesendanger, R. (2006) *Physical Review Letters*, **96**, 237203.
124 Rusponi, S., Weiss, N., Chen, T., Epple, M. and Brune, H. (2005) *Applied Physics Letters*, **87**, 162514.
125 Meier, F., von Bergmann, K., Ferriani, P., Wiebe, J., Bode, M., Hashimoto, K., Heinze, S. and Wiesendanger, R. (2006) *Physical Review B - Condensed Matter*, **74**, 195411.
126 Meier, F., von Bergmann, K., Wiebe, J., Bode, M. and Wiesendanger, R. (2007) *Journal of Physics D*, **40**, 1306.
127 Kukunin, A., Prokop, J. and Elmers, H.J. (2006) *Acta Physica Polonica A*, **109**, 371.
128 Prokop, J., Kukunin, A. and Elmers, H.J. (2007) *Physical Review B - Condensed Matter*, **75**, 144423.
129 Krause, S., Berbil-Bautista, L., Herzog, G., Bode, M. and Wiesendanger, R. (2007) *Science*, **317**, 1537.
130 Evoy, S., Carr, D.W., Sekaric, L., Suzuki, Y., Parpia, J.M. and Craighead, H.G. (2000) *Journal of Applied Physics*, **87**, 404.
131 Hertel, R. (2002) *Zeitschrift Fur Metallkunde*, **93**, 957.
132 Shinjo, T., Okuno, T., Hassdorf, R., Shigeto, K. and Ono, T. (2000) *Science*, **289**, 930.
133 Yayon, Y., Brar, V.W., Senapati, L., Erwin, S.C. and Crommie, M.F. (2007) *Physical Review Letters*, **99**, 067202.
134 Crommie, M.F., Lutz, C.P. and Eigler, D.M. (1993) *Nature*, **363**, 524.
135 Ravlić, R., Bode, M. and Wiesendanger, R. (2003) *Journal of Physics: Condensed Matter*, **15**, S2513.

136 Bode, M., Ravlić, R., Kleiber, M. and Wiesendanger, R. (2005) *Applied Physics A: Materials Science and Processing*, **80**, 907.

137 Getzlaff, M., Bansmann, J. and Schönhense, G. (1999) *Journal of Magnetism and Magnetic Materials*, **192**, 458.

138 Getzlaff, M., Bansmann, J. and Schönhense, G. (1996) *Journal of Electron Spectroscopy and Related Phenomena*, **77**, 197.

139 Getzlaff, M., Westphal, C., Bansmann, J. and Schönhense, G. (2000) *Journal of Electron Spectroscopy and Related Phenomena*, **107**, 293.

140 Getzlaff, M., Bansmann, J. and Schönhense, G. (1993) *Physics Letters A*, **182**, 153.

141 von Bergmann, K., Bode, M., Kubetzka, A., Heide, M., Blügel, S. and Wiesendanger, R. (2004) *Physical Review Letters*, **92**, 046801.

142 Berbil-Bautista, L., Krause, S., Hänke, T., Bode, M. and Wiesendanger, R. (2006) *Surface Science*, **600**, L20.

143 Getzlaff, M., Bansmann, J. and Schönhense, G. (1993) *Solid State Communications*, **87**, 467.

144 Krause, S., Berbil-Bautista, L., Hänke, T., Vonau, F., Bode, M. and Wiesendanger, R. (2006) *Europhysics Letters*, **76**, 637.

145 Hubert, A. and Schäfer, R. (1998) *Magnetic Domains*, Springer, Berlin.

146 Kubetzka, A., Pietzsch, O., Bode, M., Ravlić, R. and Wiesendanger, R. (2003) *Acta Physica Polonica B*, **104**, 259.

147 Bode, M., Vedmedenko, E.Y., von Bergmann, K., Kubetzka, A., Ferriani, P., Heinze, S. and Wiesendanger, R. (2006) *Nature Materials*, **5**, 477.

148 Nakamura, K., Takeda, Y., Akiyama, T., Ito, T. and Freeman, A.J. (2004) *Physical Review Letters*, **93**, 057202.

149 Douglass, D.C., Cox, A.J., Bucher, J.P. and Bloomfield, L.A. (1993) *Physical Review B - Condensed Matter*, **47**, 12874.

150 von Bergmann, K., Heinze, S., Bode, M., Vedmedenko, E.Y., Bihlmayer, G., Blügel, S. and Wiesendanger, R. (2006) *Physical Review Letters*, **96**, 167203.

151 Pappas, D.P., Popov, A.P., Anisimov, A.N., Reddy, B.V. and Khamna, S.N. (1996) *Physical Review Letters*, **76**, 4332.

152 Hobbs, D., Kresse, G. and Hafner, J. (2000) *Physical Review B - Condensed Matter*, **62**, 11556.

153 Kurz, Ph., Bihlmayer, G., Hirai, K. and Blügel, S. (2001) *Physical Review Letters*, **86**, 1106.

154 Heinze, S., Kurz, P., Wortmann, D., Bihlmayer, G. and Blügel, S. (2002) *Applied Physics A: Materials Science and Processing*, **75**, 25.

155 Kurz, Ph., Förster, F., Nordström, L., Bihlmayer, G. and Blügel, S. (2004) *Physical Review B - Condensed Matter*, **69**, 024415.

156 Lizárraga, R., Nordström, L., Bergqvist, L., Bergman, A., Sjöstedt, E., Mohn, P. and Eriksson, O. (2004) *Physical Review Letters*, **93**, 107205.

157 Heinze, S. (2006) *Applied Physics A: Materials Science and Processing*, **85**, 407.

158 Dzialoshinskii, I.E. (1957) *Soviet Physics JETP*, **5**, 1259.

159 Moriya, T. (1960) *Physical Review*, **120**, 91.

160 Bode, M., Heide, M., von Bergmann, K., Ferriani, P., Heinze, S., Bihlmayer, G., Kubetzka, A., Pietzsch, O., Blügel, S. and Wiesendanger, R. (2007) *Nature*, **447**, 190.

161 Rößler, U.K., Bogdanov, A.N. and Pfleiderer, C. (2006) *Nature*, **442**, 797.

2
Nanoscale Imaging and Force Analysis with Atomic Force Microscopy
Hendrik Hölscher, André Schirmeisen, and Harald Fuchs

2.1
Principles of Atomic Force Microscopy

2.1.1
Basic Concept

The direct measurement of the force interaction between distinct molecules has been a challenge for scientists for many years. In fact, only very recently was a demonstration given that these forces can be determined for a single atomic bond, by using the powerful technique of atomic force microscopy (AFM). But how is it possible to measure interatomic forces, which may be as small as one billionth of one Newton?

The answer to this question is surprisingly simple: It is the same mechanical principle used by a pair of kitchen scales, where a spring with a defined elasticity is elongated or compressed due to the weight of the object to be measured. The compression Δz of the spring (with spring constant c_z) is a direct measure of the force F exerted, which in the regime of elastic deformation obeys Hooke's law:

$$F = c_z \times \Delta z. \tag{2.1}$$

The only difference from the kitchen scale is the sensitivity of the measurement. In AFM, the 'spring' is a bendable cantilever with a stiffness of 0.01 N m^{-1} to 10 N m^{-1}. As interatomic forces are in the range of some nN, the cantilever will be deflected by 0.01 nm to 100 nm. Consequently, the precise detection of the cantilever bending is the key feature of an atomic force microscope. If a sufficiently sharp tip is directly attached to the cantilever, it would be possible to measure the interacting forces between the last atoms of the tip and the sample through the bending of the cantilever.

In 1986, Binnig, Quate and Gerber presented exactly this concept for the first atomic force microscope [1]. These authors measured the deflection of a cantilever with sub-Ångström precision by a scanning tunneling microscope [2] and used a gold foil as the spring. The tip was a piece of diamond glued to this home-made cantilever

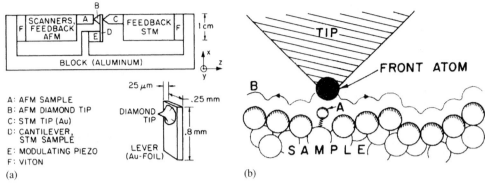

Figure 2.1 (a) The basic concept of the first atomic force microscope built in 1986 by Binnig, Quate and Gerber. A sharp diamond tip glued to a gold foil scanned the surface, while the bending of the cantilever was detected with scanning tunneling microscopy; (b) The ultimate goal was to measure the force between the front atom of the tip and a specific sample atom. (Reproduced from Ref. [1].)

(see Figure 2.1), and by using this set-up the group was able to image sample topographies down to the nanometer scale.

2.1.2
Current Experimental Set-Ups

During the past few years the experimental set-up has been modified while AFM has become a widespread research tool. Some 20 years after its invention, the commercial atomic force microscope is available from a variety of manufacturers. Although most of these instruments are designed for specific applications and environments, they are typically based on the following types of sensors detection method and scanning principles.

2.1.2.1 Sensors

Cantilevers are produced by standard microfabrication techniques, mostly from silicon and silicon nitride as rectangular or V-shaped cantilevers. Spring constants and resonance frequencies of cantilevers depend on the actual mode of operation. For contact AFM measurements these are about 0.01 to $1\,\mathrm{N\,m^{-1}}$ and 5–100 kHz, respectively. In a typical atomic force microscope, cantilever deflections ranging from 0.1 Å to a few micrometers are measured, which corresponds to a force sensitivity ranging from $10^{-13}\,\mathrm{N}$ to $10^{-5}\,\mathrm{N}$.

Figure 2.2 shows two scanning electron microscopy (SEM) images of a typical rectangular silicon cantilever. When using this imaging technique, the length (l), width (w) and thickness (t) can be precisely measured. The spring constant c_z of the cantilever can then be determined from these values [3].

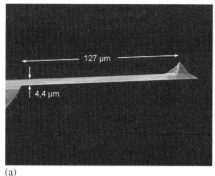 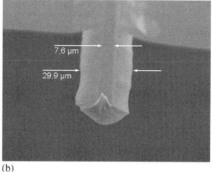

Figure 2.2 (a) Scanning electron microscopy image of a rectangular silicon cantilever with a width of 127 μm and a thickness of 4.4 μm; (b) A different view of the same cantilever reveals that the cross-section of the cantilever is trapezoidal and the cantilever has two geometric widths – a smaller one on the tip side and a broader one on the reverse side. Hence, most manufacturers provide a 'mean width', which for the cantilever shown is (7.6 + 29.9)/2 = 18.75 μm. The trapezoidal shape of the cantilever is caused by the anisotropic etching of the silicon during the microfabrication of the cantilever. (Images courtesy of Boris Anczykowski, nanoAnalytics GmbH; used with kind permission.)

$$c_z = E_{Si} \frac{w}{4} \left(\frac{t}{l}\right)^3 \tag{2.2}$$

where $E_{Si} = 1.69 \times 10^{11}\,\mathrm{N\,m^{-2}}$ is the Youngs's modulus. The typical dimensions of silicon cantilevers are as follows: lengths of 100–300 μm; widths of 10–30 μm; and thicknesses of 0.3–5 μm.

The torsion of the cantilever due to lateral forces between tip and surface depends also on the height of the tip, h. The torsional spring constant can be calculated from [3]

$$c_{tor} = \frac{G\,wt^3}{3\,lh^2} \tag{2.3}$$

where $G_{Si} = 0.68 \times 10^{11}\,\mathrm{N\,m^{-1}}$ is the shear modulus of silicon.

As the dimensions of cantilevers given by the manufacturer are only average values, the high-accuracy calibration of the spring constant requires the measurement of length, width and thickness for each individual cantilever. The length and width can be measured with sufficient accuracy using an optical microscope, but the thickness requires high-resolution techniques such as SEM. In order to avoid this time- and cost-consuming measurement one can determine the cantilever thickness from its eigenfrequency in normal direction [3–5]

$$t = \underbrace{\frac{4\sqrt{3}}{0.596861^2\pi}\sqrt{\frac{\rho_{Si}}{E_{Si}}}}_{\approx 7.23 \times 10^{-4}\,\mathrm{s/m}} l^2 f_0 \tag{2.4}$$

where the density of silicon $\rho_{Si} = 2330\,\mathrm{kg\,m^{-3}}$.

The formulas presented above are only valid for rectangular cantilevers. Equations for V-shaped cantilevers can be found in Refs. [6, 7].

2.1.2.2 Detection Methods

In addition to changes in the cantilevers, the detection methods used to measure the minute bendings have also been improved. Today, commercial AFM instruments use the so-called *laser beam deflection* scheme shown in Figure 2.3. The bending and torsion of cantilevers can be detected by a laser beam reflected from their reverse side [8, 9], while the reflected laser spot is detected with a sectioned photo-diode. The different parts are read out separately. For this, a four-quadrant diode is normally used, in order to detect the normal as well as the torsional movements of the cantilever. With the cantilever at equilibrium the spot is adjusted such that the upper and lower sections show the same intensity. Then, if the cantilever bends up or down, the spot will move and the difference signal between the upper and lower sections will provide a measure of the bending.

The sensitivity can be improved by interferometer systems adapted by several research groups (see Refs [10–13]). It is also possible to use cantilevers with integrated deflection sensors based on piezoresistive films [14–16]. As no optical parts are

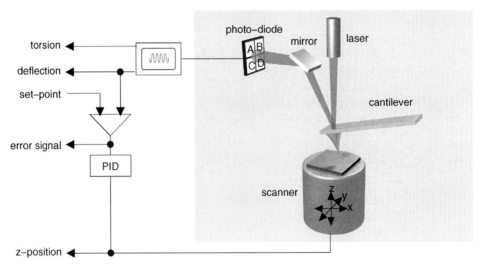

Figure 2.3 Principle of an atomic force microscope using the laser beam deflection method. Deflection (normal force) and torsion (friction) of the cantilever are measured simultaneously by measuring the lateral and vertical deflection of a laser beam while the sample is scanned in the x–y-plane. The laser beam deflection is determined using a four-quadrant photo diode. If A, B, C and D are proportional to the intensity of the incident light of the corresponding quadrant, the signal (A + B) − (C + D) is a measure for the deflection, and (A + C) − (B + D) is a measure of the torsion of the cantilever. A schematic of the feedback system is shown by the solid lines. The actual deflection signal of the photo-diode is compared with the set-point chosen by the experimentalist. The resultant error signal is fed into the PID controller, which moves the z-position of the scanner in order to minimize the deflection signal.

required in the experimental set-up of an atomic force microscope, when using these cantilevers the design can be very compact [17]. An extensive comparison of the different detection methods available can be found in Ref. [4].

2.1.2.3 Scanning and Feedback System

As the surface is scanned, the deflection of the cantilever is kept constant by a feedback system that controls the vertical movement of the scanner (shown schematically in Figure 2.3). The system functions as follows: (i) The current signal of the photo-diode is compared with a preset value; (ii) the feedback system, which includes a proportional, integral and differential (PID) controller, then varies the z-movement of the scanner to minimize the difference. As a consequence, the tip-sample force is kept practically constant for an optimal set-up of the PID parameters.

While the cantilever is moving relative to the sample in the x–y-plane of the surface by a piezoelectric scanner (see Figure 2.4), the current z-position of the scanner is recorded as a function of the lateral x–y-position with (ideally) sub-Ångström precision. The obtained data represents a map of equal forces, which is then analyzed and visualized by computer processing.

A principle of a simple laboratory class experiment – the imaging of a test grid – is shown in Figure 2.5. The comparison between the topography and the error signal shows that the PID controller needs some time at the step edges to correct the actual deflection error.

2.1.3
Tip–Sample Forces in Atomic Force Microscopy

A large variety of sample properties related to tip-sample forces can be detected using the atomic force microscope. The obtained contrast depends on the operational mode and

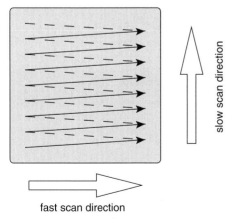

Figure 2.4 Schematic of the scan process. The cantilever scans the sample surface systematically in the x- and y-directions. Typical scan sizes ranging from less than 1 nm × 1 nm to 150 μm × 150 μm can be used.

2 Nanoscale Imaging and Force Analysis with Atomic Force Microscopy

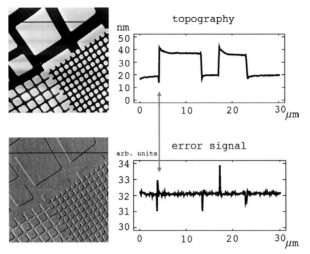

Figure 2.5 A simple laboratory class experiment demonstrating the scanning process of an atomic force microscope in contact mode. The cantilever is scanned over a simple test grid made of silicon. The feedback keeps the deflection (and therefore the force) constant, and the z-position of the scanner is interpreted as the topography of the sample. The resultant map is plotted as a gray-scale image (lighter areas correspond to higher topography, upper left graph). The simultaneously plotted error signal (lower graph) shows that the feedback fails to keep the deflection constant at the step edges.

the actual tip-sample interactions. Before discussing details of the operational modes of AFM, however, we must first specify the most important tip–sample interactions.

Figure 2.6 shows the typical shape of the interaction force curve that the tip senses during an approach towards the sample surface. Upon approach of the tip towards the

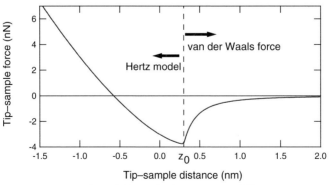

Figure 2.6 Tip–sample model force after the DMT-M model for air Equation 2.9, using the parameters described in the text. The dashed line marks the position z_0, where the tip touches the surface.

sample, the negative attractive forces (which represent, for example, van der Waals or electrostatic interaction forces) increase until a minimum is reached. This turnaround point is due to the onset of repulsive forces, caused by Pauli repulsion, which will start to dominate upon further approach. Eventually, the tip is pushed into the sample surface and elastic deformation will occur.

In general, the effective tip–sample interaction force is a sum of different force contributions, and these can be roughly divided into *attractive* and *repulsive* components. The most important forces are summarized as follows.

2.1.3.1 Van der Waals Forces

These forces are caused by fluctuating induced electric dipoles in atoms and molecules. The distance-dependence of this force for two distinct molecules follows $1/z^7$. For simplicity, solid bodies are often assumed to consist of many independent noninteracting molecules, and the van der Waals forces of these bodies are obtained by simple summation. For example, for a sphere over a flat surface the van der Waals force is given by

$$F_{vdW}(z) = -\frac{A_H R}{6z^2}, \quad (2.5)$$

where R is the radius of the sphere and A_H is the Hamaker constant, which is typically in the range of ≈ 0.1 aJ [18]. This geometry is often used to approximate the van der Waals forces between the tip and sample. Due to the $1/z^2$ dependency, van der Waals forces are considered long-range forces compared to the other forces that occur in AFM.

2.1.3.2 Capillary Forces

Capillary forces are important under ambient conditions. Water molecules condense at the sample surface (and also on the tip) and cause the occurrence of an adsorption layer. Consequently, the atomic force microscope tip penetrates through this layer when approaching the sample surface. At the tip–sample contact, a water meniscus is formed which causes a very strong attractive force [19]. For soft samples these forces often lead to unwanted deformations of the surface; however, this effect can be circumvented by measuring directly in liquids. Alternatively, capillary forces can be avoided by performing the experiments in a glovebox with dry gases, or in vacuum.

2.1.3.3 Pauli or Ionic Repulsion

These forces are the most important in conventional contact mode AFM. The Pauli exclusion principle forbids that the charge clouds of two electrons showing the same quantum numbers can have some significant overlap; first, the energy of one of the electrons must be increased, and this yields a repulsive force. In addition, an overlap of the charge clouds of electrons can cause an insufficient screening of the nuclear charge, leading to ionic repulsion of coulombic nature. The Pauli and the ionic repulsion are nearly hard-wall potentials. Thus, when the tip and sample are in intimate contact most of the (repulsive) interaction is carried by the atoms directly at the interface. The Pauli repulsion is of purely quantum mechanical origin, and semi-empirical potentials are mostly used to allow an easy and fast calculation. One well-known model is the *Lennard–Jones* potential, which combines short-range repulsive interactions with long-range attractive van der Waals interactions:

$$V_{LJ}(z) = E_0 \left(\left(\frac{r_0}{z}\right)^{12} - 2\left(\frac{r_0}{z}\right)^6 \right) \tag{2.6}$$

where E_0 is the bonding energy and r_0 the equilibrium distance. In this case, the repulsion is described by an inverse power law with $n=12$. The term with $n=6$ describes the attractive van der Waals potential between two atoms/molecules.

2.1.3.4 Elastic Forces

Elastic forces and deformations can occur if the tip is in contact with the sample. As this deformation affects the effective contact area, knowledge about the elastic forces and the corresponding deformation mechanics of the contact is an important issue in AFM. The repulsive forces that occur during the elastic indentation of a sphere into a flat surface were analyzed as early as 1881 by H. Hertz (see Refs [20, 21]),

$$F_{\text{Hertz}}(z) = \frac{4}{3} E^* \sqrt{R} (z_0 - z)^{3/2} \quad \text{for} \quad z \leq z_0, \tag{2.7}$$

where the effective elastic modulus E^*

$$\frac{1}{E^*} = \frac{(1-\mu_t^2)}{E_t} + \frac{(1-\mu_s^2)}{E_s} \tag{2.8}$$

depends on the Young's moduli $E_{t,s}$ and the Poisson ratios $\mu_{t,s}$ of the tip and surface, respectively. Here, R is the tip radius and z_0 is the point of contact.

This model does not include adhesion forces, however, which must be considered at the nanometer scale. Two extreme cases were analyzed by Johnson et al. [22] and Derjaguin et al. [23]. The model of Johnson, Kendall and Roberts (the JKR model) considers only the adhesion forces *inside* the contact area, whereas the model of Derjaguin, Muller and Toporov (the DMT model) includes only the adhesion *outside* the contact area. Various models analyzing the contact mechanics in the intermediate regime were suggested by other authors (see Ref. [24] for a recent overview).

However, in many practical cases it is sufficient to assume that the geometric shape of the tip and sample does not change until contact has been established at $z = z_0$, and that afterwards, the tip–sample forces are given by the DMT-M theory, denoting Maugis' approximation to the earlier DMT model [24]. In this approach, an offset $F_{\text{vdW}}(z_0)$ is added to the well-known Hertz model, which accounts for the adhesion force between tip and sample surface. Therefore, the DMT-M model is often also referred to as *Hertz-plus-offset model* [24]. The resulting overall force law is given by

$$F_{\text{DMT-M}}(z) = \begin{cases} -\dfrac{A_H R}{6z^2} & \text{for} \quad z \geq z_0, \\ \dfrac{4}{3} E^* \sqrt{R}(z_0-z)^{3/2} - \dfrac{A_H R}{6z_0^2} & \text{for} \quad z < z_0. \end{cases} \tag{2.9}$$

Figure 2.6 shows the resulting tip–sample force curve for the DMT-M model. The following parameters were used, representing typical values for AFM measurements under ambient conditions: $A_H = 0.2$ aJ; $R = 10$ nm; $z_0 = 0.3$ nm; $\mu_t = \mu_s = 0.3$; $E_t = 130$ GPa; and $E_s = 1$ GPa.

2.1.3.5 Frictional Forces
Frictional forces counteract the movement of the tip during the scan process, and dissipate the kinetic energy of the moving tip–sample contact into the surface or tip material. This can be due to permanent changes in the surface itself, by scratching or indenting, or also by the excitation of lattice vibration (i.e. phonons) in the material.

2.1.3.6 Chemical Binding Forces
Chemical binding forces arise from the overlap of molecular orbitals, due to specific bonding states between the tip and the surface molecules. These forces are extremely short-ranged, and can be exploited to achieve atomic resolution imaging of surfaces. As these forces are also specific to the chemical identity of the molecules, it is conceivable to identify the chemical character of the surface atoms with AFM scans.

2.1.3.7 Magnetic and Electrostatic Forces
These forces are of long-range character and might be either attractive or repulsive; they are usually measured when the tip is not in contact with the surface (i.e. 'noncontact' mode). For magnetic forces, magnetic materials must be used for tip or tip coating. Well-defined electrical potentials between tip and sample are necessary for the measurement of electrostatic forces.

More detailed information on the intermolecular and surface forces relevant for AFM measurements can be found in the monographs of Israelachvili [18] and Sarid [25]. Details of the most important forces are summarized in Figure 2.7. Although, in principle, every type of force can be measured using the atomic force microscope, the actual sensitivity to a specific force depends on the mode of operation. Hence, the most important modes are introduced in the next section.

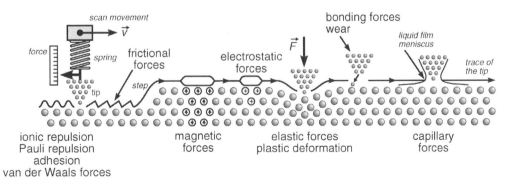

Figure 2.7 Summary of the forces relevant in atomic force microscopy. (Image courtesy of Udo D. Schwarz, Yale University; used with kind permission.)

2.2
Modes of Operation

Although an atomic force microscope can be driven in different modes of operation, we concentrate here on the two most important modes that are widely used to image sample surfaces down to the atomic scale.

2.2.1
Static or Contact Mode

The contact mode, which historically is the oldest, is used frequently to obtain nanometer-resolution images on a wide variety of surfaces. This technique also has the advantage that not only the deflection, but also the torsion of the cantilever, can be measured. As shown by Mate et al. [26], the lateral force can be directly correlated to the friction between tip and sample, thus extending AFM to friction force microscopy (FFM).

Some typical applications of an atomic force microscope driven in contact-mode are shown in Figure 2.8a and b. Here, the images represent a measurement of a L-α-

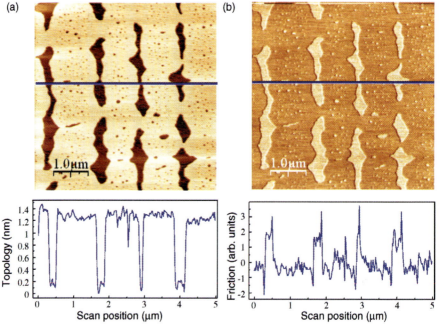

Figure 2.8 (a) Atomic force microscopy image obtained in contact mode of a monomolecular DPPC (L-α-dipalmitoyl-phophatidycholine) film adsorbed onto mica. The image is color-coded; that is, dark areas represent the mica substrate and light areas the DPPC film; (b) The simultaneously recorded friction image shows a lower friction on the film (dark areas) than on the substrate (light areas). The graphs represent single scan lines obtained at the positions marked by a dark line in the above images.

dipalmitoyl-phophatidycholine (DPPC) film adsorbed onto a mica substrate. The lateral force was simultaneously recorded with the topography, and shows a contrast between the DPPC film and the substrate. This effect can be attributed to the different frictional forces on DPPC and the mica substrate, and is frequently used to obtain a chemical contrast on flat surfaces [27, 28].

2.2.1.1 Force versus Distance Curves

So far, we have neglected one important issue for the operation of the atomic force microscope, namely the mechanical stability of the measurement. In static AFM the tip is allowed to approach very slowly towards the surface, and the attractive forces between the tip and sample must be counteracted by the restoring force of the cantilever. However, this fails if the force gradient of the tip–sample forces is larger than the spring constant of the cantilever. Mathematically speaking, an instability occurs if

$$c_z < \frac{\partial F_{ts}(z)}{\partial z}. \tag{2.10}$$

In this case the attractive forces can no longer be sustained by the cantilever and the tip 'jumps' towards the sample surface [29].

This effect has a strong influence on static-mode AFM measurements, as exemplified by a typical force-versus-distance curve shown in Figure 2.9. Here, the force

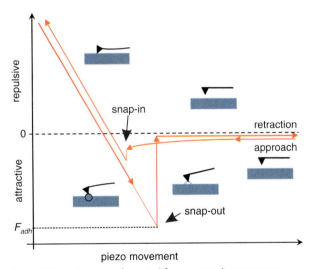

Figure 2.9 A schematic of a typical force versus distance curve obtained in static mode. The cantilever is approached towards the sample surface. Due to strong attractive forces it 'jumps' (snap-in) towards the sample surface at a specific position. During retraction, the tip is strongly attracted by the surface and the 'snap-out' point is considerably behind the 'snap-in' point. This results in an hysteresis between approach and retraction.

acting on the tip recorded during an approach and retraction movement of the cantilever is depicted. Upon approach of the cantilever towards the sample, the attractive forces acting on the tip bend the cantilever towards the sample surface. At a specific point close to the sample surface these forces can be no longer sustained by the cantilever spring, and the tip 'jumps' towards the sample surface. Now, the tip and sample are in direct mechanical contact, and a further approach towards the sample surface pushes the tip into the sample. As the spring constant of the cantilever usually is much softer than the elasticity of the sample, the bending of the cantilever increases almost linearly.

If the cantilever is now retracted from the surface, the tip stays in contact with the sample because it is strongly attracted by the sample due to adhesive forces, and the force F_{adh} is necessary to disconnect the tip from the surface. The 'snap-out' point is always at a larger distance from the surface than the 'snap-in', and this results in an hysteresis between the approach and retraction of the cantilever. This phenomenon of mechanical instability is often referred to as the *jump-to-contact*. Unfortunately, this sudden jump can lead to undesired changes of the tip and/or sample.

2.2.2
Dynamic Modes

Despite the success of contact-mode AFM, the resolution was found to be limited in many cases (in particular for soft samples) by lateral forces acting between tip and sample. In order to avoid this effect, the cantilever can be oscillated in a vertical direction near the sample surface. AFM imaging with vibrating cantilever is often denoted as dynamic force microscopy (DFM).

The historically oldest scheme of cantilever excitation in DFM imaging is the *external* driving of the cantilever at a *fixed excitation frequency* exactly at or very close to the cantilever's first resonance [30–32]. For this driving mechanism, different detection schemes measuring either the change of the oscillation amplitude or the phase shift were proposed. Over the years, the amplitude modulation (AM) or 'tapping' mode, where the oscillation amplitude is used as a measure of the tip–sample distance, has developed into the most widespread technique for imaging under ambient conditions and liquids.

In a vacuum, any external oscillation of the cantilever is disadvantageous. Standard AFM cantilevers constructed from silicon exhibit very high Q-values in vacuum, which results in very long response times of the system. Consequently, in 1991 Albrecht *et al.* [33] introduced the frequency modulation (FM) mode, which works well for high-Q systems and subsequently has developed into the dominant driving scheme for high-resolution DFM experiments in ultra-high vacuum (UHV) [34–37]. In contrast to the AM mode, this approach features a so-called *self-driven* oscillator [38, 39] which, when placed in a closed-loop set-up ('active feedback'), uses the cantilever deflection itself as the driving signal, thus ensuring that the cantilever instantaneously adapts to changes in the resonance frequency. These two driving mechanisms are discussed in more detail in the following section.

2.3
Amplitude Modulation (Tapping Mode)

2.3.1
Experimental Set-Up of AM-Atomic Force Microscopy

As an alternative to the contact mode, the cantilever can be excited to vibrate near its resonant frequency close to the sample surface. Under the influence of tip–sample forces the resonant frequency (and consequently also the amplitude and phase) of the cantilever will change and serve as the measurement parameters. This is known as the *dynamic* mode. If the tip is approached towards the surface, the oscillation parameters of amplitude and phase are influenced by the tip–surface interaction, and can therefore be used as feedback channels. A certain set-point (e.g. the amplitude) is given, whereby the feedback loop will adjust the tip–sample distance so that the amplitude remains constant. The controller parameter is recorded as a function of the lateral position of the tip with respect to the sample, and the scanned image essentially represents the surface topography.

The technical realization of dynamic-mode AFM is based on the same key components as a static AFM set-up. A sketch of the experimental set-up of an atomic force microscope driven in AM mode is shown in Figure 2.10.

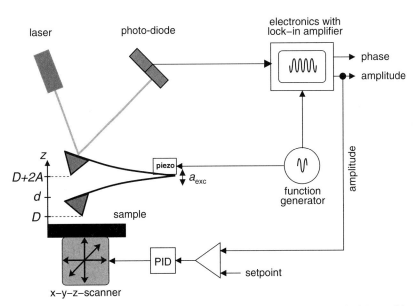

Figure 2.10 Set-up of a dynamic force microscope operated in AM or tapping mode. A laser beam is deflected by the reverse side of the cantilever, with the deflection being detected by a split photo-diode. The cantilever vibration is caused by an external frequency generator driving an excitation piezo. A lock-in amplifier is used to compare the cantilever driving with its oscillation. The amplitude signal is held constant by a feedback loop which controls the cantilever–sample distance.

The deflection of the cantilever is typically measured with the laser beam deflection method, as indicated [8, 9], but other displacement sensors such as *interferometric sensors* [12, 13, 30, 40] have also been applied. During operation in conventional tapping mode, the cantilever is driven at a fixed frequency with a constant excitation amplitude from an external function generator, while the resulting oscillation amplitude and/or the phase shift are detected by a lock-in amplifier. The function generator supplies not only the signal for the dither piezo; its signal serves simultaneously as a reference for the lock-in amplifier.

This set-up can be operated both in air and in liquids. A typical image obtained with this experimental set-up in ambient conditions is shown in Figure 2.11. For a direct comparison with the static mode, the sample is also DPPC-adsorbed onto a mica substrate. In contact mode the frictional forces are measured simultaneously with the topography, whereas in dynamic mode the phase between excitation and oscillation is acquired as an additional channel. The phase image provides information about the different material properties of DPPC and the mica substrate. It can be shown, that the phase signal is closely related to the energy dissipated in the tip–sample contact [41–43].

Due to its technical relevance the investigation of polymers has been the focal point of many studies (see Ref. [44] for a recent review). High-resolution imaging has been extensively performed in the area of materials science; for example, by using specific

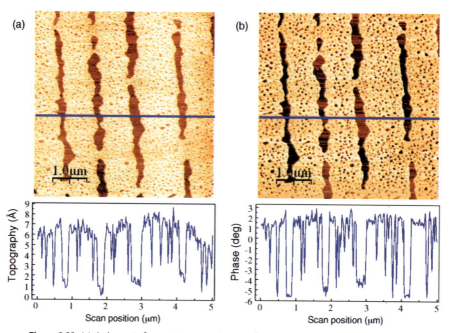

Figure 2.11 (a) A dynamic force microscopy image of a monomolecular DPPC film adsorbed onto mica; (b) The phase contrast is directly related to the topography; that is, the phase is different between the substrate and the DPPC film.

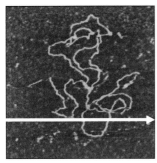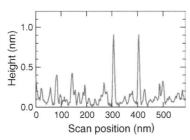

Figure 2.12 Topography of DNA adsorbed onto mica imaged in buffer solution by tapping mode AFM. The graph shows a single scan line obtained at the position marked by a white arrow in the image.

tips with additionally grown sharp spikes, Klinov *et al.* [45] obtained true molecular resolution on a polydiacetylene crystal.

Imaging in liquids opens up an avenue for the investigation of biological samples in their natural environment. For example, Möller *et al.* [46] have obtained high-resolution images of the topography of the hexagonally packed intermediate (HPI) layer of *Deinococcus radiodurans*, using tapping-mode AFM. A typical example of the imaging of DNA in liquid solution is shown in Figure 2.12.

2.3.1.1 Theory of AM-AFM

Based on the above description of the experimental set-up, it is possible to formulate the basic equation of motion describing the cantilever dynamics of AM-AFM:

$$m\ddot{z}(t) + \frac{2\pi f_0 m}{Q_0}\dot{z}(t) + c_z(z(t)-d) + \; = \; \underbrace{a_d c_z \cos(2\pi f_d t)}_{\text{external driving force}} + \underbrace{F_{ts}[z(t),\dot{z}(t)]}_{\text{tip-sample force}}. \tag{2.11}$$

Here, $\dot{z}(t)$ is the position of the tip at the time t; c_z, m and $f_0 = \sqrt{(c_z/m)}/(2\pi)$ are the spring constant, the effective mass, and the eigenfrequency of the cantilever, respectively. As a small simplification, it is assumed that the quality factor Q_0 combines the intrinsic damping of the cantilever and all influences from surrounding media, such as air or liquid (if present) in a single value. The equilibrium position of the tip is denoted as d. The first term on the right-hand side of the equation represents the external driving force of the cantilever by the frequency generator. It is modulated with the constant excitation amplitude a_d at a fixed frequency f_d. The (nonlinear) tip–sample interaction force F_{ts} is introduced by the second term.

Before discussing the solutions of this equation, some words of caution should be added with regards to the universality of the equation of motion and the various solutions discussed below. Equation 2.11 disregards two effects, which might become important under certain circumstances. First, we describe the cantilever by a spring-mass-model and neglect in this way the higher modes of the cantilever.

This is justified in most cases, as the first eigenfrequency is by far the most dominant in typical AM-AFM experiments (see Refs [41, 47–50]). Second, we assume in our model equation of motion that the dither piezo applies a sinusoidal force to the spring, but do not consider that the movement of the dither piezo simultaneously also changes the effective position of the tip at the cantilever end by $a_{exc}(t) = a_d \cos(2\pi f_d t)$ [47, 51, 52]. This effect becomes important when a_d is in the range of the cantilever oscillation amplitude.

In a first step, we assume that the cantilever vibrates far away from the sample surface. Consequently, we can neglect tip–sample forces ($F_{ts} \equiv 0$), resulting in the well-known equation of motion of a driven damped harmonic oscillator.

After some time the external driving amplitude forces the cantilever to oscillate exactly at the driving frequency f_d. Therefore, the steady-state solution is given by the ansatz

$$z(t \gg 0) = d + A \cos(2\pi f_d t + \phi), \quad (2.12)$$

where ϕ is the phase difference between the excitation and the oscillation of the cantilever. With this, we obtain two functions for the amplitude and phase curves:

$$A = \frac{a_d}{\sqrt{\left(1 - \frac{f_d^2}{f_0^2}\right)^2 + \left(\frac{1}{Q_0}\frac{f_d}{f_0}\right)^2}}, \quad (2.13a)$$

$$\tan\phi = \frac{1}{Q_0} \frac{f_d/f_0}{1 - f_d^2/f_0^2}. \quad (2.13b)$$

The features of such an oscillator are well known from introductory physics courses.

If the cantilever is brought closer towards the sample surface, the tip senses the tip–sample interaction force, F_{ts}, which changes the oscillation behavior of the cantilever. However, as the mathematical form of realistic tip–sample forces is highly nonlinear, this fact complicates the analytical solution of the equation of motion Equation 2.11. For the analysis of DFM experiments we need to focus on steady-state solutions of the equation of motion with sinusoidal cantilever oscillation. Therefore, it is advantageous to expand the tip–sample force into a Fourier series

$$\begin{aligned}F_{ts}[z(t),\dot{z}(t)] \approx &\, f_d \int_0^{1/f_d} F_{ts}[z(t),\dot{z}(t)] dt \\ &+ 2f_d \int_0^{1/f_d} F_{ts}[z(t),\dot{z}(t)] \cos(2\pi f_d t + \phi) dt \times \cos(2\pi f_d t + \phi) \\ &+ 2f_d \int_0^{1/f_d} F_{ts}[z(t),\dot{z}(t)] \sin(2\pi f_d t + \phi) dt \times \sin(2\pi f_d t + \phi) \\ &+ \ldots, \end{aligned} \quad (2.14)$$

where $z(t)$ is given by Equation 2.12.

2.3 Amplitude Modulation (Tapping Mode)

The first term in the Fourier series reflects the averaged tip–sample force over one full oscillation cycle, which shifts the equilibrium point of the oscillation by a small offset Δd from d to d_0. Actual values for Δd, however, are very small. For typical amplitudes used in AM-AFM in air (some nm to some tens of nm), the averaged tip–sample force is in the range of some pN. The resultant offset Δd is less than 1 pm for typical sets of parameters. As this is well beyond the resolution limit of an AM-AFM experiment in air, we neglect this effect in the following and assume $d \approx d_0$ and $D = d - A$.

For further analysis, we now insert the first harmonics of the Fourier series Equation 2.14 into the equation of motion (Equation 2.11), thus obtaining two coupled equations [53, 54]

$$\frac{f_0^2 - f_d^2}{f_0^2} = I_+(d, A) + \frac{a_d}{A}\cos\phi, \tag{2.15a}$$

$$-\frac{1}{Q_0}\frac{f_d}{f_0} = I_-(d, A) + \frac{a_d}{A}\sin\phi, \tag{2.15b}$$

where the following integrals have been defined:

$$I_+(d, A) = \frac{2f_d}{c_z A}\int_0^{1/f_d} F_{ts}[z(t), \dot{z}(t)]\cos(2\pi f_d t + \phi)dt$$
$$= \frac{1}{\pi c_z A^2}\int_{d-A}^{d+A}(F_\downarrow + F_\uparrow)\frac{z-d}{\sqrt{A^2 - (z-d)^2}}dz, \tag{2.16a}$$

$$I_-(d, A) = \frac{2f_d}{c_z A}\int_0^{1/f_d} F_{ts}[z(t), \dot{z}(t)]\sin(2\pi f_d t + \phi)dt$$
$$= \frac{1}{\pi c_z A^2}\int_{d-A}^{d+A}(F_\downarrow - F_\uparrow)dz \tag{2.16b}$$
$$= \frac{1}{\pi c_z A^2}\Delta E(d, A).$$

Both integrals are functions of the actual oscillation amplitude A and the cantilever–sample distance d. Furthermore, they depend on the sum and the difference of the tip–sample forces during approach ($F_\downarrow$) and retraction ($F_\uparrow$), as manifested by the labels '+' and '−' for easy distinction. The integral I_+ is a weighted average of the tip–sample forces ($F_\downarrow + F_\uparrow$). On the other hand, the integral I_- is directly connected to ΔE, which reflects the energy dissipated during an individual oscillation cycle. Consequently, this integral vanishes for purely conservative tip–sample forces, where $F_\downarrow$ and $F_\uparrow$ are identical. A more detailed discussion of these integrals can be found in Refs. [55, 56].

By combining Equations 1.13b and 1.16b we obtain a direct correlation between the phase and the energy dissipation.[1]

---

[1] The '−'-sign on the right-hand side of the equation is due to our definition of the phase ϕ in Equation 2.12.

$$\sin\phi = -\left(\frac{A f_d}{A_0 f_0} + \frac{Q_0 \Delta E}{\pi c_z A_0 A}\right). \tag{2.17}$$

This relationship can be also obtained from the conservation of energy principle [41–43].

Equation (2.15) can be used to calculate the resonance curves of a dynamic force microscope, including tip–sample forces. The results are

$$A = \frac{a_d}{\sqrt{\left(1 - \frac{f_d^2}{f_0^2} - I_+(d, A)\right)^2 + \left(\frac{1}{Q_0}\frac{f_d}{f_0} + I_-(d, A)\right)^2}}, \tag{2.18a}$$

$$\tan\phi = \frac{\frac{1}{Q_0}\frac{f_d}{f_0} + I_-(d, A)}{1 - \frac{f_d^2}{f_0^2} - I_+(d, A)}. \tag{2.18b}$$

Equation 2.18a describes the shape of the resonance curve, but it is an implicit function of the oscillation amplitude A, and cannot be plotted directly.

Figure 2.13 contrasts the solution of this equation (solid lines) with numerical solution (symbols). As pointed out by various authors (see Refs [47, 57–63]), the amplitude versus frequency curves are multivalued within certain parameter ranges. Moreover, as the gradient of the analytical curve increases to infinity at specific positions, some branches become unstable. The resulting instabilities are reflected by the 'jumps' in the simulated curves (marked by arrows in Figure 2.13), where only stable oscillation states are obtained. Obviously, they are different for increasing and decreasing driving frequencies. This well-known effect is frequently observed in nonlinear oscillators (see Refs [64, 65]).

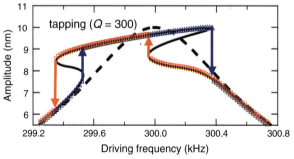

Figure 2.13 Resonance curves for tapping mode operation if the cantilever oscillates near the sample surface with $d = 8.5$ nm and $A_0 = 10$ nm, thereby experiencing the model force field given by Equation 2.9. The solid lines represent the analytical result of Equation 2.21, while the symbols are obtained from a numerical solution of the equation of motion, Equation 2.11. The dashed lines reflect the resonance curves without tip–sample force, and are shown purely for comparison. The resonance curve exhibits instabilities ('jumps') during a frequency sweep; these jumps take place at different positions (marked by arrows), depending on whether the driving frequency is increased or decreased.

2.3 Amplitude Modulation (Tapping Mode)

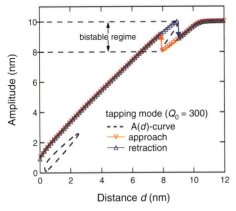

Figure 2.14 Amplitude versus distance curve for conventional ('tapping mode') AM-AFM for $A_0 = 10$ nm, $f_0 = 300$ kHz, and a tip–sample interaction force as given in Figure 2.6. The dashed lines represent the analytical result, while the symbols are obtained from a numerical solution of the equation of motion, Equation 2.11. The overall amplitudes decrease during an approach towards the sample surface in both cases, although instabilities (indicated by red and blue arrows) occur.

In AM-AFM, the cantilever might be oscillated at any frequency around the resonance peak. Here, we restrict ourselves to the situation where the driving frequency is set *exactly* to the eigenfrequency of the cantilever ($f_d = f_0$). With this choice – which is also very common in actual DFM experiments – we have defined imaging conditions leading to handy formulas suitable for further analysis. A discussion on the alternative cases of driving the cantilever slightly above or slightly below resonance can be found in the Refs [66–68].

From Equation (2.18a) we obtain the following relationship between the free oscillation amplitude A_0, the actual amplitude A, and the equilibrium tip position d:

$$A_0 = A\sqrt{1 + (Q_0 I_+[d, A])^2}. \tag{2.19}$$

In order to derive this formula, we used the approximation that the maximal value of the free oscillation amplitude at resonance is given by $A_0 \approx a_d Q_0$.

Solving this equation allows us to study amplitude versus distance curves for different effective Q-factors, as shown in Figure 2.14 for a Q-factor of 300. As observed previously in the resonance curves displayed in Figure 2.13, stable and unstable branches develop, which can be unambiguously identified by a comparison with numerical results (symbols).

Most noticeable, the tapping-mode curve exhibits jumps between unstable branches, which occur at different locations for approach and retraction. The resulting bistable regime then causes a hysteresis between approach and retraction, which has been the focus of numerous experimental and theoretical studies (see Refs [47, 61, 66, 67, 69–72]). As shown by various authors, the instability in conventional AM-AFM divides the tip–sample interaction into two regimes [47, 61, 70, 72]. Before the instability occurs, the tip interacts during an individual oscillation exclusively with the attractive part of

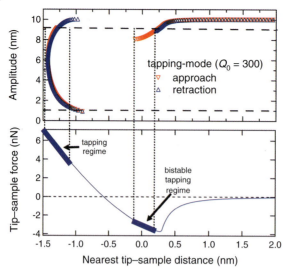

Figure 2.15 A comparison between the maximum tip–sample forces (tip–sample forces acting at the point of closest tip–sample approach/nearest tip–sample position D) experienced by conventional 'tapping mode' AM-AFM, assuming the same parameters as in Figure 2.14. The upper graph shows the nearest tip–sample position D versus the actual oscillation amplitude A for tapping mode. The lower graph shows the force regimes sensed by the tip. The maximal tip–sample forces in tapping mode are on the repulsive (tapping regime) as well as attractive (bistable tapping regime) part of the tip–sample force curve.

the tip–sample force. After jumping to the higher branch, however, the tip senses also the repulsive part of the tip–sample interaction.

In Figure 2.15 the oscillation amplitude is plotted as a function of the nearest tip–sample distance. In addition, the lower graph depicts the corresponding tip–sample force (cf. Figure 2.6). The origin of the nearest tip–sample position D is defined by this force curve. As both the amplitude curves and the tip–sample force curve are plotted as a function of the nearest tip–sample position, it is possible to identify the resulting maximum tip–sample interaction force for a given oscillation amplitude.

A closer look at the $A(D)$-curves helps to identify the different interaction regimes in AM-AFM. During the approach of the vibrating cantilever towards the sample surface, this curve shows a discontinuity for the nearest tip–sample position D (the point of closest approach during an individual oscillation) between 0 and -1 nm. This gap corresponds to the bistability and the resulting jumps in the amplitude versus distance curve. When the jump from the attractive to the repulsive regime has occurred, the amplitude decreases continuously, but the nearest tip–sample position does not reduce accordingly, remaining roughly between -0.8 nm and -1.5 nm. As a result, larger A/A_0 ratios do not necessarily translate into lower tip–sample interactions – a point which is important to bear in mind while adjusting imaging parameters in tapping mode AM-AFM imaging. In contrast, once the repulsive regime has been reached, the user's ability to influence the tip–sample interaction strength by modifying the set-point for A is limited, thus also limiting the possibilities of improving the image quality.

For practical applications, it is reasonable to assume that the set-point of the amplitude used for imaging has been set to a value between 90% (= 9 nm) and 10% (= 1 nm) of the free oscillation amplitude. With this condition, we can identify the accessible imaging regimes indicated by the horizontal (dashed) lines and the corresponding vertical (dotted) lines in Figure 2.15. In tapping mode, two imaging regimes are realized: the *tapping* regime (left) and the *bistable tapping* regime (middle). The first can be accessed by any amplitude set-point between 9 nm and 1 nm, and results in a maximum tip–sample forces well within the repulsive regime. The second regime, belonging to the bistable imaging state, is only accessible during approach; here, the corresponding amplitude set-point is between 9 nm and 8 nm. Imaging in this regime is possible with the limitation that the oscillating cantilever might jump into the repulsive regime [68, 70].

2.3.1.2 Reconstruction of the Tip–Sample Interaction

Previously, we have outlined the influence of the tip–sample interaction on the cantilever oscillation, calculated the maximum tip–sample interaction forces based on the assumption of a specific model force, and subsequently discussed possible routes for image optimization. However, during AFM imaging, the tip–sample interaction is not known *a priori*. However, several groups [52, 73–75] have suggested solutions to this inversion problem. Here, we present an approach which is based on the analysis of the amplitude and phase versus distance curves which can easily be measured with most AM-AFM set-ups.

Let us start by applying the transformation $D = d - A$ to the integral I_+ in Equation 2.16a, where D corresponds to the nearest tip–sample distance, as defined in Figure 2.10. Next we note that, due to the cantilever oscillation, the current method intrinsically recovers the values of the force that the tip experiences at its lower turning point, where $F_\downarrow$ necessarily equals $F_\uparrow$. We thus define $F_{ts} = (F_\downarrow + F_\uparrow)/2$, and Equation 2.16a subsequently reads as

$$I_+ = \frac{2}{\pi c_z A^2} \int_D^{D+2A} F_{ts} \frac{z - D - A}{\sqrt{A^2 - (z - D - A)^2}} dz. \tag{2.20}$$

The amplitudes commonly used in AM-AFM are considerably larger than the interaction range of the tip–sample force. Consequently, tip–sample forces in the integration range between $D + A$ and $D + 2A$ are insignificant. For this so-called 'large-amplitude approximation' [76, 77], the last term can be expanded at $z \to D$ to $(z - D - A)/\sqrt{A^2 - (z - D - A)^2} \approx -\sqrt{A/2(z - D)}$, resulting in

$$I_+ \approx -\frac{\sqrt{2}}{\pi c_z A^{3/2}} \int_D^{D+2A} \frac{F_{ts}}{\sqrt{z - D}} dz. \tag{2.21}$$

By introducing this equation into Equation 2.15a, we obtain the following integral equation:

$$\underbrace{\frac{c_z A^{3/2}}{\sqrt{2}} \left[\frac{a_d \cos(\phi)}{A} - \frac{f_0^2 - f_d^2}{f_0^2} \right]}_{\kappa} = \frac{1}{\pi} \int_D^{D+2A} \frac{F_{ts}}{\sqrt{z-D}} dz. \qquad (2.22)$$

The left-hand side of this equation contains only experimentally accessible data, and we denote this term as κ. The benefit of these transformations is that the integral equation can be inverted [65, 76] and, as a final result, we find

$$F_{ts}(D) = -\frac{\partial}{\partial D} \int_D^{D+2A} \frac{\kappa(z)}{\sqrt{z-D}} dz. \qquad (2.23)$$

It is now straightforward to recover the tip–sample force using Equation 2.23 from a 'spectroscopy experiment' – that is, an experiment where the amplitude and the phase are continuously measured as a function of the actual tip–sample distance $D = d - A$ at a fixed location. With this input, one first calculates κ as a function of D. In a second step, the tip–sample force is computed solving the integral in Equation 2.23 numerically.

Additional information about the tip–sample interaction can be obtained, remembering that the integral I_- is directly connected to the energy dissipation ΔE. By simply combining Equations 1.15b and 1.16b, we get

$$\Delta E = \left(\frac{1}{Q_0} \frac{f_d}{f_0} + \frac{a_d}{A} \sin\phi \right) \pi c_z A^2. \qquad (2.24)$$

The same result was found earlier by Cleveland *et al.* [41], using the conservation of energy principle. However, in a further development of Cleveland's investigations we suggest plotting the energy dissipation as a function of the nearest tip–sample distance $D = d - A$ in order to have the same scaling as for the tip–sample force.

An application of the method to experimental data obtained on a silicon wafer is shown in Figure 2.16, where only the data points before the jump were used to reconstruct the tip–sample force and energy dissipation. As a consequence, the experimental force curve showed only the attractive part of the force between tip and sample, with a minimum of -1.8 nN. This result was in agreement with previous studies which stated that the tip sensed only attractive forces before the jump [66, 69, 78].

2.3.2
Frequency-Modulation or Noncontact Mode in Vacuum

In order to obtain high-resolution images with an atomic force microscope, it is very important to prepare clean sample surfaces that are free from unwanted adsorbates. Therefore, these experiments are usually performed in ultra-high vacuum with pressures below 1×10^{-10} mbar. As a consequence, most DFM experiments in

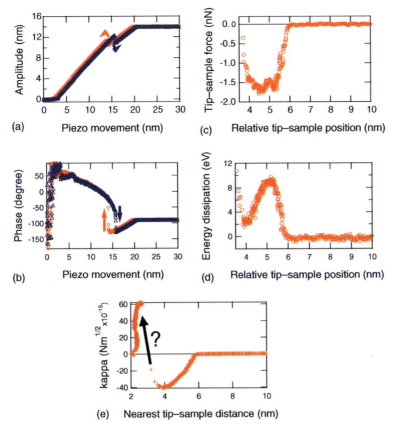

Figure 2.16 Dynamic force spectroscopy experiment on a silicon wafer in air (parameters of the cantilever: $f_d = f_0 = 328.61$ kHz, $c_z = 33.45$ N m^{-1}, $Q_0 = 537$). (a) A measurement of the oscillation amplitude as a function of the oscillation amplitude shows jumps at different positions during approach and retraction; (b) The jumps are also observed in the phase versus distance curves; (c) Using the algorithm described in the text, the tip–sample force can be reconstructed. This curve is calculated from the approach data. Only the data points before the jump are used for reconstruction of the tip–sample force; (d) The energy dissipation per oscillation cycle can be easily obtained from Equation 2.28; (e) This graph shows the $\kappa(D)$-values computed from the amplitude and phase versus distance curves plotted in panels (a) and (b). The jump in these curves results also in a jump in the κ-curve.

vacuum utilize the FM detection scheme introduced by Albrecht *et al.* [33]. In this mode, the cantilever is self-oscillated, in contrast to the AM- or tapping-mode discussed in Section 2.2.3. The FM technique enables the imaging of single point defects on clean sample surfaces in vacuum, and its resolution is comparable with that of the scanning tunneling microscope, while not restricted to conducting surfaces. During the years after the invention of the FM technique the term noncontact atomic force microscopy (NC-AFM) was established, because it is commonly believed that a repulsive, destructive contact between the tip and sample

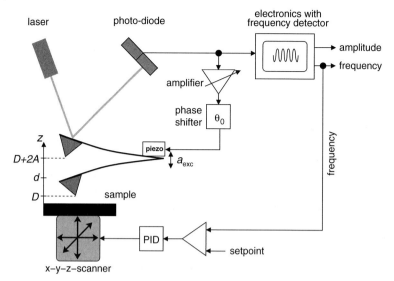

Figure 2.17 The schematic set-up of a dynamic force microscope using the frequency modulation technique. This experimental set-up is often used in UHV. A significant feature is the positive feedback of the self-driven cantilever. The detector signal is amplified and phase-shifted before being used to drive the piezo. The measured quantity is the frequency shift due to the tip–sample interaction, which serves as the control signal for the cantilever–sample distance.

is prevented by this technique. In the following subsection we introduce the basic principles of the experimental set-up, explain the origin and calculation of the detected frequency shift, and present applications of this mode.

2.3.2.1 Set-Up of FM-AFM

In vacuum applications, the Q-factor of silicon cantilevers is in the range of 10 000 to 30 000. High Q-factors, however, limit the acquisition time (bandwidth) of DFM, as the oscillation amplitude of the cantilever requires a long time to adjust. This problem is avoided by the FM-detection scheme based on the specific features of a self-driven oscillator.

The basic set-up of a dynamic force microscope utilizing this driving mechanism is shown schematically in Figure 2.17. The movement of the cantilever is measured with a displacement sensor, after which this signal is fed back into an amplifier with an automatic gain control (AGC); the signal is subsequently used to excite the piezo oscillating the cantilever. The time delay between the excitation signal and cantilever deflection is adjusted by a time ('phase') shifter to a value $t_0 = 1/(4f_0)$, corresponding to $\approx 90°$, as this ensures an oscillation at resonance. Two different modes have been established: (i) the *constant-amplitude mode* [33], where the oscillation amplitude A is kept at a constant value by the AGC; and (ii) the *constant excitation mode* [79], where the excitation amplitude is kept constant. In the following, however, we focus on the constant amplitude mode.

The key feature of the described set-up is the positive feedback loop which oscillates the cantilever always at its resonance frequency, f [39]. The reason for this behavior is that the cantilever serves as the frequency-determining element. This is in contrast to an external driving of the cantilever by a frequency generator, where the driving frequency f_d is not necessarily the resonant frequency of the cantilever.

If the cantilever oscillates near the sample surface, the tip–sample interaction alters its resonant frequency, which is then different from the eigenfrequency f_0 of the free cantilever. The actual value of the resonant frequency depends on the nearest tip–sample distance and the oscillation amplitude. The measured quantity is the *frequency shift* Δf, which is defined as the difference between both frequencies ($\Delta f = f - f_0$). The detection method received its name from the frequency demodulator (FM-detector). The cantilever driving mechanism, however, is independent of this part of the set-up. Other set-ups use a phase-locked loop (PLL) to detect the frequency and to oscillate the cantilever exactly with the frequency measured by the PLL [80].

For imaging, the frequency shift Δf is used to control the cantilever sample distance. Thus, the frequency shift is constant and the acquired data represents planes of constant Δf, which can be related to the surface topography in many cases. The recording of the frequency shift as a function of the tip–sample distance, or alternatively the oscillation amplitude can be used to determine the tip–sample force with high resolution (see Section 2.2.4.5).

2.3.2.2 Origin of the Frequency Shift

Before presenting experimental results obtained in vacuum, we will analyze the origin of the frequency shift. A good insight into the cantilever dynamics is provided by examining the tip potential displayed in Figure 2.18. If the cantilever is far away from the sample surface, the tip moves in a symmetric, parabolic potential (dotted line), and its oscillation is harmonic. In such a case, the tip motion is sinusoidal and the resonance

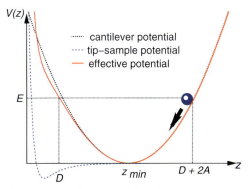

Figure 2.18 The frequency shift in dynamic force microscopy is caused by the tip–sample interaction potential (dashed line), which alters the harmonic cantilever potential (dotted line). Therefore, the tip moves in an anharmonic and asymmetric effective potential (solid line). z_{min} is the minimum position of the effective potential.

frequency is given by the eigenfrequency f_0 of the cantilever. If, however, the cantilever approaches the sample surface, the potential – which determines the tip oscillation – is modified to an effective potential V_{eff} (solid line) given by the sum of the parabolic potential and the tip–sample interaction potential V_{ts} (dashed line). This effective potential differs from the original parabolic potential and shows an asymmetric shape.

As a result of this modification of the tip potential the oscillation becomes anharmonic, and the resonance frequency of the cantilever depends now on the oscillation amplitude A. Since the effective potential experienced by the tip changes also with the nearest distance D, the frequency shift is a functional of both parameters ($\Rightarrow \Delta f := \Delta f(D, A)$).

Figure 2.19 displays some experimental frequency shift versus distance curves for different oscillation amplitudes. These experiments were carried out with

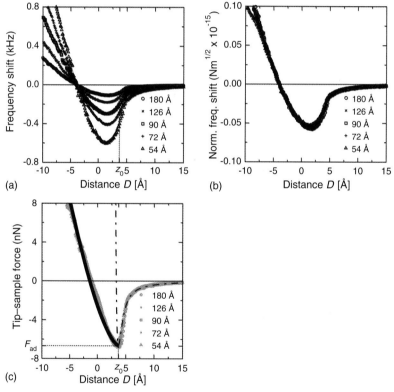

Figure 2.19 (a) Experimental frequency shift versus distance curves acquired with a silicon cantilever ($c_z = 38$ N m^{-1}; $f_0 = 171$ kHz) and a graphite sample for different amplitudes (54–180 Å) in UHV at low temperature ($T = 80$ K). The curves are shifted along the x-axes to make them comparable; (b) Transformation of all frequency shift curves shown in (a) to one universal curve using Equation 2.32. The normalized frequency shift $\gamma(D)$ is nearly identical for all amplitudes; (c) The tip–sample force calculated with the experimental data shown in (a) and (b), using the formula in Equation 2.33. The force F_{ts} (Equation 2.34) is plotted using a dashed-dotted line; the best fit using the force law F_c is displayed by a solid line. The border between 'contact' and 'noncontact' force is marked by the position z_0.

an atomic force microscope designed for operation in UHV and at low temperatures [10].

The obtained experimental frequency shift versus distance curves show a behavior expected from the simple model explained above. All curves show a similar overall shape, but differ in magnitude depending on the oscillation amplitude and the nearest tip–sample distance. During the approach of the cantilever towards the sample surface, the frequency shift decreases and reaches a minimum. With a further reduction of the nearest tip–sample distance, the frequency shift increases again and becomes positive. For smaller oscillation amplitudes, the minimum of the $\Delta f(z)$-curves is deeper and the slope after the minimum is steeper than for larger amplitudes – that is, the overall effect is larger for smaller amplitudes.

This can also be explained by the simple potential model: A decrease in the amplitude A for a fixed nearest distance D moves the minimum of the effective potential closer to the sample surface. Therefore, the relative perturbation of the harmonic cantilever potential increases, which increases also the absolute value of the frequency shift.

2.3.2.3 Theory of FM-AFM

As described in the previous subsection, it is a specific feature of the FM technique that the cantilever is 'self-driven' by a positive feedback loop. Due to this experimental set-up, the corresponding equation of motion is different from the case of the externally driven cantilever discussed in Section 2.2.3. The external driving term must be replaced in order to describe the self-driving mechanism correctly; therefore, the equation of motion is given by

$$m^*\ddot{z}(t) + \frac{2\pi f_0 m^*}{Q}\dot{z}(t) + c_z(z(t)-d) + \underbrace{gc_z(z(t-t_0)-d)}_{\text{driving}} = F_{ts}[z(t), \dot{z}(t)] \quad (2.25)$$

where $z := z(t)$ represents the position of the tip at the time t; and c_z, m and Q are the spring constant, the effective mass and the quality factor of the cantilever, respectively. $F_{ts} = -(\partial V_{ts})/(\partial z)$ is the tip–sample interaction force. The last term on the left describes the active feedback of the system by the amplification of the displacement signal by the *gain factor g* measured at the retarded time $t - t_0$.

The frequency shift can be calculated from the above equation of motion with the ansatz

$$z(t) = d + A\cos(2\pi f t) \quad (2.26)$$

describing the stationary solutions of Equation 2.25. As described in Section 2.2.3, it is assumed that the cantilever oscillations are more or less sinusoidal, such that the tip–sample force F_{ts} is developed into a Fourier-series, as in Equation 2.14. This procedure results in a set of two coupled trigonometric equations [38, 39]:

$$g\cos(2\pi f t_0) = \frac{f^2 - f_0^2}{f_0^2} + I_+ \quad (2.27)$$

$$g \sin(2\pi f t_0) = \frac{1}{Q} \frac{f}{f_0} + I_- \qquad (2.28)$$

where the two integrals I_+ Equation 2.16a and I_- Equation 2.16b were defined in accordance to Section 2.2.3.2. These two coupled equations can be solved numerically, if one is interested in the exact dependency of the tip–sample interaction force F_{ts} and the time delay t_0 on the oscillation frequency f and the gain factor g.

Fortunately, a detailed analysis shows that the results of a FM-AFM experiment are mainly determined by the tip–sample force, and only very slightly by the time delay, if t_0 is set to an optimal value before approaching the tip towards the sample surface. These values of the time delay are specific resonance values corresponding to 90° (i.e. $t_0 = 1/4f_0$), and can be easily found by minimizing the gain factor as a function of the time delay. Therefore, it can be assumed that $\cos(2\pi f t_0) \approx 0$ and $\sin(2\pi f t_0) \approx 1$ and the two coupled Equations 1.27 and 1.28 can be decoupled. As a result, an equation for the frequency shift is obtained:

$$\Delta f \cong -\frac{f_0}{2} I_+ = \frac{1}{\pi c_z A^2} \int_{d-A}^{d+A} (F_\downarrow + F_\uparrow) \frac{z-d}{\sqrt{A^2 - (z-d)^2}} dz \qquad (2.29)$$

and the energy dissipation

$$\Delta E = \left(g - \frac{1}{Q} \frac{f}{f_0} \right) \pi c_z A^2. \qquad (2.30)$$

As no assumptions were made about the specific force law of the tip–sample interaction F_{ts}, these equations are valid for any type of interaction as long as the resulting cantilever oscillations remain nearly sinusoidal.

As the amplitudes in FM-AFM are often considerably larger than the distance range of the tip–sample interaction, we can again make the 'large amplitude approximation' [76, 77] and introduce the approximation Equation 2.21 for the integral I_-. This yields the formula

$$\Delta f = \frac{1}{\sqrt{2\pi}} \frac{f_0}{c_z A^{3/2}} \int_D^{D+2A} \frac{F_{ts}(z)}{\sqrt{z-D}} dz \qquad (2.31)$$

It is interesting to note that the integral in this equation is virtually independent of the oscillation amplitude. The experimental parameters (c_z, f_0 and A) appear as prefactors. Consequently, it is possible to define the *normalized frequency shift* [77]

$$\gamma(z) := \frac{c_z A^{3/2}}{f_0} \Delta f(z). \qquad (2.32)$$

This is a very useful quantity to compare experiments obtained with different amplitudes and cantilevers. The validity of Equation 2.32 is nicely demonstrated by the application of this equation to the frequency shift curves already presented in Figure 2.19a. As shown in Figure 2.19b, all curves obtained for different amplitudes result in one universal γ-curve, which depends only on the actual tip–sample distance, D.

2.3.2.4 Applications of FM-AFM

The initial excitement surrounding the NC-AFM technique in UHV was driven by the first results of Giessibl [81], who was able to image the true atomic structure of the Si(1 1 1)-7 × 7-surface with this technique in 1995. In the same year, Sugawara *et al.* [82] observed the motion of single atomic defects on InP with true atomic resolution. However, imaging on conducting or semi-conducting surfaces is also possible by using scanning tunneling microscopy (STM), and these first NC-AFM images provided no new information on surface properties. The true potential of NC-AFM lies in the imaging of nonconducting surface with atomic precision, which was first demonstrated by Bammerlin *et al.* [83] on NaCl. A longstanding question about the surface reconstruction of the technological relevant material aluminum oxide could be answered by Barth *et al.* [84], who imaged the atomic structure of the high-temperature phase of α-Al_2O_3(0 0 0 1).

The high-resolution capabilities of NC-AFM are nicely demonstrated by the images shown in Figure 2.20. Allers *et al.* [85] resolved atomic steps and defects with atomic resolution on nickel oxide. Today, such a resolution is routinely obtained by various research groups (for an overview, see Refs [3, 34, 35, 86]). Recent efforts have also been concentrated on the analysis of functional organic molecules, since in the field of nanoelectronics it is anticipated that organic molecules in particular will play an important role as the fundamental building blocks of nanoscale electronic device elements. For example, atomic resolution on the highly curved surface of a nanotube [87] was achieved. The analysis of growth properties of thin films [88–90] with respect to their electronic properties has been investigated, while the intramolecular contrast of individual molecules has also been resolved [91], which might be directly related to the internal charge density distribution inside the molecules.

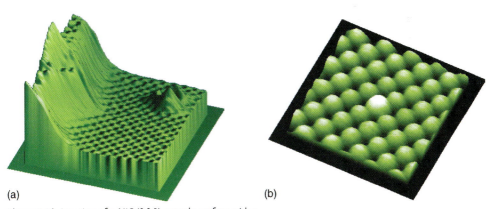

Figure 2.20 Imaging of a NiO(0 0 1) sample surface with a noncontact AFM. (a) Surface step and an atomic defect. The lateral distance between two atoms is 4.17 Å; (b) A dopant atom is imaged as a light protrusion about 0.1 Å higher than the other atoms. (Images courtesy of W. Allers and S. Langkat, University of Hamburg; used with kind permission.)

2.3.2.5 Dynamic Force Spectroscopy

Since its invention in 1986, the atomic force microscope has been used extensively to study tip–sample interactions for various material combinations. Unfortunately, in contact mode such investigations were often hindered close to the sample surface by a 'jump to contact' (see Section 2.2.1.1). In tapping mode, on the other hand, the force analysis is limited due to the instabilities in the amplitude and phase versus distance curves (see Section 2.2.3.3). Such problems, however, are avoided by using large oscillation amplitudes in the FM technique.

In Section 2.2.4.3 it was shown how the frequency shift can be calculated for a given tip–sample interaction law. The inverse problem, however, is even more interesting: *How can the tip-sample interaction be determined from frequency shift data?* Various mathematical solutions to this question have been presented by many research groups [76, 92–97], and this has led to the dynamic force spectroscopy (DFS) technique, which is a direct extension of the FM-AFM mode.

Here, we present the approach of Dürig [76], which is based on the inversion of the integral Equation 2.31 already presented in Section 2.2.4.3. This can be transformed to

$$F_{ts}(D) = \sqrt{2} \frac{c_z A^{3/2}}{f_0} \frac{\partial}{\partial D} \int_D^{\infty} \frac{\Delta f(z)}{\sqrt{z-D}} dz, \qquad (2.33)$$

which allows a direct calculation of the tip–sample interaction force from the frequency shift versus distance curves.

An application of this formula to the experimental frequency shift curves already presented in Section 2.2.4.2 is shown in Figure 2.19c. The obtained force curves are almost identical, despite being obtained with different oscillation amplitudes. As the tip–sample interactions can be measured with high resolution, DFS opens a direct way to compare experiments with theoretical models and predictions.

Giessibl [77] suggested a description of the force between the tip and the sample by combining a long-range (van der Waals) and a short-range (Lennard–Jones) term (see Section 2.1.3). Here, the long-range part describes the van der Waals interaction of the tip, modeled as a sphere with a specific radius, with the surface. The short-range Lennard–Jones term is a superposition of the attractive van der Waals interaction of the last tip apex atom with the surface and the coulombic repulsion. For a tip with radius R, this assumption results in the tip–sample force:

$$F_{nc}(z) = -\frac{A_H R}{6z^2} + \frac{12 E_0}{r_0} \left(\left(\frac{r_0}{z}\right)^{13} - \left(\frac{r_0}{z}\right)^7 \right). \qquad (2.34)$$

As this approach does not explicitly consider elastic contact forces between tip and sample, we call this the 'noncontact' force law in the following sections.

A fit of this equation to the experimental tip–sample force curve is shown in Figure 2.19c by a solid line; the obtained parameters are $A_H R = 2.4 \times 10^{-27}$ Jm, $r_0 = 3.4$ Å, and $E_0 = 3$ eV [98]. The regime on the right from the minimum fits well to the experimental data, but the deep and wide minimum of the experimental curves cannot be described accurately with the noncontact force. This is caused by the steep increase in the Lennard–Jones force in the repulsive regime ($\Rightarrow F_{ts} \propto 1/r^{13}$ for $z < r_0$).

The elastic contact behavior can be described with the assumption of the above-described DMT-M model (see Section 2.1.3), that the overall shape of tip and sample changes only slightly until point contact is reached and that, after the formation of this point contact, the tip–sample forces are described by the Hertz theory. A fit of the Hertz model to the experimental data is shown in Figure 2.19 by a solid line. The experimental force curves agree quite well with the contact force law for distances $D<z_0$. This shows that the overall behavior of the experimentally obtained force curves can be described by a combination of long-range (van der Waals), short-range (Lennard–Jones), and contact (Hertz/DMT) forces.

As Equation 2.31 was derived under the assumption that the resonance amplitude is considerably larger than the decay length of the tip–sample interaction, the same restriction applies for Equation 2.33. However, by using more advanced algorithms it is also possible to determine forces from DFS experiments without the large amplitude restriction. The numerical approach of Gotsmann et al. [94], as well as the semi-empirical methods of Dürig [92], Giessibl [93] and Sader and Jarvis [97], are applicable in all regimes.

The resolution of DFS can be driven down to the atomic scale. Lantz et al. [99] measured frequency shift versus distance curves at different lattice sites of the Si (1 1 1)-(7 × 7) surface, and in this way were able to distinguish differences in the bonding forces between inequivalent adatoms of the 7 × 7 surface reconstruction of silicon.

The concept of DFS can be also extended to three-dimensional (3-D)-force spectroscopy by mapping the complete force field above the sample surface [100]. A schematic of the measurement principle is shown in Figure 2.21a. Frequency shift versus distance curves are recorded on a matrix of points perpendicular to the sample

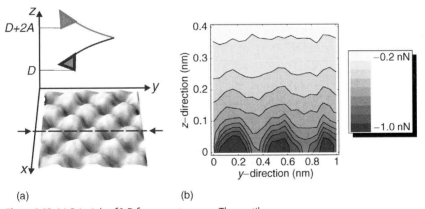

(a) (b)

Figure 2.21 (a) Principle of 3-D force spectroscopy. The cantilever oscillates near the sample surface and measure the frequency shift in a x–y–z-box. The 3-D surface shows the topography of the sample (image size: 10 Å × 10 Å) obtained immediately before recording of the spectroscopy field; (b) The reconstructed force field of NiO (0 0 1) shows atomic resolution. The data are recorded along the dotted line shown in (a).

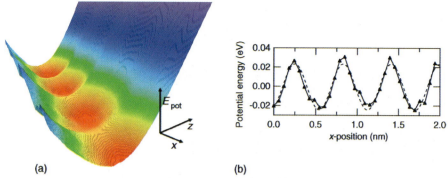

Figure 2.22 (a) A 3-D representation of the interaction energy map determined from 3-D force spectroscopy experiments on a NaCl(1 0 0) crystal surface. The red circular depressions represent the local energy minima; (b) Potential energy profile obtained from (a) by collecting the energy minimum values along the x-axis. This curve thus directly reveals the potential energy barrier of $\Delta E_{barrier} = 48$ meV which separates the local energy minima.

surface. By using Equation 2.33, the complete 3-D force field between the tip and sample can be recovered with atomic resolution. Figure 2.21b shows a cut through the force field as a two-dimensional (2-D) map.

The 3D-force technique has been applied also to a NaCl(1 0 0) surface, where not only conservative but also the dissipative tip–sample interaction could be measured in full space [101]. Initially, the forces were measured in the attractive as well as repulsive regime, allowing for the determination of the local minima in the corresponding potential energy curves (Figure 2.22). This information is directly related to the atomic energy barriers responsible for a multitude of dynamic phenomena in surface science, such as diffusion, faceting and crystalline growth. The direct comparison of conservative with the simultaneously acquired dissipative processes furthermore allowed determining atomic-scale mechanical relaxation processes.

If the NC-AFM is capable of measuring forces between single atoms with sub-nN precision, why should it not be possible to also exert forces with this technique? In fact, the new and exciting field of nanomanipulation would be driven to a whole new dimension, if defined forces could be reliably applied to single atoms or molecules. In this respect, Loppacher *et al.* [102] were able to exert pressure on different parts of an isolated Cu–TBBP molecule, which is known to possess four rotatable legs. Here, the force–distance curves were measured while one of the legs was pushed by the AFM tip and turned by 90°, and hence were able to measure the energy which was dissipated during the 'switching' of this molecule between different conformational states. The manipulation of single silicon atoms with NC-AFM was demonstrated by Oyabu *et al.* [103], who removed single atoms from a Si(1 1 1)-7 × 7 surface with the AFM tip and were able subsequently to re-deposit atoms from the tip onto the surface. This approach was driven to its limits by Sugimoto *et al.*, who manipulated single Sn-

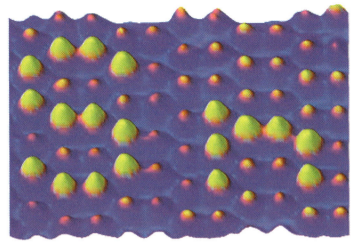

Figure 2.23 Final topographic NC-AFM image of the process of rearranging single atoms at room temperature. The image was acquired with a cantilever oscillation amplitude of 15.7 nm, using a Si cantilever. (Reproduced from Ref. [104].)

atoms on the Ge(1 1 1)-c(2 × 8) semiconductor surface. By pushing single Sn-atoms from one lattice site to another, they finally succeeded in writing the term 'Sn' with single atoms (Figure 2.23).

2.4 Summary

In summary, we have presented an overview over the basic principles and modern applications of AFM. This versatile technique can be categorized into two operational modes, static and dynamic. The static mode allows the simultaneous measurement of normal and lateral forces, thus yielding direct information about friction mechanisms of nanoscale contacts. The main advantage of the dynamic mode is the possibility to control tip–sample distances while avoiding the undesirable and destructive 'jump-to-contact' phenomenon. Two different excitation schemes for dynamic force microscopy were introduced, where the amplitude-modulation or tapping mode are in particular well-suited to high-resolution imaging under ambient or liquid conditions. The ultimate 'true' atomic resolution, however, is limited to vacuum conditions using FM or noncontact techniques. Nonetheless, the impact of AFM reaches far beyond the high-resolution imaging of surface topography. DFS allows the quantification of tip–sample forces, through the systematic acquisition of parameters such as amplitude, phase and oscillation frequency as a function of the relative tip–sample distance. Based on this approach, not only the bonding force of single interatomic chemical bonds can be measured, but also the full 3-D force field can be determined, at atomic resolution. Finally, the finding that atomic forces can not only be measured but also exerted with atomic precision will open up the new and exciting field of nanomanipulation.

Acknowledgments

The authors would like to thank all colleagues who contributed to these studies with their images and experimental results, including Boris Anczykowski, Daniel Ebeling, Jan-Erik Schmutz, Domenique Weiner (University of Münster), Wolf Allers, Shenja Langkat, Alexander Schwarz (University of Hamburg) and Udo D. Schwarz (Yale University).

References

1 Binnig, G., Quate, C.F. and Gerber, Ch. (1986) Atomic force microscopy. *Physical Review Letters*, **56**, 930–933.
2 Binnig, G., Rohrer, H., Gerber, C. and Weibel, E. (1982) Surface studies by scanning tunneling microscopy. *Physical Review Letters*, **49**, 57–61.
3 Meyer, E., Hug, H.-J. and Bennewitz, R. (2004) *Scanning Probe Microscopy – The Lab on a Tip*, Springer-Verlag.
4 Bhushan, B. and Marti, O.(eds) (2005) Scanning probe microscopy – principle of operation, instrumentation, and probes, in *Nanotribology and Nanomechanics – An Introduction*, (ed. B. Bhushan) Springer-Verlag, Berlin Heidelberg, pp. 41–115.
5 Lüthi, R., Meyer, E., Haefke, H., Howald, L., Gutmannsbauer, W., Guggisberg, M., Bammerlin, M. and Güntherodt, H.-J. (1995) Nanotribology: an UHV-SFM study on thin films of C60 and AgBr. *Surface Science*, **338**, 247–260.
6 Neumeister, J.M. and Ducker, W.A. (1994) Lateral, normal, and longitudinal spring constants of atomic force microscopy cantilevers. *Review of Scientific Instruments*, **65**, 2527–2531.
7 Sader, J.E. (1995) Parallel beam approximation for V-shaped atomic force microscope cantilevers. *Review of Scientific Instruments*, **66**, 4583–4587.
8 Alexander, S., Hellemans, L., Marti, O., Schneir, J., Elings, V. and Hansma, P.K. (1988) An atomic-resolution atomic-force microscope implemented using an optical lever. *Journal of Applied Physics*, **65**, 164–167.
9 Meyer, G. and Amer, N.M. (1988) Novel optical approach to atomic force microscopy. *Applied Physics Letters*, **53**, 1045–1047.
10 Allers, W., Schwarz, A., Schwarz, U.D. and Wiesendanger, R. (1998) A scanning force microscope with atomic resolution in ultrahigh vacuum and at low temperatures. *Review of Scientific Instruments*, **69**, 221–225.
11 Kawakatsu, H., Kawai, S., Saya, D., Nagashio, M., Kobayashi, D., Toshiyoshi, H. and Fujita, H. (2002) Towards atomic force microscopy up to 100 MHz. *Review of Scientific Instruments*, **73** (6), 2317–2320.
12 Moser, A., Hug, H.-J., Jung, T., Schwarz, U.D. and Guntherodt, H.-J. (1993) A miniature fibre optic force microscope scan head. *Measurement Science and Technology*, **4**, 769–775.
13 Rugar, D., Mamin, H.J. and Guethner, P. (1989) Improved fiber-optic interferometer for atomic force microscopy. *Applied Physics Letters*, **55**, 2588–2590.

14 Linnemann, R., Gotzalk, T., Rangelow, I.W., Dumania, P. and Oesterschulze, E. (1996) Atomic force microscopy and lateral force microscopy using piezoresistive cantilevers. *Journal of Vacuum Science & Technology B*, **14** (2), 856–860.

15 Tortonese, M., Barrett, R.C. and Quate, C.F. (1993) Atomic resolution with an atomic force microscope using piezoresistive detection. *Applied Physics Letters*, **62**, 834–836.

16 Yuan, C.W., Batalla, E., Zacher, M., de Lozanne, A.L., Kirk, M.D. and Tortonese, M. (1994) Low temperature magnetic force microscope, utilizing a piezoresistive cantilever. *Applied Physics Letters*, **65**, 1308–1310.

17 Stahl, U., Yuan, C.W., de Lozanne, A.L. and Tortonese, M. (1994) Atomic force microscope using piezoresistive, cantilevers and combined with a scanning electron microscope. *Applied Physics Letters*, **65**, 2878–2880.

18 Israelachvili, J.N. (1992) *Intermolecular and Surface Forces*, Academic Press, London.

19 Stifter, Th., Marti, O. and Bhushan, B. (2000) Theoretical investigation of the distance dependence of capillary and van der Waals forces in scanning force microscopy. *Physical Review B - Condensed Matter*, **62**, 13667–13673.

20 Johnson, K.L. (1985) *Contact Mechanics*, Cambridge University Press, Cambridge, UK.

21 Landau, L.D. and Lifschitz, E.M. (1991) *Lehrbuch der theoretischen Physik VII: Elastizitätstheorie*, Akademie-Verlag, Berlin.

22 Johnson, K.L., Kendall, K. and Roberts, A.D. (1971) Surface energy and contact of elastic solids. *Proceedings of the Royal Society of London Series A - Mathematical, Physical and Engineering Sciences*, **324**, 301.

23 Derjaguin, B.V., Muller, V.M. and Toporov, Y.P. (1975) Effect of contact deformations on the adhesion of particles. *Journal of Colloid and Interface Science*, **53**, 314–326.

24 Schwarz, U.D. (2003) A generalized analytical model for the elastic deformation of an adhesive contact between a sphere and a flat surface. *Journal of Colloid and Interface Science*, **261**, 99–106.

25 Sarid, D. (1994) *Scanning Force Microscopy – With Applications to Electric, Magnetic, and Atomic Forces*, Oxford University Press.

26 Mate, C.M., McClelland, G.M., Erlandsson, R. and Chiang, S. (1987) Atomic-scale friction of a tungsten tip on a graphite surface. *Physical Review Letters*, **59**, 1942–1945.

27 McKendry, R., Theoclitou, M.-E., Rayment, T. and Abell, C. (1998) Chiral discrimination by chemical force microscopy. *Nature*, **391**, 566–568.

28 Overney, R.M., Meyer, E., Frommer, J., Brodbeck, D., Lüthi, R., Howald, L., Güntherodt, H.-J., Fujihira, M., Takano, H. and Gotoh, Y. (1992) Friction measurements on phase-separated thin films with a modified atomic force microscope. *Nature*, **359**, 133–135.

29 Burnham, N.A. and Colton, R.J. (1989) Measuring the nanomechanical properties and surface forces of materials using an atomic force microscopy. *Journal of Vacuum Science and Technology A - Vacuum Surfaces and, Films*, **7**, 2906.

30 Martin, Y., Williams, C.C. and Wickramasinghe, H.K. (1987) Atomic force microscope–force mapping and profiling on a sub 100-Å scale. *Journal of Applied Physics*, **61**, 4723–4729.

31 Putman, C.A.J., Vanderwerf, K.O., Degrooth, B.G., Vanhulst, N.F. and Greve, J. (1994) Tapping mode atomic force microscopy in liquid. *Applied Physics Letters*, **64**, 2454–2456.

32 Zhong, Q.D., Inniss, D., Kjoller, K. and Elings, V.B. (1993) Fractured polymer/silica fiber surface studied by tapping mode atomic force microscopy. *Surface Science Letters*, **290**, L688–L692.

33 Albrecht, T.R., Grütter, P., Horne, D. and Rugar, D. (1991) Frequency modulation detection using high-Q cantilevers for enhanced force microscope sensitivity. *Journal of Applied Physics*, **69**, 668–673.

34 Garcia, R. and Pérez, R. (2002) Dynamic atomic force microscopy methods. *Surface Science Reports*, **47**, 197–301.

35 Giessibl, F.-J. (2003) Advances in atomic force microscopy. *Reviews of Modern Physics*, **75**, 949–983.

36 Hölscher, H. and Schirmeisen, A. (2005) Dynamic force microscopy and spectroscopy, in *Advances in Imaging and Electron Physics*, (ed. P.W. Hawkes), Academic Press Ltd, London, pp. 41–101.

37 Morita, S., Wiesendanger, R. and Meyer, E.(eds) (2002) *Noncontact Atomic Force Microscopy*, Springer-Verlag, Berlin.

38 Hölscher, H., Gotsmann, B., Allers, W., Schwarz, U.D., Fuchs, H. and Wiesendanger, R. (2001) Measurement of conservative and dissipative tip-sample interaction forces with a dynamic force microscope using the frequency modulation technique. *Physical Review B - Condensed Matter*, **64**, 075402.

39 Hölscher, H., Gotsmann, B., Allers, W., Schwarz, U.D., Fuchs, H. and Wiesendanger, R. (2002a) Comment on "damping mechanism in dynamic force microscopy". *Physical Review Letters*, **88**, 019601.

40 Schönenberger, C. and Alvarado, S.F. (1989) A differential interferometer for force microscopy. *Review of Scientific Instruments*, **60**, 3131–3134.

41 Cleveland, J.P., Anczykowski, B., Schmid, A.E. and Elings, V.B. (1998) Energy dissipation in tapping-mode atomic force microscopy. *Applied Physics Letters*, **72**, 2613.

42 Garcia, R., Gómez, C.J., Martinez, N.F., Patil, S., Dietz, C. and Magerle, R. (2006) Identification of nanoscale dissipation processes by dynamic atomic force microscopy. *Physical Review Letters*, **97**, 016103.

43 Tamayo, J. and García, R. (1998) Relationship between phase shift and energy dissipation in tapping-mode scanning force microscopy. *Applied Physics Letters*, **73**, 2926–2928.

44 Maganov, S. (2004) Visualization of polymer structures with atomic force microscopy, in *Applied Scanning Probe Methods*, (eds H. Fuchs M. Hosaka and B. Bhushan) Springer-Verlag, pp. 207–250.

45 Klinov, D. and Maganov, S. (2004) True molecular resolution in tapping-mode atomic force microscopy with high-resolution probes. *Applied Physics Letters*, **84**, 2697–2698.

46 Möller, C., Allen, M., Elings, V., Engel, A. and Müller, D.J. (1999) Tapping-mode atomic force microscopy produces faithful high-resolution images of protein surfaces. *Biophysical Journal*, **77**, 1150–1158.

47 Lee, S.I., Howell, S.W., Raman, A. and Reifenberger, R. (2002) Nonlinear dynamics of microcantilevers in tapping mode atomic force microscopy: a comparison between theory and experiment. *Physical Review B - Condensed Matter*, **66**, 115409.

48 Rodríguez, T.R. and García, R. (2002) Tip motion in amplitude modulation (tapping-mode) atomic-force microscopy: Comparison between continuous and point-mass models. *Applied Physics Letters*, **80**, 1646–1648.

49 Stark, R.W. and Heckl, W. (2000) Fourier transformed atomic force microscopy: tapping mode atomic force microscopy beyond the hookian approximation. *Surface Science*, **457**, 219–228.

50 Stark, R.W., Schitter, G., Stark, M., Guckenheimer, R. and Stemmer, A. (2004) State-space model of freely vibrating and surface-coupled cantilever dynamics in atomic force microscopy. *Physical Review B - Condensed Matter*, **69**, 085412.

51 Legleiter, J. and Kowalewski, T. (2005) Insight into fluid tapping-mode atomic force microscopy provided by numerical

simulations. *Applied Physics Letters*, **87**, 163120.

52 Legleiter, J., Park, M., Cusick, B. and Kowalewski, T. (2006) Scanning probe acceleration microscopy (SPAM) in fluids: Mapping mechanical properties of surfaces at the nanoscale. *Proceedings of the National Academy of Sciences of the United States of America*, **103**, 4813–4818.

53 Hölscher, H., Ebeling, D. and Schwarz, U.D. (2006) Theory of Q-controlled dynamic force microscopy in air. *Journal of Applied Physics*, **99**, 084311.

54 Sahin, O., Quate, C.F., Solgaard, O. and Atalar, A. (2004) Resonant harmonic response in tapping-mode atomic force microscopy. *Physical Review B - Condensed Matter*, **69**, 165416.

55 Dürig, U. (2000a) Interaction sensing in dynamic force microscopy. *New Journal of Physics*, **2**, 5.1.

56 Sader, J.E., Uchihashi, T., Farrell, A., Higgins, M.J., Nakayama, Y. and Jarvis, S.P. (2005) Quantitative force measurements using frequency modulation atomic force microscopy – theoretical foundations. *Nanotechnology*, **16**, S94–101.

57 Aimé, J.P., Boisgard, R., Nony, L. and Couturier, G. (1999) Nonlinear dynamic behavior of an oscillating tip-microlever system and contrast at the atomic scale. *Physical Review Letters*, **82**, 3388–3391.

58 Gleyzes, P., Kuo, P.K. and Boccara, A.C. (1991) Bistable behavior of a vibrating tip near a solid surface. *Applied Physics Letters*, **58**, 2989–2991.

59 Kühle, A., Sorensen, A. and Bohr, J. (1998a) Role of attractive forces in tapping tip force microscopy. *Journal of Applied Physics*, **81**, 6562–6569.

60 Nony, L., Boisgard, R. and Aimé, J.-P. (2001) Stability criterions of an oscillating tip-cantilever system in dynamic force microscopy. *European Physical Journal B*, **24**, 221–229.

61 San Paulo, A. and García, R. (2002) Unifying theory of tapping-mode atomic-force microscopy. *Physical Review B - Condensed Matter*, **66**, 041406.

62 Sasaki, Naruo and Tsukada, Masaru (1999) Theory for the effect of the tip-surface interaction potential on atomic resolution in forced vibration system of noncontact AFM. *Applied Surface Science*, **140**, 339–343.

63 Wang, L. (1998) Analytical descriptions of the tapping-mode atomic force microscopy response. *Applied Physics Letters*, **73**, 3781–3783.

64 Hoppenstaedt, Frank C. (1993) *Analysis and Simulation of Chaotic Systems*, Springer-Verlag, New York.

65 Landau, L.D. and Lifschitz, E.M. (1990) *Lehrbuch der Theoretischen Physik: Mechanik*, Akademie-Verlag, Berlin.

66 Anczykowski, B., Krüger, D. and Fuchs, H. (1996) Cantilever dynamics in quasinoncontact force microscopy: Spectroscopic aspects. *Physical Review B - Condensed Matter*, **53**, 15485–15488.

67 Haugstad, G. and Jones, R.R. (1999) Mechanisms of dynamic force microscopy on polyvinyl alcohol: region-specific non-contact and intermittent contact regimes. *Ultramicroscopy*, **67**, 77–86.

68 Stark, R.W., Schitter, G. and Stemmer, A. (2003) Tuning the interaction forces in tapping mode atomic force microscopy. *Physical Review B - Condensed Matter*, **68**, 085401.

69 García, R. and Paulo, A.S. (2000) Dynamics of a vibrating tip near or in intermittent contact with a surface. *Physical Review B - Condensed Matter*, **61**, R13381–R13384.

70 San Paulo, A. and García, R. (2000) High-resolution imaging of antibodies by tapping-mode atomic force microscopy: Attractive and repulsive tip-sample interaction regimes. *Biophysical Journal*, **78**, 1599–1605.

71 Tamayo, J. and García, R. (1996) Deformation, contact time, and phase contrast in tapping mode scanning force microscopy. *Langmuir*, **12**, 4430–4435.

72 Zitzler, L., Herminghaus, S. and Mugele, F. (2002) Capillary forces in tapping mode atomic force microscopy. *Physical Review B - Condensed Matter*, **66**, 155436.

73 Hölscher, H. (2006) Quantitative measurement of tip-sample interactions in amplitude modulation atomic force microscopy. *Applied Physics Letters*, **89**, 123109.

74 Lee, M. and Jhe, W. (2006) General solution of amplitude-modulation atomic force microscopy. *Physical Review Letters*, **97**, 036104.

75 Stark, M., Stark, R.W., Heckl, W.M. and Guckenberger, R. (2002) Inverting dynamic force microscopy: From signals to time,-resolved interaction forces. *Proceedings of the National Academy of Sciences of the United States of America*, **99**, 8473–8478.

76 Dürig, U. (1999) Relations between interaction force and frequency shift in large-amplitude dynamic force microscopy. *Applied Physics Letters*, **75**, 433–435.

77 Giessibl, F.J. (1997) Forces and frequency shifts in atomic-resolution dynamic-force microscopy. *Physical Review B - Condensed Matter*, **56**, 16010–16015.

78 Kühle, A., Sørensen, A.H., Zandbergen, J.B. and Bohr, J. (1998b) Contrast artifacts in tapping tip atomic force microscopy. *Applied Physics A: Materials Science & Processing*, **66**, S329–S332.

79 Ueyama, H., Sugawara, Y. and Morita, S. (1998) Stable operation mode for dynamic noncontact atomic force microscopy. *Applied Physics A: Materials Science & Processing*, **66**, S295–S297.

80 Loppacher, Ch., Bammerlin, M., Battiston, F., Guggisberg, M., Müller, D., Lüthi, R., Hidber, H.R., Meyer, E. and Güntherodt, H.-J. (1998) Fast digital electronics for application in dynamic force microscopy using high-q cantilevers. *Applied Physics A: Materials Science & Processing*, **66**, S215–S218.

81 Giessibl, F.-J. (1995) Atomic resolution of the silicon (111)-(7 × 7) surface by atomic force microscopy. *Science*, **267**, 68.

82 Sugawara, Y., Otha, M., Ueyama, H. and Morita, S. (1995) Defect motion on an InP (110) surface observed with noncontact atomic force microscopy. *Science*, **270**, 1646.

83 Bammerlin, M., Lüthi, R., Meyer, E., Baratoff, A., Lü, J., Guggisberg, M., Gerber, Ch., Howald, L. and Güntherodt, H.-J. (1997) True atomic resolution on the surface of an insulator via ultrahigh vacuum dynamic force microscopy. *Probe Microscopy*, **1**, 3.

84 Barth, C. and Reichling, M. (2001) Imaging the atomic arrangement on the high-temperature reconstructed α-Al_2O_3(0001) surface. *Nature*, **414**, 54–57.

85 Allers, W., Langkat, S. and Wiesendanger, R. (2001) Dynamic low-temperature scanning force microscopy on nickel oxide (001). *Applied Physics A: Materials Science & Processing*, **72**, S27.

86 Morita, S. and Sugawara, Y. (2002) *Noncontact Atomic Force Microscopy*, (eds S. Morita, R. Wiesendanger and E. Meyer), Springer-Verlag, Heidelberg, Germany. pp. 47–77.

87 Ashino, M., Schwarz, A., Behnke, T. and Wiesendanger, R. (2004) Atomic-resolution dynamic force microscopy and spectroscopy of a single-walled carbon nanotube: Characterization of interatomic van der Waals forces. *Physical Review Letters*, **93**, 136101.

88 Burke, S.A., Mativetsky, J.M., Hoffmann, R. and Grütter, P. (2005) Nucleation and submonolayer growth of C60 on KBr. *Physical Review Letters*, **94**, 096102.

89 Kunstmann, T., Schlarb, A., Fendrich, M., Wagner, Th., Möller, R. and Hoffmann, R. (2005) Dynamic force microscopy study of 3,4,9,10-perylenetetracarboxylic dianhydride on KBr(001). *Physical Review B - Condensed Matter*, **71**, 121403.

90 Loppacher, Ch., Zerweck, U., Eng, L.M., Gemming, S., Seifert, G., Olbrich, C., Morawetz, K. and Schreiber, M. (2006)

Adsorption of PTCDA on a partially KBr covered Ag(111) substrate. *Nanotechnology*, **17**, 1568.

91 Such, B., Weiner, D., Schirmeisen, A. and Fuchs, H. (2006) Influence of the local adsorption environment on the intramolecular contrast of organic molecules in non-contact atomic force microscopy. *Applied Physics Letters*, **89**, 093104.

92 Dürig, U. (2000b) Extracting interaction forces and complementary observables in dynamic probe microscopy. *Applied Physics Letters*, **76**, 1203–1205.

93 Giessibl, F.-J. (2001) A direct method to calculate tip-sample forces from frequency shifts in frequency-modulation atomic force microscopy. *Applied Physics Letters*, **78**, 123–125.

94 Gotsmann, B., Ancykowski, B., Seidel, C. and Fuchs, H. (1999) Determination of tip-sample interaction forces from measured dynamic force spectroscopy curves. *Applied Surface Science*, **140**, 314–319.

95 Hölscher, H., Gotsmann, B. and Schirmeisen, A. (2003) On dynamic force spectroscopy using the frequency modulation technique with constant excitation. *Physical Review B - Condensed Matter*, **68**, 153401.

96 Hölscher, H., Schwarz, U.D. and Wiesendanger, R. (1999) Calculation of the frequency shift in dynamic force microscopy. *Applied Surface Science*, **140**, 344–351.

97 Sader, J.E. and Jarvis, S.P. (2004) Accurate formulas for interaction force and energy in frequency modulation force spectroscopy. *Applied Physics Letters*, **84**, 1801–1803.

98 Hölscher, H., Schwarz, A., Allers, W., Schwarz, U.D. and Wiesendanger, R. (2000) Quantitative analysis of dynamic force spectroscopy data on graphite(0001) in the contact and non-contact regime. *Physical Review B - Condensed Matter*, **61**, 12678–12681.

99 Lantz, M.A., Hug, H., Hoffmann, R., van Schendel, P.J.A., Kappenberger, P., Martin, S., Baratoff, A. and Güntherodt, H.-J. (2001) Quantitative measurement of short-range, chemical bonding forces. *Science*, **291**, 2580–2583.

100 Hölscher, H., Langkat, S.M., Schwarz, A. and Wiesendanger, R. (2002b) Measurement of three-dimensional force fields with atomic resolution using dynamic force spectroscopy. *Applied Physics Letters*, **81**, 4428–4430.

101 Schirmeisen, A., Weiner, D. and Fuchs, H. (2006) Single-atom contact mechanics: From atomic scale energy barrier to mechanical relaxation hysteresis. *Physical Review Letters*, **97**, 136101.

102 Loppacher, Ch., Guggisberg, M., Pfeiffer, O., Meyer, E., Bammerlin, M., Luthi, R., Schlittler, R., Gimzewski, J.K., Tang, H. and Joachim, C. (2003) Direct determination of the energy required to operate a single molecule switch. *Physical Review Letters*, **90**, 066107.

103 Oyabu, N., Custance, O., Yi, I., Sugawara, Y. and Morita, S. (2003) Mechanical vertical manipulation of selected single atoms by soft nanoindentation using near contact atomic force microscopy. *Physical Review Letters*, **90**, 176102.

104 Sugimoto, Y., Abe, M., Hirayama, S., Oyabu, N., Custance, O. and Morita, S. (2005) Atom inlays performed at room temperature using atomic force microscopy. *Nature Materials*, **4**, 156–159.

3
Probing Hydrodynamic Fluctuations with a Brownian Particle
Sylvia Jeney, Branimir Lukic, Camilo Guzman, and László Fórró

3.1
Introduction

The observation of Brownian motion has been a subject of interest since the invention of optical microscopy during the seventeenth century [1]. From then on, the understanding of its origin remained a subject of debate until 1905, when Einstein described a convincing model which, assumed that the fluctuations of a small-sized particle floating in a fluid were caused by momentum transfer from thermally excited fluid molecules. Einstein identified the mean square displacement of the particle as *the* characteristic experimental observable of Brownian motion, and showed that it grows linearly with time as $\langle \Delta x^2(t) \rangle = 2Dt$, thereby introducing the diffusion coefficient D [2]. In 1908, Langevin reformulated Newton's force balance equation by adding to the instantaneous Stokes' friction [3] a stochastic force term, representing the random impacts of surrounding medium molecules on the Brownian particle [4]. At the same time, Henri pointed out the limited nature of Stokes' formula for the friction force, when applied to neutrally buoyant, micron-sized particles [5], which is the case for most Brownian particles used in experiments. In such cases, correlations between friction and velocity are non-instantaneous, and the positions of the particle are expected to be correlated up to longer times. The origin of this effect comes from the non-negligible inertia of the fluid, and this must be accounted for in the description of Brownian motion. The expression for the mean square displacement using the noninstantaneous friction force was introduced by Vladimirsky and Terletzky in 1945 [6], but remained largely ignored, as their contribution was published in Russian. In 1967, Alder and Wainwright discovered, in numerical simulations, that the particle's velocity correlations (another characteristic observable of Brownian motion) display a power-law decay [7] instead of an exponential relaxation, as expected for instantaneous friction. These simulations led theoreticians during the 1970s to reconsider the contribution of fluid mechanics to Brownian motion [8–14], and to address its relevance in experiments. The idea of using a particle subjected to Brownian motion as a reporter of its local environment was settled. With this approach, any deviation from the normal diffusive behavior of the particle could be interpreted as a response to the material properties of its complex environment [15, 16]. To measure such behavior, a high spatial resolution down to the nanometer scale is needed. Experiments using dynamic light scattering in colloidal suspensions confirmed that the diffusion of colloidal particles is influenced by fluid mechanics, and hence is time-dependent [17–21]. However, in order to achieve a sufficiently high resolution, averaging over an ensemble of different particles was necessary. Nowadays, tracking a *single* particle in a fluid on time scales sufficiently short to detect hydrodynamic contributions can be realized by using optical tracking interferometry (OTI). This allows a *direct* measurement of Brownian motion at the same resolution as techniques averaging over many particles, and an individual particle comes to be the local Brownian probe. OTI utilizes a weak optical trap [22] and interferometric particle position detection. The trapping laser provides a light source for the position detection of the particle, and at the same time ensures that the particle remains within the detector range.

In this chapter we provide a complete picture of the measurements of a Brownian particle immersed in a viscous fluid and held by an optical trap. First, relevant theoretical insights are exposed and the different timescales of Brownian motion are summarized. Next, the technical aspects of OTI are described, and methodologies on data acquisition, analysis and interpretation provided. The influences of experimentally relevant parameters, such as the trapping force constant, the fluid properties and the Brownian particle itself, are presented. Finally, the overlap of the different measurable time scales of Brownian motion is quantified, and the consequences are discussed.

3.2
Theoretical Model of Brownian Motion in an Optical Trap

In general, the Brownian motion of a particle in a fluid is the result of thermal fluctuations of the surrounding fluid molecules. Collisions between a bath of fluid molecules at temperature T and the particle lead to an exchange of energy that allows the establishment of a thermal equilibrium between the particle and its environment. A simple model system to describe the phenomenon quantitatively in terms of statistical mechanics *as well as* hydrodynamics consists of a micrometer-sized sphere immersed in a viscous, so-called Newtonian fluid. Such a system can typically be observed in OTI. In this section, theoretical predictions on the motion of a Brownian sphere in a harmonic potential are discussed, starting from the Langevin equation, and including effects arising from hydrodynamic interactions with the viscous fluid. The section ends with an overview of different relaxation times related to the Brownian particle, the fluid and the optical trap.

3.2.1
The General Langevin Equation for a Brownian Sphere in an Incompressible Fluid

There are principally two, counteracting forces that govern the motion of a Brownian particle. First, the particle is driven through the thermal force $F_{th}(t)$, that arises from random fluctuations of the fluid molecules excited by the thermal energy $k_B T$. Second, $F_{th}(t)$ is resisted by the friction force $F_{fr}(t)$, which (over-)damps the motion of the fluctuating particle. $F_{fr}(t)$ is the force exerted on the Brownian sphere by the surrounding viscous fluid, when the fluid is perturbed through the particle's fluctuations. Following from Newton's second law, the equation of motion can be written as the generalized Langevin equation:

$$m_s \ddot{x}(t) = F_{th}(t) + F_{fr}(t) + F_{ex}(t), \quad (3.1)$$

where m_s is the inertia of the Brownian sphere (s referring to the sphere's parameters), and $F_{ex}(t)$ represents the sum of any external forces, such as gravity or the force of the optical trap. For simplicity, we will discuss only the one-dimensional case for the axis x, even though OTI measurements give access to all three directions x, y and z.

3.2.1.1 The Random Thermal Force $F_{th}(t)$

The random force $F_{th}(t)$ represents the collisions of the fluid molecules on the particle. Its contribution in a homogeneous and isotropic medium varies so rapidly compared to the observable time scales, that $F_{th}(t)$ should be, on average, zero; $\langle F_{th}(t) \rangle = 0$.

Furthermore, as a very large number of collisions occurs during two successive measurements at times t and t', the correlation time of $F_{th}(t)$ is much shorter than the time interval between the two measurements [10]:

$$\langle F_{th}(t) F_{th}(t') \rangle = 2 k_B T \gamma \langle \xi(t) \xi(t') \rangle$$

where k_B is the Boltzmann constant, γ is the viscous drag of the fluid on the particle and $\zeta(t)$ is a white noise term with no finite correlation time: $\langle \xi(t) \xi(t') \rangle = \delta(t - t')$.

$F_{th}(t)$ obeys the fluctuation–dissipation theorem, and has the expression:

$$F_{th}(t) = \sqrt{2 k_B T \gamma} \xi(t) \quad (3.2)$$

In real systems the correlation time will not typically vanish instantaneously, because of the finite-size and finite-scale interactions which also exist between the fluid molecules themselves. Viscous and thermal forces will then become spatially and temporally correlated, through a time-dependent viscous drag γ (this is discussed next). It is worthy of note that $F_{th}(t)$ can only be described in terms of its statistical properties, and as its effect has already vanished on experimentally accessible time scales, $F_{th}(t)$ has never been measured [23].

3.2.1.2 The Friction Force $F_{fr}(t)$

An incompressible isotropic fluid with a viscosity $\eta_f(t)$ and density ρ_f (f refers to fluid parameters) generates a viscous drag $\gamma(t)$ on the thermally excited Brownian particle as it moves through the fluid, giving rise to the friction force $F_{fr}(t)$. A correct expression of $\gamma(t)$ and the resulting $F_{fr}(t)$ is given by solving the Navier–Stokes equation, describing the hydrodynamic properties of the fluid [24]. Here, the molecular character of the viscous fluid is neglected, and the fluid is treated as a continuum, which is valid when the Brownian sphere with radius a_s and density ρ_s is much larger than the fluid molecules. Then, the average free path length of the molecules which compose the fluid is small compared to the dimension of, for example, the sphere immersed in it. Furthermore, we will only consider the sphere moving far away from any boundary, like an obstacle placed in its trajectory, which would make the fluid anisotropic. Two experimentally relevant solutions of the Navier–Stokes equation can then be distinguished, the first being a commonly used approximation of the second:

(i) $\rho_s \gg \rho_f$: If the sphere has a high inertia m_s, and hence a density much higher than the fluid's density ρ_f, it will move steadily and at very slow speeds through the fluid. The fluid's response to the particle's presence can then be considered as instantaneous, and the solution of the Navier–Stokes equation is simply the constant Stokes' drag [3]:

$$\gamma = 6\pi\eta_f a_s \qquad (3.3)$$

and the friction force F_{fr} follows Stokes' law, which states that it is instantaneously linear with the sphere's velocity $\dot{x}$:

$$F_{fr} = -6\pi\eta_f a_s \dot{x} \qquad (3.4)$$

It must be noted here that η_f, the dynamic viscosity of the fluid, is considered as time-independent, thus implying that correlations between successive collisions of the fluid molecules on the Brownian particle have already vanished. Motion can hence be observed as a Markovian process – that is, a random walk.

(ii) $\rho_s \approx \rho_f$: As noted by Lorentz [25], Equation 3.1, which includes Stoke's law (Equation 3.4), is only consistent with hydrodynamics when $m_s \gg m_f = (4/3)\pi a_s^3 \rho_f$. When the sphere has a density similar to its surrounding medium – which is usually the case for the neutrally buoyant particles used in optical trapping – the sphere's motion will be determined not only by its own inertia but also by the inertia m_f of the surrounding fluid. Then, Brownian motion theory needs to include frequency-dependent effects, and the time dependence of $\dot{x}$ should be accounted for when computing the drag γ.

As the particle receives momentum from the fluctuating fluid molecules, it displaces the fluid in its immediate vicinity. Although the fluid can still be considered as a continuum, and even with the conditions of a low Reynold's number, the fluid's flow field will be perturbed. The non-negligible inertia of entrained fluid $m_f = (4/3)\pi a_s^3 \rho_f$ will act back on the sphere. As a consequence, correlations in the fluid's fluctuations will persist up to time scales observable by OTI and become experimentally relevant. The Brownian sphere will move with a non-constant velocity and perform a non-random walk, which will depend heavily on the nature of the surrounding medium. This phenomenon is commonly called hydrodynamic memory. Such perturbations give rise to the Stokes–Boussinesq friction force that is derived from the Navier–Stokes equation accounting for the inertia of the fluid, and given as [9]:

$$F_{fr} = -6\pi\eta_f a_s \dot{x} - \frac{2}{3}\pi a_s^3 \rho_f \ddot{x} - 6a_s^2 \sqrt{\pi\eta_f \rho_f} \int_0^t (t-t')^{-1/2} \ddot{x}(t') dt' \quad (3.5)$$

The first term is the ordinary Stokes' friction from Equation 3.4, while the second term is connected to the mass m_f of the incompressible fluid displaced by the Brownian particle. In principle, this term defines an effective mass $M = m_s + m_f/2$ that should replace m_s on the left-hand side of the Langevin equation (Equation 3.1). The third term is time-dependent, stating that the friction force at time t is determined by the penetration depth of the viscous, unsteady flow around the sphere at all preceding times. Equation 3.5 confirms that, for a fluid with a density similar to the density of the Brownian particle, the terms containing ρ_f cannot be neglected, as they are of the same order of magnitude as the inertial term.

Consequently, an instantaneous disturbance of the fluid from the thermally excited Brownian sphere will spread, and its initial momentum will be shared with all of the molecules in a small volume around the sphere. The velocity field of this moving volume then grows by vorticity diffusion. In an incompressible liquid, this enforces a back flow at short times, which creates a vortex ring in three dimensions [12]. The diffusive spreading of this vortex carries the momentum into the fluid on a time scale $\tau_f = a_s^2 \rho_f / \eta_f$ – the time needed for vorticity to diffuse over the distance of one particle radius. Figure 3.1 shows a simplified scheme of the characteristic double-vortex structure of the velocity field caused by the initial displacement of the Brownian sphere in a simple liquid.

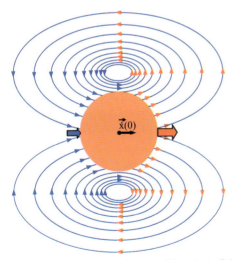

Figure 3.1 Schematic visualization of the velocity field of the fluid after the colloidal particle has been set in motion. (Inspired from computer simulations by Ref. [26].)

3.2.1.3 The External Force $F_{ex}(t)$

Only two different cases for $F_{ex}(t)$ will be considered for deriving the solutions of the Langevin equation:

(i) $F_{ex}(t) = 0$: When no additional force acts on the sphere and its motion is considered as free.

(ii) $F_{ex}(t) = -kx(t)$: Corresponding to the harmonic trapping potential with a force constant k created by the optical trap, which retains the sphere within the observation volume of the detector. The sphere's motion is then qualified as optically confined.

3.2.2
Solutions to the Different Langevin Equations for Cases Observable by OTI

From the solution of the Langevin equation for a Brownian sphere, the following measurable quantities of physical interest are derived for further studies in experiments (see Sections 3.3 and 3.4) the velocity autocorrelation function (VAF) $\langle \dot{x}(t)\dot{x}(0) \rangle$; the mean square displacement (MSD) $\langle \Delta x^2(t) \rangle$, which is related to the velocity autocorrelation function through:

$$\langle \Delta x^2(t) \rangle = 2\int_0^t (t-t')\langle \dot{x}(t')\dot{x}(0)\rangle dt'; \tag{3.6}$$

and also the power spectral density (PSD) $\langle |\tilde{x}(f)|^2 \rangle$, which mirrors the MSD through its Fourier transform as

$$\langle \Delta x^2(t) \rangle = 4\int_0^\infty \cos(2\pi f t)\langle |\tilde{x}(f)|^2 \rangle df. \tag{3.7}$$

The following listing of all three measurables derived from the four discussed Langevin equations is meant to provide a summarizing overview of the theoretical models of Brownian motion derived in the literature by various authors. Each model can be picked accordingly to fit the data acquired by OTI, as discussed in Section 3.4. The most accurate expressions are also the most complex; however, a good understanding of the problem of Brownian motion in a viscous fluid is already gained by only considering the characteristic limiting behaviors in each situation.

3.2.2.1 Free Brownian Motion

Solving the Langevin Equation using Stokes Friction Solving the Langevin equation using the Stokes friction of Equation 3.4 and $F_{ex}(t) = 0$ results in a VAF:

$$\langle \dot{x}(t)\dot{x}(0)\rangle_{free} = \frac{k_B T}{m_s} e^{-t/\tau_s} \tag{3.8}$$

and in a MSD:

$$\langle \Delta x^2(t)\rangle_{free} = 2Dt\left[1 + \frac{\tau_s}{t}(e^{-t/\tau_s}-1)\right] \quad (3.9)$$

that both decay exponentially with a characteristic time constant $\tau_s = m_s/\gamma = 2a_s^2\rho_s/9\eta_f$. This implies that, for a larger/heavier particle and/or a less viscous fluid, correlations will last longer. If $D = k_BT/\gamma$ is the diffusion constant, then the PSD will be:

$$\langle|\tilde{x}(f)|^2\rangle_{free} = \frac{D}{\pi^2 f^2[(\gamma f/2\pi m_s)^2 + 1]} = \frac{D}{\pi^2 f^2[(f/\phi_s)^2 + 1]} \quad (3.10)$$

with $\phi_s = 1/2\pi\tau_s$, the corresponding characteristic frequency.

For short times ($t \to 0$), and respectively, high frequencies ($f \to \infty$), the particle moves with its initial velocity:

$$\dot{x}(0) = \sqrt{k_BT/m_s} \quad (3.11)$$

Then, the motion is ballistic with:

$$\langle \Delta x^2(t)\rangle_{free} = (k_BT/m_s)t^2 \quad (3.12)$$

and:

$$\langle|\tilde{x}(f)|^2\rangle_{free} = (D\phi_s^2/\pi^2)f^{-4} \quad (3.13)$$

At long times ($t \to \infty$), and respectively, low frequencies ($f \to 0$), velocity correlations vanish exponentially with τ_s, the relaxation time of the particle's initial momentum. The particle has then lost all information about its initial velocity, and diffuses randomly with

$$\langle \Delta x^2(t)\rangle_{free} = 2Dt \quad (3.14)$$

and respectively

$$\langle|\tilde{x}(f)|^2\rangle_{free} = (D/\pi^2)f^{-2} \quad (3.15)$$

Solving the Langevin Equation using the Stokes–Boussinesq Friction Force Solving the Langevin equation using the Stokes–Boussinesq friction force of Equation 3.5 and $F_{ex}(t) = 0$ results in more complex expressions [6, 11]:

$$\langle \dot{x}(t)\dot{x}(0)\rangle_{free} = \frac{k_BT}{2\pi a_s^3\rho_f\sqrt{5-8\rho_s/\rho_f}}\left[\alpha_+ e^{\alpha_+^2 t}\mathrm{erfc}(\alpha_+\sqrt{t}) - \alpha_- e^{\alpha_-^2 t}\mathrm{erfc}(\alpha_-\sqrt{t})\right]$$

$$\text{with} \quad \alpha_\pm = \frac{3}{2}\frac{3\pm\sqrt{5-36\tau_s/\tau_f}}{\sqrt{t}(1+9\tau_s/\tau_f)} \quad (3.16)$$

now also including the hydrodynamic effect, decaying with a fluid-dependent time constant $\tau_f = a_s^2\rho_f/\eta_f$, which represents the time needed by the perturbed fluid flow field to diffuse over the distance of one particle radius. For an incompressible fluid, the initial velocity is now given by $\dot{x}(0) = \sqrt{k_BT/(m_s+m_f/2)} = \sqrt{k_BT/M}$, and

τ_s should in principle be expressed as $\tau_s = (m_s + m_f/2)/\gamma = a_s^2(2\rho_s + \rho_f)/9\eta_f$. The MSD is given by:

$$\langle \Delta x^2(t) \rangle_{free} = 2Dt \left[1 - 2\sqrt{\frac{\tau_f}{\pi t}} + 4\frac{\tau_f}{t} - \frac{\tau_s}{t} + \Xi\left(\frac{\tau_s}{\tau_f}, \frac{t}{\tau_f}\right) \right]$$

with $\Xi\left(\frac{\tau_s}{\tau_f}, \frac{t}{\tau_f}\right) = \frac{3}{t\sqrt{5 - 36\tau_s/\tau_f}} \left[\frac{1}{\alpha_+^3} e^{\alpha_+^2 t} erfc(\alpha_+ \sqrt{t}) - \frac{1}{\alpha_-^3} e^{\alpha_-^2 t} erfc(\alpha_- \sqrt{t}) \right].$

(3.17)

Equation 3.17 depends on the particle's inertia through τ_s and on the fluid's inertia through τ_f. The two times are connected by the relationship $\frac{\tau_s}{\tau_f} \approx \frac{2\rho_s}{9\rho_f}$.

The corresponding characteristic frequency $\phi_f = 1/2\pi\tau_f$ appears in the PSD as [27]:

$$\langle |\tilde{x}(f)|^2 \rangle_{free} = \frac{D}{\pi^2 f^2} \frac{\left(1 + \sqrt{f/2\phi_f}\right)}{\left[(f/\phi_s) + \sqrt{f/2\phi_f} + (f/9\phi_f)\right]^2 + \left(1 + \sqrt{f/2\phi_f}\right)^2} \quad (3.18)$$

The behaviors in the time and frequency limits of all three functions remain similar to the previously discussed case, meaning that, for very short times, the motion is ballistic and, for very long times, the motion is diffusive. However, the transition between the two regimes is algebraic and delayed to significantly longer times compared to the case of simple Stokes' friction. This translates into a slow algebraic decay in the MSD (Equation 3.12) and results in a VAF which is governed by a power-law rather than by an exponential tail [7]:

$$\langle \dot{x}(t)\dot{x}(0) \rangle_{free} \propto (t/\tau_f)^{-3/2}, \quad \text{for} \quad t \geq \tau_f \quad (3.19)$$

This power law is usually referred to in the literature as the 'long-time tail', and arises from the fluid vortices observed around the colloidal particle, as sketched in Figure 3.1.

The log-log plot in Figure 3.2a compares the VAF given by Equation 3.8 (red line) with that given by Equation 3.16 (blue line), both normalized by their respective initial velocity, for a sphere with radius $a_s = 1\,\mu m$, density $\rho_s = 1.51\,kg\,dm^{-3}$ immersed in water with viscosity $\eta_f = 10^{-3}\,Pa\cdot s$ at $T = 22\,°C$. It can be seen that the exponential relaxation resulting from the Stokes' friction changes to a power-law decay when the fluid's inertia is accounted for. In the same way, the log-log representations in Figures 3.2b and 3.2c show a comparison between Equations 3.9 and 3.17, as well as between Equations 3.10 and 3.18. Differences in the MSD and PSD are less visible in this representation, but the respective common limiting behaviors, translating to characteristic slopes are indicated. In Figure 3.2c, the discrepancies visible at high frequencies above 2 MHz (arrow) between both functions, arise from the differences in the displaced masses; m_s, and respectively M. The green bars on the abscissa highlight the time, and respectively, frequency regions accessible by OTI, as introduced in Section 3.3.

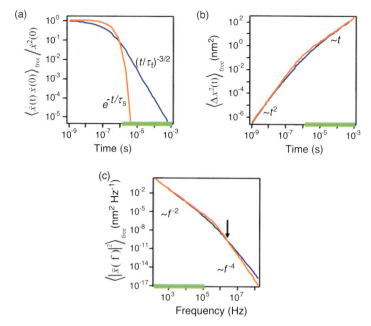

Figure 3.2 Log-log plots of (a) the normalized VAF given by Equation 3.8 (red line) and Equation 3.16 (blue line); (b) the MSD given by Equation 3.9 (red line) and Equation 3.17 (blue line); and (c) the PSD given by Equation 3.10 (red line) and Equation 3.18 (blue line) for a 2 µm-sized sphere in water. The chosen time span ranges from the short-time to the long-time limits of each function, which are indicated in each graph. The green bars on the abscissa highlight the regions accessible by OTI.

3.2.2.2 Optically Confined Brownian Motion

The Case of a Nonfree Particle In the case of a nonfree particle the situation becomes more complex. The diffusion of such a particle in an unbounded fluid was first described by Ornstein and Uhlenbeck [28]. The Langevin equation using the Stokes friction of Equation 3.4 gives then a velocity autocorrelation function:

$$\langle \dot{x}(t)\dot{x}(0) \rangle = \frac{k_B T}{m_s(\varsigma_+ - \varsigma_-)} [\varsigma_+ e^{-\varsigma_+ t} - \varsigma_- e^{-\varsigma_- t}]$$

$$\text{with} \quad \varsigma_\pm = \frac{1}{2\tau_s}\left(1 \pm \sqrt{1 - 4\tau_s/\tau_k}\right) \quad (3.20)$$

The VAF now has a positive part which decays exponentially to zero as $\langle \dot{x}(t)\dot{x}(0) \rangle \propto e^{-t/\tau_s}$ for $t \approx \tau_s$, and a negative part which decays exponentially as:

$$\langle \dot{x}(t)\dot{x}(0) \rangle \propto -e^{-t/\tau_k} \quad (3.21)$$

for $t \approx \tau_s$, a new characteristic time constant $\tau_k = 6\pi\eta_f a_s/k$ determined by the trap stiffness k. The MSD follows as:

$$\langle \Delta x^2(t) \rangle = \frac{2k_BT}{k}\left[1 + \frac{\varsigma_-}{\varsigma_+ - \varsigma_-}e^{-\varsigma_+ t} + \frac{\varsigma_+}{\varsigma_- - \varsigma_+}e^{-\varsigma_- t}\right] \quad (3.22)$$

and the PSD becomes Lorentzian [29]:

$$\langle |\tilde{x}(f)|^2 \rangle = \frac{D}{\pi^2 f^2 [1 + (\phi_k/f)^2]} \quad (3.23)$$

showing that the motion is also influenced by the trapping potential with the characteristic frequency $\phi_k = 1/2\pi\tau_k = k/12\pi^2\eta_f a_s$, corresponding to the corner frequency of the Lorentzian function.

In the short time limits, the behavior remains the same as introduced above, but in the long time limit it is now governed by the confining potential – that is, the optical trapping constant. Then, the velocity correlations still disappear, but Equation 3.22 approaches the time-independent limit:

$$\langle \Delta x^2(t) \rangle_{t \to \infty} = 2k_BT/k \quad (3.24)$$

and Equation 3.23 the limit:

$$\langle |\tilde{x}(f)|^2 \rangle_{f \to 0} = 4k_BT\gamma/\pi k^2 \quad (3.25)$$

The sphere's motion is now governed by the potential's restoring force with stiffness k.

Using the Stokes–Boussinesq Friction Force The most complete solution considered in this work is when the Stokes–Boussinesq friction force of Equation 3.5 and $F_{ex}(t) = -kx(t)$ are used to set up the Langevin equation. Then, the VAF is [14]:

$$\langle \dot{x}(t)\dot{x}(0) \rangle = \frac{\zeta_1^3 e^{\zeta_1^2 t} erfc(\zeta_1\sqrt{t})}{(\zeta_1-\zeta_2)(\zeta_1-\zeta_3)(\zeta_1-\zeta_4)} + \frac{\zeta_2^3 e^{\zeta_2^2 t} erfc(\zeta_2\sqrt{t})}{(\zeta_2-\zeta_1)(\zeta_2-\zeta_3)(\zeta_2-\zeta_4)}$$

$$+ \frac{\zeta_3^3 e^{\zeta_3^2 t} erfc(\zeta_3\sqrt{t})}{(\zeta_3-\zeta_1)(\zeta_3-\zeta_2)(\zeta_3-\zeta_4)} + \frac{\zeta_4^3 e^{\zeta_4^2 t} erfc(\zeta_4\sqrt{t})}{(\zeta_4-\zeta_1)(\zeta_4-\zeta_2)(\zeta_4-\zeta_3)} \quad (3.26)$$

where the coefficients $\zeta_1, \zeta_2, \zeta_3$ and ζ_4 are the four roots of the equation:

$$\left(\tau_s + \frac{1}{9}\tau_f\right)\zeta^4 + \sqrt{\tau_f}\zeta^3 + \zeta^2 + \frac{1}{\tau_k} = 0.$$

Correspondingly, the MSD results in:

$$\langle \Delta x^2(t) \rangle = 2D\tau_k + \frac{2D}{\tau_s + \frac{\tau_f}{9}} \left[\frac{e^{\zeta_1^2 t} erfc(\zeta_1 \sqrt{t})}{\zeta_1(\zeta_1-\zeta_2)(\zeta_1-\zeta_3)(\zeta_1-\zeta_4)} + \frac{e^{\zeta_2^2 t} erfc(\zeta_2 \sqrt{t})}{\zeta_2(\zeta_2-\zeta_1)(\zeta_2-\zeta_3)(\zeta_2-\zeta_4)} \right]$$

$$+ \frac{2D}{\tau_s + \frac{\tau_f}{9}} \left[\frac{e^{\zeta_3^2 t} erfc(\zeta_3 \sqrt{t})}{\zeta_3(\zeta_3-\zeta_1)(\zeta_3-\zeta_2)(\zeta_3-\zeta_4)} + \frac{e^{\zeta_4^2 t} erfc(\zeta_4 \sqrt{t})}{\zeta_4(\zeta_4-\zeta_1)(\zeta_4-\zeta_2)(\zeta_4-\zeta_3)} \right]$$

(3.27)

and the PSD is given by [27]:

$$\langle |\tilde{x}(f)|^2 \rangle = \frac{D}{\pi^2 f^2} \left[\frac{\left(1+\sqrt{f/2\phi_f}\right)}{\left[(\phi_k/f)-(f/\phi_s)-\sqrt{f/2\phi_f}-\left(f/9\phi_f\right)\right]^2 + \left(1+\sqrt{f/2\phi_f}\right)^2} \right]$$

(3.28)

which is by far the most convenient formula of all three quantities to use for fitting data acquired in OTI experiments. Fortunately, despite their complexity, the limiting behaviors of all three expressions simplify in the same way as in the above-discussed cases.

3.2.3
Time Scales of Brownian Motion

All of the above considerations show that many different time scales (as outlined in Table 3.1) govern the physics of Brownian motion. To better appreciate these times, we can consider a sphere with radius $a_s = 1\,\mu m$ in water. The sphere is set into motion by random collisions with fluid molecules within $\tau_{col} = \bar{d}_{mol}\bar{v}_{mol} \approx 0.1\,ps$, the correlation time of $F_{th}(t)$, estimated by the ratio of the mean solvent particle separation $\bar{d}_{mol}$ and fluctuation speed $\bar{v}_{mol}$ [30]. The momentum is then transferred from the particle to the fluid on very different time scales. If compressibility effects of the fluid are taken into account, one-third of the initial momentum is carried off

Table 3.1 Overview of the characteristic time scales of a Brownian particle in a harmonic potential.

Time constant	Origin	Typical value for our model system
$\tau_{col} = \bar{d}_{mol}/\bar{v}_{mol}$	Random molecular collisions	$\sim 10^{-13}$ s
$\tau_{sw} = a_s/c$	Sound wave propagation over the distance of a sphere's radius	$\sim 10^{-9}$ s
$\tau_s = m_s/\gamma = 2a_s^2 \rho_s/9\eta_f$	Inertia of the Brownian particle	$\sim 10^{-6}$ s
$\tau_f = a_s^2 \rho_f/\eta_f$	Inertia of the perturbed fluid surrounding the Brownian particle	$\sim 10^{-6}$ s
$\tau_k = 6\pi\eta_f a_s/k$.	Harmonic potential of the optical trap	$\sim 10^{-4}$ s

Typical values are calculated for a resin sphere ($a_s = 1\,\mu m$, $\rho_s = 1.51\,kg\,dm^{-3}$) in water ($\rho_f = 1\,kg\,dm^{-3}$, $\eta = 10^{-3}\,Pa\cdot s$).

rapidly by a spherical sound wave, the front of which leaves the sphere after a time $\tau_{sw} = a_s/c \approx 0.7$ ns, where c is the speed of sound in the fluid [31]. Equation 3.5, which was set for an incompressible fluid, has simply to be corrected to include a rapid change of the particle's inertial mass m_s to the combined mass $M = m_s + m_f/2$ [13]. Consequently, the velocity correlation function starts with the initial value given by the equipartition theorem, $\langle \dot{x}(0)\dot{x}(0) \rangle = k_B T/m_s$ and, after a short time, on the order of τ_{sw}, decays from $k_B T/m_s$ to $k_B T/M$ due to acoustic damping of the particle's velocity. When the sound wave has separated and the particle has relaxed with $\tau_s = m_s/\gamma \approx 0.9$ μs or $\tau_{s+f} = (m_s + m_f/2)/\gamma \approx 0.7$ μs, the vortex ring develops around the particle. The region of vorticity (see Figure 3.1) grows diffusively, as the remaining momentum is distributed over increasingly larger volumes, with the disturbance taking a time of order $\tau_f = a_s^2 \rho_f/\eta_f$ to leave the particle. Finally, the optical trapping force, $F_{ex}(t)$, sets in after a time $\tau_k = 6\pi\eta_f a_s/k$ and slows down the sphere, confining its motion around the potential minimum. The stronger the trap, the earlier optical confinement will reduce the particle's free motion.

For the model of a sphere in a harmonic potential with a typical spring constant $k = 10$ μN m^{-1}, τ_s and τ_f are in the microsecond range, whereas τ_k is in the millisecond range. The relaxation time of the optical trapping potential is thus well separated from the others, and its separation can be adjusted by choosing suitable experimental parameters, that is, a_s, ρ_s, ρ_f, η_f and k. The diffusion constant D is on the order of 1 μm^2 s^{-1}, and hence the time for the sphere to diffuse over its radius is on the order of 1 s. Correspondingly, within 1 μs, the sphere will have diffused about 1 nm, which is the time and distance range accessible by OTI. This will allow study of the interplay between τ_f and τ_k, as shown in detail in Section 3.4.

3.3
Experimental Aspects of Optical Trapping Interferometry

In OTI, optical trapping is combined with high-resolution interferometric position detection. The optical trap provides the linear external force $F_{ex}(t) = -kx(t)$ in the Langevin equation, which will maintain the studied microsphere within the detection range of the system. In general, the principles of optical trapping, as well as instrumental aspects, are well known and have been reported extensively in the literature [29, 32]. Therefore, at this point we will present only briefly the opto-mechanical and electronic components of the set-up. However, the data acquisition and analysis strategy deserves greater emphasis, as it allows the fine characterization of interactions between the Brownian probe and its environment.

3.3.1
Experimental Set-Up

The apparatus consists mainly of a custom-built inverted light microscope with a 3-D sample positioning stage, an infrared laser for trapping, and a quadrant photodiode for high-resolution 3-D and time-resolved position detection. The two main custom-made

Figure 3.3 Photograph of the experimental set-up on a vibration isolation table. The mechanical frame is made from titanium, steel and aluminum to achieve maximal mechanical and temperature stability. For details of abbreviations, see text and Figure 3.4. (Photograph courtesy of Daniel Gutierrez.)

circular base plates are made from the titanium alloy Ti-6AL-4V, on the basis of its high tensile strength, light weight, low thermal conductivity, low thermal expansion coefficient and corrosion resistance compared to steel or aluminum. The commercially available mechanical pieces are either made of aluminum or steel. In order to minimize mechanical vibrations, the whole set-up is mounted on a table with tuned damping (RS-4000, Newport, UK), supported by pneumatic isolators (I-2000, Newport). The main features of the instrument are shown in Figure 3.3.

3.3.1.1 Optical Trapping Interferometry and Microscopy Light Path

The optical paths can be divided into an infrared (IR) trapping and detection light path, and a visible illumination and imaging light path, as shown in Figure 3.4. The trapping beam is emitted by a diode-pumped, ultra-low-noise Nd:YaG laser with a wavelength of $\lambda = 1064$ nm (IRCL-500-1064-S; CrystaLaser, USA) and a maximal light power of 500 mW in continuous-wave mode. The choice of the near-IR wavelength satisfies the requirement of minimal water absorption to avoid heating of the sample in the laser focus [29]. A high-intensity gradient for good trapping efficiency is achieved by over-illuminating the high numerical back-aperture of the focusing lens (OBJ). Therefore, the effective laser beam diameter is expanded 20-fold by a telecentric lens system (EXP; Beam Expander, Sill Optics, Germany). In order to

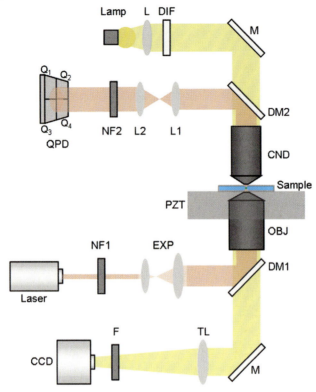

Figure 3.4 Schematic layout of the infrared (IR; red) and visible (yellow) light paths. The laser beam is expanded 20-fold by a beam expander (EXP), attenuated if necessary by a neutral density filter (NF1), and then reflected by a dichroic mirror (DM1) and focused by the objective lens (OBJ) into the sample chamber, which is mounted on a piezostage (PZT). The scattered IR light is collected by a condenser (CND), and directed by a second dichroic mirror (DM2) onto the quadrant photodiode (QPD). A second neutral density filter (NF2) is placed in front of the QPD, to avoid possible saturation of the detector. A 50 W halogen light source (Lamp) illuminating the object plane, through a lens (L) and diffuser (DIF), is reflected by the first mirror M, but transmitted through both dichroic mirrors (DM1 and 2). The image created by the CND and OBJ is reflected by the second lower mirror (M) and the 180 mm tube lens (TL) onto the charge-coupled device camera (CCD).

minimize noise, the laser is operated at high power, and its intensity is adapted after expansion by neutral density filters (NF1) with variable transmission coefficients ($T = 0.25$, 0.1, 0.01 or 0.001; OWIS, Germany). Increasing the transmission coefficient of NF1 will decrease the trapping stiffness, as this depends linearly on the laser power [33]. Next, the IR-beam is reflected by a dichroic mirror (DM1; AHF analysentechnik AG, Germany) into the high numerical aperture (NA = 1.2) of a ×60 water-immersion objective (OBJ; UPLapo/IR, Olympus, Japan), which focuses the laser down to its refraction limit into the object plane of the microscope and creates the optical trap. The choice of a water-immersion objective lens offers a longer working distance of up to 280 μm compared to oil-immersion lenses. Such a long working distance guarantees a stable space-invariant trap through the entire sample chamber. This is particularly essential when studies on Brownian particles far away from any surface boundary are of interest (as will be the case for the experiments discussed below).

The sample is mounted onto an *xyz*-piezo scanning table (PZT; P-561, Physikalishe Instrumente, Germany) for manipulation and positioning. The PZT with controller (E-710 Digital PZT Controller; Physikalishe Instrumente, Germany) has a travel range of 100 µm along all three dimensions, with a precision of 1 nm. The laser light focused by the objective lens is collected with a condenser (CND; 63X, Achroplan, NA = 0.9, water-immersion; Zeiss, Germany), and projected by a second dichroic mirror (DM2) and two lenses (L1 and L2, with focal lengths $f_1 = 30$ mm and $f_2 = 50$ mm) with a magnification of $f_1/f_2 \approx 1.67$ onto an InGaAs quadrant photodiode (QPD; G6849, Hamamatsu Photonics, Japan). The QPD is placed in the back focal plane of the condenser and fixed on an x–y translation stage (OWIS, Germany) for manual centering of the detector relative to the IR-beam. In order to avoid possible saturation of the QPD, a second neutral density filter (NF2) can be optionally placed in front of the QPD. For illumination in the visible range, a 50 W halogen lamp is diffused (DIF) and projected by a mirror (M) through the condenser, objective and a 180 mm tube lens (TL) that creates an image of the object plane onto a charge-coupled device camera (CCD; ORCA ER S5107, Hamamatsu Photonics, Japan). The image of the object plane is digitized (Hamamatsu Digital Controller ORCA ER), and can be further processed (HiPic, Hamamatsu Photonics, Japan).

3.3.1.2 Sample Preparation

The sample chamber consists of a custom-made flow cell. A coverslip (thickness ∼130 µm) is stuck to a standard microscope slide using two pieces of double-sided tape, arranged in such a way as to form a ∼5 mm-wide and ∼70 µm-thick channel with a volume of ∼20 µl. After loading with a suspension of microspheres, the flow-cell is mounted upside down on the 3-D piezo-stage. In the experiments presented in Section 3.4, either polystyrene ($\rho_s = 1.05$ kg dm^{-3}; Bangs Laboratories, USA), melamine resin ($\rho_s = 1.51$ kg dm^{-3}; Sigma-Aldrich, USA) or silica spheres ($\rho_s = 1.96$ kg dm^{-3}; Bangs Laboratories, USA) were used with radii (a_s) varying from 0.27 to 2 µm. To guarantee the manipulation and analysis of exclusively one particle, a particle concentration of 10^6 spheres per milliliter was used to maximize the average distance between two neighboring particles and minimize their mutual influence on their motions [34].

3.3.2
Position Signal Detection and Acquisition

When following the 3-D Brownian motion of the trapped particle, the scattering of the strongly focused trapping laser on the particle is measured by the QPD. The InGaAs Quadrant Photodiode (G6849, Hamamatsu Photonics, Japan) is 2.0 mm in diameter with a dead zone of 0.1 mm between the quadrants. The photosensitivity is 0.67 A W^{-1} at 1064 nm. The QPD signals are fed into a custom-built preamplifier (Pre-AMP;

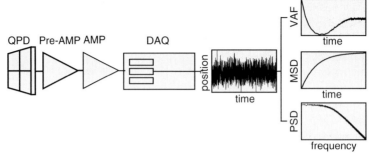

Figure 3.5 Position signal acquisition and data processing. Intensity fluctuations are recorded on the quadrant photodiode (QPD) and converted to volts (Pre-AMP). The signal is amplified (AMP) and digitized using the acquisition card (DAQ). The VAF, MSD and PSD are then calculated from the recorded position time trace.

Öffner MSR-Technik, Germany) which provides two differential signals between the segments and one signal that is proportional to the total light intensity (Figure 3.5). Preamplification of the quadrant photodiode signals at $20\,\text{V}\,\text{mA}^{-1}$ with $0.67\,\text{A}\,\text{W}^{-1}$ photosensitivity leads to a voltage of $13.4\,\text{V}\,\text{mW}^{-1}$. Subsequently, differential amplifiers (AMP; Öffner MSR-Technik) adjust the preamplifier signals for optimal digitalization by the data acquisition board (DAQ) with a dynamic range of 12 bits (NI-6110, National Instruments, USA). Amplification of the QPD signal is chosen to span the maximal dynamic range of the acquisition card. The amplifier (Öffner MSR-Technik), with a maximal gain of 500, has a cut-off frequency around 1 MHz. The particle's position can be detected in all three dimensions. On the QPD, scattered and unscattered light generate an interference pattern that corresponds to the probe's position. A displacement of the particle near the beam focus modulates the optical power collected by the QPD. When the sphere moves perpendicularly to the optical axis – that is, along the x-direction – the current signal $S_x = (Q_1 + Q_2) - (Q_3 + Q_4)$ measured between both top segments (Q_1, Q_2) and both bottom segments (Q_3, Q_4) of the QPD changes correspondingly. The same holds for movements in the y-direction. Displacements along the z-axis instead affect the sum-signal $S_z = Q_1 + Q_2 + Q_3 + Q_4$ of the QPD, and so z displacements can be determined by the changes in total intensity. The full 3-D position information of the probe is thus encoded in the interference pattern of the forward-scattered and transmitted laser light recorded by the QPD. A detailed analysis of the detector response is given for fluctuations perpendicular to the optical axis in Ref. [35], and along the optical axis in Refs [36] and [37]. For small displacements from the trap center, the differential signals from the QPD are proportional to the lateral displacements of the particle in the optical trap, and the sum signal to the axial displacement.

The scanning stage, CCD camera and data acquisition are controlled and coordinated by a custom-made program written in VEE (Agilent, USA). Data are saved in binary format and analyzed with Igor 6.0 (WaveMetrics, USA).

The 3-D position–time traces of the Brownian motion of the probe can be acquired up to maximally $N = 10^7$ points (this limitation is set by the working memory of the

VEE). Data analysis also becomes very slow when the data files exceed 3.10^7 points per channel, and therefore the combination between data acquisition rate f_{acq} and total recording time t_{tot}, must be adjusted according to $N = f_{acq} t_{tot}$. A schematic overview of the signal acquisition and data processing schemes is shown in Figure 3.5.

3.3.3
Position Signal Processing

The three quantities of VAF, MSD and PSD, which were introduced in Section 3.2 can be calculated from the same experimental time trace $x(t)$, which is recorded in volts and converted to nanometers after fitting to the suitable theory of Brownian motion (as described in Section 3.4.1). An example of such a time trace, as well as the resulting VAF, MSD and PSD is shown on the right-hand side of Figure 3.5 for a 2 μm resin bead. For the sake of clarity, only one-dimensional time traces $x(t)$ will be presented through the remainder of this chapter, even though the developed data analysis strategy also applies to the y and z directions.

The velocity $\dot{x}(t) = \Delta x(t)/\Delta t$ of the studied sphere is derived from the steps $\Delta x(t) = x(t + \Delta t) - x(t)$ it performed, where Δt is the lag time related to the acquisition frequency as $f_{acq} = 1/\Delta t$. For the total number of acquired points N, the total recording time is expressed as $t_{tot} = N\Delta t$.

The velocity autocorrelation function $\langle \dot{x}(t)\dot{x}(0) \rangle$ is then defined as:

$$\langle \dot{x}(t)\dot{x}(0) \rangle = \frac{1}{(\Delta t)^2} \langle \Delta x(t) \Delta x(0) \rangle = \frac{f_{acq}^2}{N} \sum_i \Delta x(t + i\Delta t) \Delta x(i\Delta t) \quad (3.29)$$

with N the total number of acquired points.

$\langle \dot{x}(t)\dot{x}(0) \rangle$ can be normalized by its initial value in an incompressible fluid:

$\langle \dot{x}^2(0) \rangle = k_B T/(m_s + m_f/2)$ at $t = 0$.

The MSD is calculated from $x(t)$ as:

$$\langle \Delta x^2(t) \rangle = \frac{1}{N} \sum_i [x(t + i\Delta t) - x(i\Delta t)]^2 \quad (3.30)$$

The discrete Fourier transform of $x(t)$ is: $\tilde{x}(f) = (1/f_s) \sum_i e^{i2\pi t f/N} x(t)$, where $t = i\Delta t$ and $i = 0, \pm 1, \ldots, \pm N/2$ and the power spectral density follows as:

$$\langle |\tilde{x}(f)|^2 \rangle = \frac{|\tilde{x}(f)|^2}{t_{tot}} \quad (3.31)$$

3.3.4
Temporal and Spatial Resolution of the Instrument

Apart from the signal arising from the particle's thermal fluctuations, anything that changes the intensity recorded by the quadrant photodiode will limit the resolution of the system. The main noise sources are the mechanical instabilities of the microscope, and electronic noise combined with laser intensity fluctuations and pointing instabilities. Mechanical instabilities mainly cause low-frequency noise (drift). Laser intensity fluctuations may lead to temporal variations in the spring constants of the optical trapping potential, and pointing instabilities to unwanted drifting of the trapping focus in the specimen plane. To quantify the contribution of laser noise to the measured signal $x(t)$, it is decomposed in:

$$x(t) = x_s(t) + x_n(t)$$

where $x_s(t)$ is the particle's thermal fluctuations and $x_n(t)$ is the noise contribution to the signal. A subtraction of its contribution can increase the quality of the signal. With the assumption that position fluctuations of the sphere and the noise are uncorrelated – that is, $\langle x_s(t) x_n(t') \rangle = 0$ – the velocity autocorrelation function can be written as:

$$\langle \dot{x}(t)\dot{x}(0) \rangle = \langle \dot{x}_s(t)\dot{x}_s(0) \rangle + \langle \dot{x}_n(t)\dot{x}_n(0) \rangle$$

the MSD of the acquired signal as:

$$\langle \Delta x^2(t) \rangle = \langle \Delta x_s^2(t) \rangle + \langle \Delta x_n^2(t) \rangle$$

and similarly,

$$|\tilde{x}(f)|^2 = |\tilde{x}_s(f)|^2 + |\tilde{x}_n(f)|^2,$$

where $\tilde{x}_s(f)$ and $\tilde{x}_n(f)$ are the Fourier transforms of $x_s(t)$ and $x_n(t)$ respectively.

The calibrated (see Section 3.4.1) position fluctuations as a function of time of a trapped sphere (resin, radius $a_s = 1\,\mu\text{m}$ in water, $k \approx 5\,\mu\text{N m}^{-1}$, $f_{acq} = 5\,\text{MHz}$, $t_{tot} = 2\,\text{s}$) are shown in Figure 3.6a (red line). The blue line represents the recordings of the empty trap's noise signal $x_n(t)$ after the sphere has been released. $x_n(t)$ can be minimized by optimizing the illumination pattern of the incident laser spot on the QPD. The comparison between $x_s(t)$ and $x_n(t)$ indicates that the laser noise contribution in this configuration is very small, and its influence on the VAF, MSD and PSD are shown in Figure 3.6b–d, respectively. As expected, the velocity fluctuations of the laser spot without a scattering particle are uncorrelated and fluctuate around zero (Figure 3.6b, blue line). Correspondingly, $\langle \Delta x^2(t) \rangle_n$ is small and constant (Figure 3.6c, blue line), whereas $\langle \Delta x^2(t) \rangle$ increases with time (Figure 3.6c, red line). The resolution of the MSD of the sphere's position fluctuations can then be enhanced by subtracting $\langle \Delta x^2(t) \rangle_n$ from $\langle \Delta x^2(t) \rangle$. The spatial resolution of $\sim 8\,\text{Å}$ achieved by the apparatus can be read from the first time points of the MSD (inset of Figure 3.6c, indicated by brackets).

In the frequency domain (Figure 3.6d), we define f_N as the frequency at which the power spectrum of the trapped bead (red line) drops to the noise level given by the power spectrum of the empty trap (blue line). The amplifier has a Butterworth-type low-pass filter with a cut-off frequency at 1 MHz, which is slightly above the frequency $f_N \approx 0.8\,\text{MHz}$. Therefore, in the following section we will analyze data in the frequency range up to a maximum of 0.5 MHz, setting the time resolution

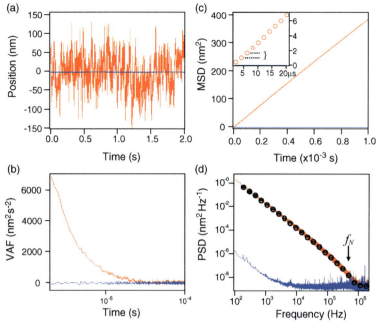

Figure 3.6 (a) Position signal of the sphere ($a_s = 2\,\mu m$) in the trap (red) and of the empty trap (blue) acquired during $t_{tot} = 2\,s$ at $f_{acq} = 5\,MHz$; (b) VAF, (c) MSD and (d) PSD calculated from the position signal (the black circles represent the PSD blocked in five bins per decade). The frequency f_N, where the signal reaches the noise floor, is indicated by an arrow.

of the OTI system at $2\,\mu s$. Additionally, in order to avoid aliasing artifacts [38] from the data acquisition card, the acquired signal is oversampled by a factor of 2; $f_{acq} > 2 f_N$.

When plotting the VAF and PSD on a log-log scale, further data processing can be made to enhance noisy data. As both functions are distributed exponentially, the number of points in a log-log plot will increase with, respectively, time or frequency. Therefore, data are commonly averaged from a 'block' of consecutive data points [39], resulting in equidistantly distributed points on the logarithmic scale. In the example discussed above, data were blocked in five bins per decade (Figure 3.6d, black circles). The data scattering around their average value gives the standard error, which remains within the size of the black circles. Together with improving the visibility of data, blocking allows fitting the data by the least-squares method, which assumes that the analyzed data points are statistically independent and conform to a Gaussian distribution [27]. Whilst the second assumption is satisfied by the VAF and PSD (as defined in Equations 3.29 and 3.31), the first assumption is satisfied only after blocking.

3.4
High-Resolution Analysis of Brownian Motion

After having characterized the performance of the OTI set-up, we can now present measurements on details in the Brownian motion of the model sphere. We demonstrate that the theory presented in Section 3.2 may be used for fitting and calibrating the data. The influence of experimentally relevant parameters on Brownian motion will be demonstrated, and the consequences of the presence of inertial effects in the data discussed.

3.4.1
Calibration of the Instrument

Position signal calibration consists of determining the detector sensitivity β and the spring constant k of the optical trapping potential. The term β has units of V nm^{-1}, as the acquired position signal is recorded in volts, and the position of the particle is expressed in nanometers. Both, experimental and theoretical investigations [40] have shown that, close to the trap center, the optical trapping forces are well approximated by three orthogonal forces derived from a harmonic trapping potential. For a given wavelength of the laser beam, the spring constant along each direction depends mainly on the particle's size, the difference between refractive indices of particle and fluid, and the intensity of the trapping laser light. When the sphere's radius is known, the physics of Brownian motion in a harmonic potential (see Section 3.2.2.2) can be exploited to calibrate the optical trap [38]. Either of three quantities presented in Section 3.3.3 – namely the VAF, MSD or PSD – can be used to obtain β and k simultaneously. An overview is provided in Figure 3.7 for measurements of a resin sphere with radius $a_s = 0.5\,\mu m, f_{acq} = 0.5$ MHz, $t_{tot} = 20$ s. For least-square fitting, the VAF (top graph) is normalized by its initial value, blocked in 10 bins per decade, and represented in a linear-log plot, whereas the PSD (bottom graph) is blocked in 10 bins per decade and plotted on a log-log scale. As can be seen by comparing Figures 3.2 and 3.6, the bandwidth of OTI allows the measurement of Brownian motion within a time range, during which it is greatly influenced by the hydrodynamic memory effect. Hence, the calibration must be made by using a theory that accounts for the fluid's inertia – that is, using Equation 3.26 instead of Equation 3.8 for fitting the VAF, Equation 3.27 instead of Equation 3.9 for fitting the MSD, and Equation 3.28 instead of Equation 3.10 for fitting the PSD. The black continuous line in each graph of Figure 3.7 therefore corresponds to Equations 3.26, 3.27 and 3.28, respectively, being fitted to the data (red circles) with the two fitting parameters β and k. All three fits generally provide an accuracy of better than 6%, depending on the acquisition frequency and total acquisition time. The relative difference for all fitted values of β acquired by either of each equation is less than a small percentage [41]. The trapping force constant k (see Table 3.2 and next section) is obtained from the long-time and respectively low-frequency, limits of each function, according to Equations 3.21, 3.24 and 3.25, or from the corner frequency ϕ_k in Equation 3.23. The two main sources of error are uncertainties in the determination of the bead size and of the temperature

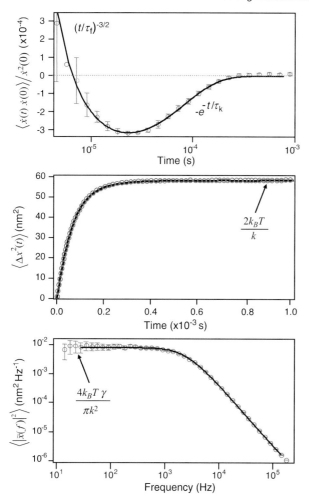

Figure 3.7 Measured VAF (raw data as red line, blocked values as red circles), MSD (circles, data points were removed for clarity) and PSD (only blocked values are shown as circles). The respective fits calculated from Equations 3.26, 3.27 or 3.28 are plotted as black lines.

inside the laser focus. The latter can lead, in particular, to unwanted fluctuations in the fluid's viscosity and density [42].

3.4.2
Influence of Different Parameters on Brownian Motion

In this section we describe the influences of the trapping potential, the surrounding fluid's properties and the particle's properties on Brownian motion.

Table 3.2 Comparison of the three values k_1, k_2 and k_3 obtained from fitting the exponential relaxation time, the corner frequency, or the long time limits of the MSD and PSD.

	k_1	k_2	k_3
$\tau_k = \frac{6\pi\eta_f a_s}{k}$	69 µs → 136 µN m^{-1}	293 µs → 33 µN m^{-1}	798 µs → 12 µN m^{-1}
$\phi_k = \frac{k}{12\pi^2\eta_f a_s}$	2361 Hz → 140 µN m^{-1}	547 Hz → 32 µN m^{-1}	197 Hz → 12 µN m^{-1}
$\langle \Delta x^2(t) \rangle_{t\to\infty} = \frac{2k_B T}{k}$	58.6 nm^2 → 139 µN m^{-1}	243.3 nm^2 → 33 µN m^{-1}	998 nm^2 → 8 µN m^{-1}
$\langle |\tilde{x}(f)|^2 \rangle_{f\to 0}$	10^{-2} nm^2 Hz^{-1}	1.6×10^{-2} nm^2 Hz^{-1}	10^{-2} nm^2 Hz^{-1}
$= \frac{96 k_B T \eta_f a_s}{k^2}$	→ 135 µN m^{-1}	→ 35 µN m^{-1}	→ 7 N m^{-1}

3.4.2.1 Changing the Trap Stiffness

The influence of k on each quantity is shown in Figure 3.8 for the example of a resin sphere, with $a_s = 0.5$ µm, held in three different potentials $k_1 = 140$ µN m^{-1} (red line), $k_2 = 32$ µN m^{-1} (blue line) and $k_2 = 12$ µN m^{-1} (green line). The variation of the trapping potential was achieved experimentally by changing the neutral density filter in front of the laser (NF1 in Figure 3.4). The acquisition frequency and time were chosen as: $f_{acq} = 0.5$ MHz, $t_{tot} = 20$ s. Changing k varies only τ_k, while τ_p and τ_f remain constant, as the particle's and the fluid's properties are fixed ($\tau_p = 0.084$ µs, $\tau_p = 0.25$ µs).

In the VAF (top graph, Figure 3.8), a distinction can be made between the three regimes in which the velocity correlations are either positive ($t < 10^{-5}$ s), negative (10^{-5} s $< t < 10^{-3}$ s), or vanish ($t > 10^{-3}$ s). As stated by Equation 3.16 and shown in Figure 3.2 (top, blue line), the velocity correlations of a free particle are always positive. However, a particle in an optical trap experiences an additional drift towards the trap center due to $F_{ext}(t)$. As the direction of the potential's restoring force $F_{ext}(t) = -kx(t)$ is opposite to the direction of the initial velocity $\dot{x}(0)$, the harmonic potential eventually introduces negative correlations, so-called anti-correlations in the VAF. As expected, the anti-correlations increase with k. The relaxation time τ_k describes the time scale for which $F_{ext}(t)$ makes the particle return from a displaced position to the trap center. For the stiffer trap in this example we find, from fitting, $\tau_{k_1} = 69$ µs, while for the softer trap $\tau_{k_3} = 798$ µs. As seen in the three data sets in Figure 3.8, $\langle \Delta x^2(t) \rangle_{t\to\infty}$ and, respectively, $\langle |\tilde{x}(f)|^2 \rangle_{f\to 0}$ approach a constant value that is inversely proportional to k (Equation 3.24) or correspondingly k^2 (Equation 3.25). In Table 3.2 it can be seen that the values for k_1, k_2 and k_3 obtained from each relationship are consistent with each other, and vary by a maximum of 10%, except for k_3. A better precision for characterizing such soft trapping potentials can be obtained by acquiring data over longer times, as this increases the number of points in the time range of τ_k (data not shown).

The data in Figure 3.8 show that for $t < 10^{-5}$ s and, respectively, $f > 10^{-4}$ Hz, there is a time range during which the particle is free from any influence of the potential. As this time range is longer for weaker traps, in order to study the influences of the fluid and the particle on Brownian motion separately from the influence of an external

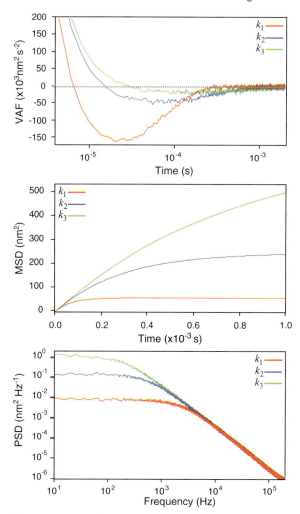

Figure 3.8 Measured VAF, MSD and PSD for three different force constants: $k_1 = 140\,\mu\text{N m}^{-1}$ (red line), $k_2 = 32\,\mu\text{N m}^{-1}$ (blue line) and $k_2 = 32\,\mu\text{N m}^{-1}$ (green line).

force, the trapping stiffness is minimized and the OTI configuration is used solely as a position detector for single particle tracking.

3.4.2.2 Changing the Fluid

In order to study the influence of the fluid's inertia and detect the long-time tail in the normalized VAF (see Figure 3.1 and Section 3.2.2.1), we track the Brownian motion of a larger resin sphere ($a_s = 1.5\,\mu\text{m}$, $\rho_s = 1.51\,\text{kg dm}^{-3}$, $f_{acq} = 0.5\,\text{MHz}$, $t_{tot} = 20\,\text{s}$) in three different solvents having different viscosities η_f and, unavoidably, different densities ρ_f (Figure 3.9). The resin beads are suspended either in a more viscous

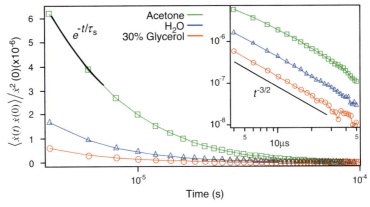

Figure 3.9 Log-linear representation of the measured normalized VAF of a resin sphere ($a_s = 1.5\,\mu m$, $\rho_s = 1.51\,kg\,dm^{-3}$, $f_{acq} = 0.5$ MHz, $t_{tot} = 20$ s) with $k \approx 10\,\mu N\,m^{-1}$, for three different fluids; 30% glycerol in H_2O ($\rho_f = 1.18\,kg\,dm^{-3}$, $\eta_{30\%glycerol} = 2.11 \times 10^{-3}$ Pa·s, ○), H_2O ($\rho_f = 1.118\,kg\,dm^{-3}$, $\eta_{H2O} = 10^{-3}$ Pa·s, △) and acetone ($\rho_f = 0.790\,kg\,dm^{-3}$, $\eta_{acetone} = 0.306 \times 10^{-3}$ Pa·s, □). Fits correspond to the continuous line with the respective color. Inset: log-log plot of the same VAFs. The $t^{-3/2}$ power-law is indicated by the black line.

solution of 30% glycerol in water ($\rho_{30\%Glycerol} = 1.18\,kg\,dm^{-3}$, $\eta_{30\%Glycerol} = 2.11 \times 10^{-3}$ Pa·s, ○), in pure water with an intermediate viscosity ($\rho_{H_2O} = 1\,kg/dm^3$, $\eta_{H_2O} = 10^{-3}$ Pa·s, △), or in pure acetone ($\rho_{acetone} = 0.790\,kg\,dm^{-3}$, $\eta_{acetone} = 0.306 \times 10^{-3}$ Pa·s, □), the fluid with the lowest viscosity we could find. Decreasing η_f results in increasing not only τ_f as: $\tau_{30\%Glycerol} = 1.26\,\mu s$, $\tau_{H_2O} = 2.25\,\mu s$ and $\tau_{acetone} = 5.8\,\mu s$, but also τ_s as: $\tau_{s_{30\%Gl}} = 0.36\,\mu s$, $\tau_{s_{H_2O}} = 0.76\,\mu s$ and $\tau_{s_{ac}} = 2.48\,\mu s$. As τ_f and τ_s depend on a_s^2, the use of a larger Brownian particle increases both time scales and places them closer to (or even within) the detection bandwidth of the instrument. The choice of a resin sphere in this particular experiment is motivated by the fact that resin, unlike polystyrene, is not soluble in acetone. Furthermore, the difference between the refractive indices of resin ($n = 1.68$) and acetone ($n = 1.36$) is still high enough to provide a good contrast and allow visualization of the particle by optical microscopy, which is not the case for example, with silica ($n = 1.37$).

In order to obtain anticorrelations that are negligible compared to the detection limit, the trap stiffness is reduced to $k \approx 10\,\mu N\,m^{-1}$, and hold this constant from one fluid to the other. However, this does not prevent τ_k from changing, as it is dependent on η_f as: $\tau_{k_{30\%Glyc}} \approx 5.9$ ms, $\tau_{k_{H2O}} \approx 2.84$ ms, $\tau_{k_{aceton}} \approx 950\,\mu s$. With such a weak trap the persistent positive correlations arising from the hydrodynamic back-flow introduced in Section 3.2.1.2 can be readily identified up to 50 µs. The data are best fitted by the normalized Equation 3.26 (continuous lines). In Figure 3.9, it can be seen that, as expected, for lower viscosities the correlations last longer (green line). Furthermore, the log-log representation in the inset allows recognizing of the $t^{-3/2}$ power law decay (black line) that is followed by all three fluids. In acetone, as $\tau_{s_{ac}} \approx 3\,\mu s$ is also within the detection range, it is possible to detect between 2 and 6 µs an exponential tendency arising from the combined fluid's and particle's inertia (Figure 3.9, green line).

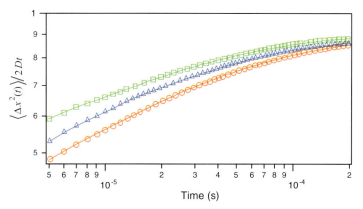

Figure 3.10 Comparison of the motion of particles with the same radius but different densities. Log-log representation of the measured normalized MSD for a silica sphere ($a_s = 1.97\,\mu\text{m}$, $\rho_s = 1.96\,\text{kg dm}^{-3}$, ○), of a resin sphere ($a_s = 2\,\mu\text{m}$, $\rho_s = 1.51\,\text{kg dm}^{-3}$, △) and a polystyrene sphere ($a_s = 1.94\,\mu\text{m}$, $\rho_s = 1.05\,\text{kg dm}^{-3}$, □). Fits correspond to the continuous line with the respective color.

3.4.2.3 Changing the Particle Density

The direct influence of the particle's inertia can be determined by comparing the motion of particles with approximately the same size but different densities. The MSD was calculated from the measured position fluctuations ($f_{acq} = 1\,\text{MHz}$, $t_{tot} = 10\,\text{s}$) of a silica sphere ($a_s = 1.97\,\mu\text{m}$, $\rho_s = 1.96\,\text{kg dm}^{-3}$, red line), of a resin sphere ($a_s = 2\,\mu\text{m}$, $\rho_s = 1.51\,\text{kg dm}^{-3}$, blue line) and a polystyrene sphere ($a_s = 1.94\,\mu\text{m}$, $\rho_s = 1.05\,\text{kg dm}^{-3}$, green line) in water with $\tau_f \approx 3.9\,\mu\text{m}$. The influence of the trap was again kept minimal at $k \approx 14\,\mu\text{N m}^{-1}$ for the three different beads. Hence, the MSD can be normalized by its long-time limit $\langle \Delta x^2(t) \rangle_{free} = 2Dt$ in the free regime, when $F_{ex}(t) \approx 0$, as shown in Figure 3.10. Equation 3.27 (continuous lines) corresponds to the best fit of the data (silica ○, resin △, polystyrene □). Here, the inertia of the perturbed fluid does not change, so that τ_f stays constant, while the inertias of the three particles are different and lead to different values of τ_s; $\tau_{s_{silica}} = 1.69\,\mu\text{s}$, $\tau_{s_{resin}} = 1.34\,\mu\text{s}$ and $\tau_{s_{PS}} = 0.88\,\mu\text{s}$. Even though differences in τ_s are in fact small compared to the time resolution of $2\,\mu\text{s}$, the contribution expected from the particle's mass can still be detected.

Changing the particle's radius a_s is equivalent to varying both τ_f and τ_s, while keeping τ_s/τ_f constant (data not shown) [34].

3.4.3
Implications of the Existence of Long-time Tails in Nanoscale Experiments

Having demonstrated agreement between Brownian motion theory and OTI data, and studied the influence of parameters which can be varied experimentally, we can derive a rule of thumb to estimate the time range during which the particle's motion can be considered as effectively free from the influence of the trap. Furthermore, we can determine for how long the inertia of a Newtonian fluid will influence the Brownian probe's motion and prevent it from performing a diffusive random walk inside an optical trapping potential. This sets the bandwidth of OTI for high-resolution single particle tracking to probe locally many different media.

3.4.3.1 Single Particle Tracking by OTI

We define the time τ_{free} starting from which the optically confined MSD given by Equation 3.27 begins to deviate by at least 2% from the free MSD described by Equation 3.17. We record the motion of different polystyrene spheres of sizes $a_s = 0.27$, 0.33, 0.50, 0.74, 1.00 and 1.25 μm confined by potentials with a spring constant ranging from 1 to 100 μN m^{-1}. For stronger traps ($\tau_k < 1$ ms), the data are recorded at 500 kHz during 20 s and calibrated in the range between 2 μs and 1 ms. For softer traps ($\tau_k > 1$ ms), the data are recorded at 200 kHz for 50 s and calibrated between 5 μs and 10 ms. In Figure 3.11, τ_{free} is represented as a function of τ_k in a log-log scale, which allows us to formulate an approximate empirical relation between both time scales [41]:

$$\tau_{free} = \tau_k/20 \qquad (3.32)$$

Hence, in the time range [2 μs, τ_{free}], which is limited on one side by the noise floor (see Figure 3.6) and on the other side by $\tau_k/20$, OTI can track the motion of a single free Brownian particle. Under these conditions, the sphere probes only its local environment, for up to three decades in time.

3.4.3.2 Diffusion in OTI

The next issue to address is the question of whether or not the Brownian probe has time to reach a purely diffusive behavior before it becomes confined by the potential of the trap. This is equivalent to studying the influence of inertial effects on the particle's motion in the optical trap at times shorter than τ_{free}. Therefore, we investigate the diffusion of a small sphere, which is expected to perturb the fluid less, and hence τ_f and the hydrodynamic memory effect are smaller. Furthermore, we adjust the trapping potential to be the softest possible, as the time region where the particle's motion is free from the influence of the trap lasts longer for high τ_k.

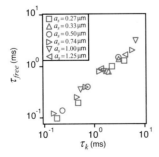

Figure 3.11 Log-log representation of τ_{free} versus τ_k for polystyrene spheres of different radii $a_s = 0.27$, 0.33, 0.50, 0.74, 1.00 and 1.25 μm.

3.4 High-Resolution Analysis of Brownian Motion | 115

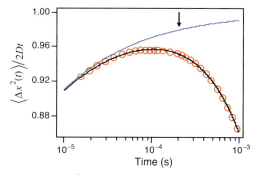

Figure 3.12 $\langle \Delta x^2(t) \rangle / 2Dt$ for the sphere with $a_s = 0.27\,\mu$m and $k = 1.5\,\mu$N m^{-1}, $f_{acq} = 0.2$ MHz, $t_{tot} = 50$ s, fitted by Equation 3.34 (black line). The theory for the free particle is given by the blue line corresponding to Equation 3.24. The arrow indicates the time when $\langle \Delta x^2(t) \rangle_{free}/2Dt$ reaches diffusive motion within 2%.

We introduce the dimensionless representation $\langle \Delta x^2(t) \rangle / 2Dt$ to distinguish between the free diffusive motion when $\langle \Delta x^2(t) \rangle / 2Dt = 1$ (see Section 3.2.2.1) and the motion influenced by inertial effects when the particle is either free or optically confined (see Section 3.2.2.1, parts (i) and (ii), respectively). In both cases motion is nondiffusive as $\langle \Delta x^2(t) \rangle / 2Dt < 1$.

The measured $\langle \Delta x^2(t) \rangle / 2Dt$ for a polystyrene sphere ($a_s = 0.27\,\mu$m, $k = 1.5\,\mu$N m^{-1}) is shown in Figure 3.12 (red circles), fitted to Equation 3.27 (black line) and compared to $\langle \Delta x^2(t) \rangle_{free}/2Dt$ given by Equation 3.17 (blue line). Here, it can be seen that $\langle \Delta x^2(t) \rangle / 2Dt$ reaches a maximum of ~ 0.96, but never 1.

For the free particle, $\langle \Delta x^2(t) \rangle_{free}/2Dt = 1$ would occur within 2% error after approximately 200 µs (Figure 3.12, arrow). Thus, in order to observe free diffusive Brownian motion, the optical trap would have to be so weak that $\tau_{free} > 0.2$ ms is satisfied, or equivalently $\tau_k > 4$ ms according to Equation 3.32. However, for all the combinations of particle sizes and spring constants studied, we could never adjust such a long relaxation time. In the particular case of the sphere with $a_s = 0.27\,\mu$m, a spring constant $k < 1$ nN/m would be needed to observe free diffusive motion for at least one decade in time. However, such a low spring constant does not allow us to trap and observe the particle for a sufficiently long period of time. Hence, in experiments using optical traps, the motion of a particle is influenced by either memory effects and/or by the harmonic potential [41]. This is in contradiction with assumptions commonly made in optical trapping experiments, where a normal diffusive behavior of the trapped particle is assumed and inertial effects from the fluid are ignored [38, 43].

The time-dependent diffusion coefficient can be derived from the VAF or the MSD as:

$$D(t) = \int_0^t \langle \dot{x}(t')\dot{x}(0) \rangle dt' = \frac{1}{2}\frac{d}{dt}\langle \Delta x^2(t) \rangle \tag{3.33}$$

and approaches the diffusion constant D in the infinite time limit when $F_{ex}(t) = 0$:

$$D = \int_0^\infty \langle \dot{x}(t')\dot{x}(0)\rangle dt' = \frac{k_B T}{6\pi \eta_f a_s} \tag{3.34}$$

3.4.3.3 Thermal Noise Statistics

According to our findings, the process of optically confined Brownian motion as observed by OTI is non-Markovian up to times $t > \tau_k$, when it becomes dominated by the trapping potential. Only from then on does motion become uncorrelated, $\langle \dot{x}(t)\dot{x}(0)\rangle = 0$, and can position data points be considered as statistically independent. It has been proposed that optical trapping data can be calibrated by thermal noise analysis using Boltzmann statistics [44]:

$$p(x)dx = ce^{-E(x)/k_B T} \tag{3.35}$$

which describes the probability $p(x)dx$ of finding the Brownian particle in the potential $E(x)$ (c normalizes the probability distribution).

From the calibrated time traces acquired with $f_{acq} = 0.5$ MHz, the probability distribution is represented in Figure 3.13 for two different resin spheres with $a_s = 0.5\,\mu\text{m}$ (red histogram) and $a_s = 2\,\mu\text{m}$ (green histogram) trapped in a similar potential with $k \approx 12.5\,\mu\text{N m}^{-1}$ but obviously different τ_k: $\tau_{k_{small}} = 69\,\mu\text{s}$ and $\tau_{k_{big}} = 798\,\mu\text{s}$. The position histograms, with a bin width of 1 nm, are compared for both spheres, and contain either $N = 400\,000$ data points, corresponding to an acquisition time $t_{tot} = 0.08$ s (top graph), $N = 400\,000$ points, corresponding to an acquisition time $t_{tot} = 0.8$ s (middle graph) or $N = 4\,000\,000$ points, corresponding to an acquisition time $t_{tot} = 8$ s (bottom graph).

Even though in each histogram there are apparently enough data points to perform statistical analysis, these points are clearly not statistically independent. Indeed, the upper histogram features only ~ 100 statistically independent points for the larger sphere and ~ 1000 for the smaller sphere. The smaller sphere will sample the trapping potential well more rapidly than the larger, which is translated in the differences between $\tau_{k_{small}}$ and $\tau_{k_{big}}$; therefore, the larger sphere will take ~ 10-fold longer to explore the potential and, consequently, for statistical analysis the trajectory should be acquired over longer times. The temporal resolution of the Boltzmann statistics method is determined by the time required to record uncorrelated data, and is heavily dependent on the particle's size and τ_k. The high acquisition rates used throughout these studies are not needed in this case.

3.5
Summary and Outlook

In this chapter we have shown how OTI can be used to study Brownian motion down to hydrodynamic time scales, where the response of the surrounding fluid becomes dominant. This is only possible due to the high bandwidth (up to 500 kHz) and

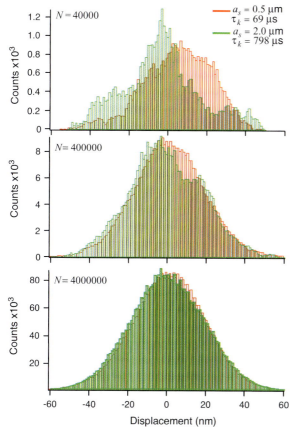

Figure 3.13 Histogram of position fluctuations (bin width of 1 nm) acquired with $f_{acq}=0.5$ MHz for a sphere with $a_s=0.5$ μm and $a_s=2.0$ μm. The number of data points N in the histograms increases from top to bottom from 40 000 ($t_{tot}=0.08$ s) to 4 000 000 ($t_{tot}=0.08$ s). The trapping stiffness $k\approx 12.5$ μN m^{-1} is similar for both bead sizes.

subnanometer spatial resolution of the position detection configuration. The precision achieved allows us to detect not only the effects of the fluid's inertia but also the more subtle effects of the sphere's inertia.

The details in the motion of the trapped model sphere provide insight into the interplay between inertial effects and the optical trapping potential, as summarized by Figure 3.14. This shows in particular, the overlap of τ_f, the characteristic time of the

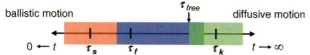

Figure 3.14 Overview of the characteristic times of a Brownian particle confined by a harmonic potential.

viscous fluid, with τ_k, the relaxation time of the restoring force of the trapping potential. The time τ_{free}, which separates τ_f from τ_k, was determined empirically and corresponds to $\sim \tau_k/20$. Below this time, OTI can be used solely as a position detector for *single particle tracking with unprecedented spatial and temporal resolution.*

At these time scales, the sphere performs a non-random walk, dominated by the nature of the surrounding medium [45]. The presented method is capable of providing new insights into the behavior of media that are more complex than just a simple viscous fluid, thus extending the bandwidth of microrheology by two decades in time [46]. For example, the high-frequency response of a viscoelastic polymer solution should provide information on the nanomechanical properties of the polymer molecules. In particular, highly dynamic polymers, such as those encountered in a living cell, should transmit their mechanical and dynamic signatures to the Brownian particle. Furthermore, an obstacle in the particle's trajectory such as a surface, with for example, various chemical functionalities, or a cell membrane, should influence Brownian motion in a characteristic way. Such studies for a variety of biopolymers and surfaces are currently in progress in our laboratory.

Acknowledgments

The authors are grateful to J. Lekki for help in data acquisition, to P. De Los Rios, H. Flyvbjerg, T. Franosch for discussions, and to D. Alexander for reading the manuscript. B.L. and C.G. acknowledge the financial support of the Swiss National Science Foundation and its NCCR. S.J. acknowledges the support of the Gebert Rüf Foundation. The authors also thank EPFL for funding the experimental equipment.

References

1 Haw, M.D. (2002) *Journal of Physics - Condensed Matter*, **14**, 7769.
2 Einstein, A. (1905) *Annalen Der Physik*, **17**, 549.
3 Stokes, G.G. (1851) *Transactions of the Cambridge Philosophical Society*, **9**, 8.
4 Langevin, P. (1908) *Comptes Rendus Hebdomadaires des Seances de L'Academie des Sciences*, **146**, 530.
5 Henri, V. (1908) *Comptes Rendus Hebdomadaires des Seances de L' Academie des Sciences*, **146**, 1024.
6 Vladimirsky, V. and Terletzky, Y. (1945) *Zhurnal Eksperimentalnoi i Teoreticheskoi Fiziki (in Russian)*, **15**, 259.
7 Alder, B.J. and Wainwright, T.E. (1967) *Physical Review Letters*, **18**, 988.
8 Widom, A. (1971) *Physical Review A*, **3**, 1394.
9 Zwanzig, R. and Bixon, M. (1970) *Physical Review A*, **2**, 2005.
10 Bedeaux, D. and Mazur, P. (1974) *Physica*, **76**, 247.
11 Hinch, E.J. (1975) *Journal of Fluid Mechanics*, **72**, 499.
12 Pomeau, Y. and Resibois, P. (1975) *Physics Reports*, **19**, 63.
13 Zwanzig, R. and Bixon, M. (1975) *Journal of Fluid Mechanics*, **69**, 21.
14 Clercx, H.J.H. and Schram, P. (1992) *Physical Review A*, **46**, 1942.

15 Frey, E. and Kroy, K. (2005) *Annalen Der Physik*, **14**, 20.
16 Gittes, F., Schnurr, B., Olmsted, P.D., MacKintosh, F.C. and Schmidt, C.F. (1997) *Physical Review Letters*, **79**, 3286.
17 Boon, J.P. and Bouiller, A. (1976) *Physics Letters A*, **55**, 391.
18 Paul, G.L. and Pusey, P.N. (1981) *Journal of Physics A - Mathematical and General*, **14**, 3301.
19 Ohbayashi, K., Kohno, T. and Utiyama, H. (1983) *Physical Review A*, **27**, 2632.
20 Weitz, D.A., Pine, D.J., Pusey, P.N. and Tough, R.J.A. (1989) *Physical Review Letters*, **63**, 1747.
21 Kao, M.H., Yodh, A.G. and Pine, D.J. (1993) *Physical Review Letters*, **70**, 242.
22 Ashkin, A., Dziedzic, J.M., Bjorkholm, J.E. and Chu, S. (1986) *Optics Letters*, **11**, 288.
23 Berg-Sorensen, K. and Flyvbjerg, H. (2005) *New Journal of Physics*, **7**, 38.
24 Landau, L.D. and Lifshitz, E.M. (1987) *Fluid Mechanics*, Vol. 6, 2nd edn, Butterworth-Heinemann, Oxford.
25 Lorentz, H.A. (1921) *Lessen over Theoretische Natuurkunde*, Vol. V, E.J. Brill, Leiden.
26 Vanderhoef, M.A., Frenkel, D. and Ladd, A.J.C. (1991) *Physical Review Letters*, **67**, 3459.
27 Berg-Sorensen, K. and Flyvbjerg, H. (2004) *Review of Scientific Instruments*, **75**, 594.
28 Uhlenbeck, G.E. and Ornstein, L.S. (1930) *Physical Review*, **36**, 0823.
29 Svoboda, K. and Block, S.M. (1994) *Annual Review of Biophysics and Biomolecular Structure*, **23**, 247.
30 Reif, F. (1985) *Fundamentals of Statistical and Thermal Physics*, McGraw-Hill, Singapore.
31 Henderson, S., Mitchell, S. and Bartlett, P. (2002) *Physical Review Letters*, **88**, 088302.
32 Sterba, R.E. and Sheetz, M.P. (1998) *Methods in Cell Biology*, **55**, 29.
33 Rohrbach, A., Tischer, C., Neumayer, D., Florin, E.L. and Stelzer, E.H.K. (2004) *Review of Scientific Instruments*, **75**, 2197.
34 Lukic, B., Jeney, S., Tischer, C., Kulik, A.J., Forro, L. and Florin, E.L. (2005) *Physical Review Letters*, **95**, 160601.
35 Gittes, F. and Schmidt, C.F. (1998) *Optics Letters*, **23**, 7.
36 Pralle, A., Prummer, M., Florin, E.L., Stelzer, E.H.K. and Horber, J.K.H. (1999) *Microscopy Research and Technique*, **44**, 378.
37 Rohrbach, A. and Stelzer, E.H.K. (2002) *Journal of Applied Physics*, **91**, 5474.
38 Neuman, K.C. and Block, S.M. (2004) *Review of Scientific Instruments*, **75**, 2787.
39 Press, W.H., Flannery, B.P., Teukolsky, S.A. and Vetterling, W.T. (1992) *Numerical Recipes in C*, Cambridge University Press, Cambridge.
40 Rohrbach, A. (2005) *Physical Review Letters*, **95**, 168102.
41 Lukic, B., Jeney, S., Sviben, Z., Kulik, A.J., Florin, E.L. and Forro, L. (2007) *Physical Review E*, **76**, 011112.
42 Guzmán, C., Flyvbjerg, H., Köszali, R., Ecoffet, C., Forró, L. and Jeney, S. (2008) *Applied Physics Letters*, **93**, 184102.
43 Gittes, F. and Schmidt, C.F. (1998) *Methods in Cell Biology*, **55**, 129.
44 Florin, E.L., Pralle, A., Stelzer, E.H.K. and Horber, J.K.H. (1998) *Applied Physics A: Materials Science and Processing*, **66**, S75.
45 Liverpool, T.B. and MacKintosh, F.C. (2005) *Physical Review Letters*, **95**, 208303.
46 Mason, T.G., Ganesan, K., vanZanten, J.H., Wirtz, D. and Kuo, S.C. (1997) *Physical Review Letters*, **79**, 3282.

4
Nanoscale Thermal and Mechanical Interactions Studies using Heatable Probes

Bernd Gotsmann, Mark A. Lantz, Armin Knoll, and Urs Dürig

4.1
Introduction

Thermal properties such as thermal conductivity and diffusivity, although rather difficult to measure, are important properties for many applications. For example, in microelectronics, where power densities can be very high, the local generation of heat and its conduction away from the heated region are a major design issue. In general, the interplay between thermal and mechanical properties of solids is a fascinating topic of research, and is of immediate practical relevance in numerous applications. For example, the mechanical properties of soft matter – namely organic polymers – exhibit such a strong temperature dependence that in standard analysis techniques, such as dynamic mechanical thermal analysis (DMTA), both mechanical stress and temperature are varied. Often, the time scale is also varied, making the analysis rather complicated [1]. The local heating of materials is also used as a micromanufacturing technique. In all of these fields, there is a clear trend to extend research down to the nanoscale, and this is further nurtured by the necessity to understand nanoscale properties in order to tailor materials for nanoscale applications. The trend towards nanoscale opens up research fields and applications beyond conventional materials science. The very definition of temperature, which is a thermodynamic (i.e. statistical) concept based on local equilibrium, becomes vague on length scales smaller than the mean free path of heat carriers (typically in the range of 10–100 nm) [2–4]. On the nanoscale, it is commonly observed that interfaces become more predominant in determining materials properties. This is also true of thermal properties in general, and leads to fascinating concepts such as phonon engineering that promise tailored thermal conductance in nanostructures [5].

In order to study and exploit microscale and nanoscale thermal phenomena, a variety of scanning probe microscopy (SPM) -based techniques have been developed, most of which are based on contact scanning force microcopy (SFM). Two examples of exciting technological applications of these techniques are in the areas of nanoscale data storage [6, 7] and lithography [8]. In this chapter, experimental procedures and results from the broad field of heated-probe SFM are addressed. Although the applications of these techniques are rather diverse, two common elements are the use of a sharp, temperature-sensitive tip and the use of SFM techniques to scan this tip over the surface and simultaneously measure the surface topography. Many of these applications also require a means to heat the tip, in turn to heat the sample, on a highly local scale. In the following sections we first describe the various types of probe that have been developed for thermal scanning probe microscopy, and outline the basics of probe-based imaging of thermal properties. Later, we analyze the various heat-loss mechanisms that play a role in the interpretation of thermal SPM data. Specific applications are discussed thereafter, including thermomechanical nanoindentation, data storage and nanopatterning and lithography.

4.2
Heated Probes

At the heart of all the techniques described in this chapter is a heatable probe with an integrated means of sensing the temperature of the probe tip. As with all scanning probe techniques, the resolution is limited at least in part by the geometry of the tip apex and the area of contact between the tip and the surface. The most widely used heated probe is a Wollaston wire probe. In this technique, a thin, bent platinum/rhodium wire is used to produce heat and detect temperature. For SPM-based applications, the wire is bent into the shape of a cantilever, as illustrated in Figure 4.1. Often, also a mirror is glued onto the back of the cantilever to improve optical detection of the cantilever bending. The temperature of the wire can be determined by measuring its electrical resistance and using the

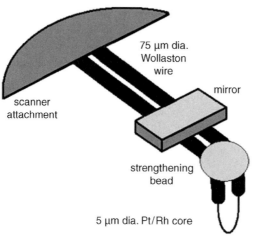

Figure 4.1 Schematic diagram of a thermal probe made from a Wollaston wire process. From Ref. [9].

known temperature dependence of the electrical resistance of the material. Although such probes are accurate and easy to handle, their spatial resolution is limited by the dimensions of the bent wire at the end that acts as the probing tip. To date, the spatial resolution reported using such probes is limited to about ∼100 nm (see for example Figure 4.5).

The spatial resolution can be improved by using microfabrication techniques to produce cantilevers with very sharp, temperature-sensitive tips. Shi *et al.* [10] have made such probes using silicon nitride for the cantilever body and microfabricating a platinum tip with a junction to chromium near the tip apex (see Figure 4.2). This junction acts as a thermoelement and can be used to measure the temperature of the tip as it is being scanned over a heated surface, or to investigate local heat sources on a sample. Majumdar *et al.* have used such tips to image hot spots in a very-large-scale integration (VLSI) chip [11] and to image the heat generated by the current flowing through a carbon nanotube [12]. In the latter experiment, an impressive lateral resolution of 50 nm was demonstrated.

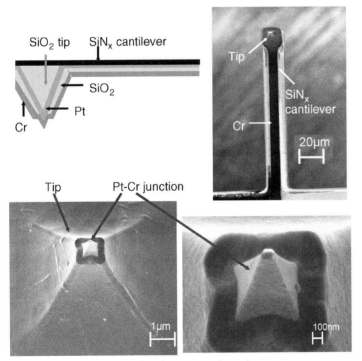

Figure 4.2 Microfabricated probes for scanning thermal microscopy (SThM). Cross-section (upper left) and scanning electron microscopy images of a SThM probe (upper right), the probe tip (lower left) and the Pt-Cr junction (lower right) at the end of the tip. (Reprinted from Ref. [10]; © 2002, American Society of Mechanical Engineering.)

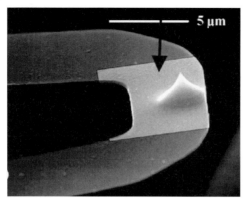

Figure 4.3 Scanning electron microscopy image of a cantilever with integrated heater (artificially contrasted) and tip. (Similar probes were used to obtain the image in Figure 4.7b). (Reprinted from Ref. [14]; © Springer-Verlag.)

Another approach, developed by IBM, uses an all-silicon microfabricated cantilever with an integrated heater and tip (see Figure 4.3) [13]. Here, the largest part of the two-legged cantilever is made from highly doped silicon (10^{20} cm^{-3} As), whereas the part of the cantilever that supports the tip is lower-doped (5×10^{17} cm^{-3}) and serves as both heater and sensor. With dimensions of 4 μm × 6 μm and a thickness of ∼200 nm, the heater has a resistance of few kΩ. The known temperature dependence of the resistivity of doped silicon can be exploited to sense the heater temperature. This type of thermal probe has been used in the majority of the experiments described in this chapter.

The temperature calibration of the heater can be carried out by measuring a current–voltage response curve (I–V curve) of the cantilever (Figure 4.4). For this, a resistor is typically placed in series with the cantilever. The measured voltage drop

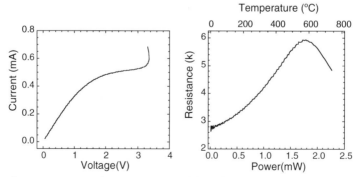

Figure 4.4 Current–voltage (I–V) curve and derived power–resistance (P–R) and temperature–resistance (T–R) curves of an integrated resistive heater.

across the resistor is used to determine the current flowing through the cantilever and, in combination with the measured voltage drop across the cantilever, can be used to calculate the electrical power dissipated in the cantilever. Initially, as the current flowing through the heater is increased, the dissipated power results in an increase in resistance due to increased scattering of the carriers. As the temperature rises, the number of thermally generated carriers also increases, which tends to reduce the rate at which the resistance increases with temperature. When the number of thermally generated carriers equals the number of dopants, the resistance reaches its maximum value, and begins to decrease with further increases in power and temperature. The temperature at which the maximum resistance occurs is a function of the doping density and is known from the literature [15]. The power needed to reach the maximum resistance, P_{Rmax}, is determined from the measured I–V data. It is assumed that all of the power dissipated in the cantilever contributes to heating of the heater structure, and that the temperature change of the heater is a linear function of the dissipated power. We can then rescale a plot of resistance versus power to a plot of resistance versus temperature using two known values, namely, the resistance at room temperature measured at low very low power, and the temperature at which the maximum resistance occurs. For the doping values used here, the maximum resistance occurs at 550 °C, and the heater temperature can thus be calculated using

$$T_{heater} = RT + P(550°C - RT)/P_{Rmax} = RT + R_{th}P,$$

where RT is room temperature.

The implicit assumption here is that the thermal resistance of the system, R_{th}, is independent of the heater temperature, T_{heater}. We find that R_{th} is typically on the order of 10^5 K W^{-1} under ambient conditions. When checking all of these assumptions by measuring the temperature of the heater by independent means, we found that the resulting systematic error is far below the measurement errors, fabrication tolerances and scatter. Using this calibration method, we estimate an absolute error of about 30% for the temperature difference $\Delta T = T_{heater} - RT$. Relative measurements and temperature changes, however, can be conducted with a temperature resolution of ~0.1 K.

The time scale that these heaters are able to probe is related to the thermal equilibration time of the cantilevers. Although the dynamics is not a simple RC-type exponential [14, 16], it can be approximated by a simple exponential and a single time constant. For the cantilevers used in most of the examples discussed below, the time constant is on the order of 7 to 10 µs; this is then the time constant that limits fast thermal sensing. For a rapid application of heat, however, the heater can be operated in nonequilibrium on timescales down to 1 µs and lower [14]. The transient temperature can be sensed reliably at any time scale by means of the electrical resistance. By design, the time constant can be decreased to values below 1 µs, but there are trade-offs between power consumption, sensitivity, time constant, ease of fabrication and the mechanical stability of the cantilever [16, 17]. Therefore, the optimum design depends critically on the application envisaged. For a detailed study of cantilever designs and time constants, the reader is referred to Ref. [14].

4.3
Scanning Thermal Microscopy (SThM)

The thermal conductivity/diffusivity of a sample can be measured by scanning a heated tip on the sample surface of interest and monitoring the heat flow between probe and surface. The heater/tip is usually mounted on a cantilever so that the surface topography can be measured simultaneously with the thermal signal by applying atomic force microscopy (AFM) techniques. This so-called scanning thermal microscopy (SThM) method has recently been reviewed [10, 18–22].

In general, the heater is also used as a temperature sensor so that changes in the heat flow can be measured during scanning. In the Wollaston wire approach, this is achieved by measuring a thermo-voltage, whereas in the IBM approach the temperature-dependent heater resistance is measured; when the tip is in contact with the surface, the contact forms a heat-loss path. The corresponding thermal conductivity can be visualized by measuring changes in the tip temperature that result from changes in the thermal conductivity as the tip is scanned over the surface. However, to be able to sense this local heat loss, the conductivity must be large enough to produce a measurable signal if compared with the electrical noise in the transducer. An example of a local thermal conductivity map of a polymer blend by Reading *et al.* [18] is shown in Figure 4.5. In such a SThM image the color contrast depends on the local thermal conductivity.

Using this technique, it is relatively straightforward to produce a qualitative image of relative differences in thermal conductivity. Quantitative measurements, however, are significantly more challenging and require knowledge of both the tip–sample contact geometry and all of the various thermal resistances and parasitic heat-loss paths in the system (see Section 4.4). In contrast to the electronic case, heat paths

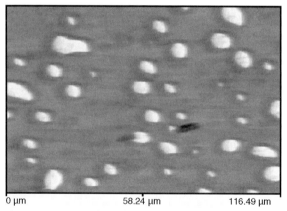

Figure 4.5 Scanning thermal microscopy image of a polymer blend. The two phases can be clearly distinguished in the thermal signal. (Reprinted from Ref. [18]; © 1998; International Scientific Communications, Inc.)

cannot easily be insulated, and therefore a heater typically has several heat-loss paths. The main heat-loss paths, which do not vary with the local sample conductivity, are conduction through the cantilever legs, nonlocal air conduction between heater and sample surface, and radiation cooling. Another potential heat-loss path is through a water meniscus that can form at the point of contact between tip and surface. This will effectively increase the heat-conduction cross-section, leading to a reduced lateral resolution. After performing experiments under ambient conditions, Shi et al. [10] have concluded that heat transfer between heater and sample is dominated by conduction through a water meniscus. The heat transfer paths are discussed in more detail in Section 4.4.

The method described above can be refined by modulating the heater drive voltage, which results in a modulation of the heat flow to the sample. The resulting ac component of the heater temperature can be measured using a lock-in amplifier and used to produce an ac thermal image. This ac heat-loss signal changes depending on how the modulation period compares with the diffusion time of heat in the tip–surface contact region – that is, lower-frequency signals diffuse further than do high-frequency signals. Thus, by varying the modulation frequency of the heater, the probing depth can be controlled. This can be seen in Figure 4.6, which shows two ac thermal images taken at 1 and 30 kHz. The sample consists of islands of high-thermal-conductivity material surrounded by low-thermal-conductivity material, both covered by a polymer layer. In the image taken at 1 kHz, the ac signal probes below the polymer layer and strong material contrast is observed, whereas at 10 kHz both probing depth and contrast are significantly reduced.

The limits of the SThM technology are not easily defined. There is a trade-off between time and temperature resolution on the one hand, and spatial resolution on the other hand. For example, increasing the contact area between the tip and sample increases the thermal conductivity, resulting in larger signals and therefore improved sensitivity – but at the expense of reduced lateral resolution. As will be shown below (Section 4.4), the high thermal impedance of the tip and the tip–surface interface make working on samples with high thermal conductivity challenging. Ideally, the

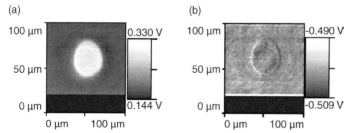

Figure 4.6 Two ac thermal images of a sample with an island of high-thermal-conductivity material within a matrix of low-thermal-conductivity material, both covered by a polymer coating. Image (a) was taken at 1 kHz and image (b) at 30 kHz. (Reprinted from Ref. [18]; © 1998, International Scientific Communications, Inc.)

thermal impedance of the tip and tip–surface interface should be comparable to, or smaller than, those of the sample. The tip and tip–sample interface impedances increase as the tip is made sharper and the contact area reduced, making high-spatial-resolution experiments challenging. Nevertheless, a spatial resolution in the range of some tens of nanometers is feasible on some samples. For example, Shi *et al.* have demonstrated a resolution better than 100 nm on metallic wires [10] and better than 50 nm when imaging a carbon nanotube [12]. A challenge for the quantitative analysis of SThM images is the unknown interaction volume under the probing tip [19]. Nevertheless, the method is very successful in the study of polymers and biological samples. For a recent review, see Ref. [19].

An example of high-lateral-resolution SThM is given in Figure 4.7. Here, the very small contrast between two materials of similar thermal conductivity (silicon oxide and hafnium oxide) is observed. The sample consisted of 2 nm-thick islands of SiO_2 surrounded by a 3 nm-thick film of HfO_2 on a single-crystal silicon substrate. Note that both materials have a considerably higher thermal conductivity than polymers. The measurements were performed in a high-vacuum environment using silicon probes with integrated silicon heaters (as described in Section 4.2). A lateral resolution of ~25 nm was achieved, and the previously unknown thermal conductivity of the 3 nm-thick HfO_2 film was determined. This example nicely demonstrates the potential of using SThM for quantitative measurements, even at high spatial resolution.

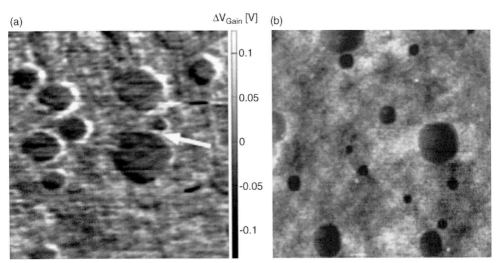

Figure 4.7 (a) Scanning thermal microscopy image (1.6 μm × 1.8 μm) and (b) topography image (2 μm × 2 μm × 5 nm) of a HfO_2 film on a Si substrate. The round holes are filled with 2 nm-thick SiO_2. From the image contrast between the HfO_2 and the SiO_2 regions, the thermal conductivity of HfO_2 can be determined quantitatively. (Images reproduced from Ref. [23].)

4.4
Heat-Transfer Mechanisms

Most of the applications described in this chapter use a sharp, heated probe to deliver heat to a surface on a highly localized scale. However, this process is often rather inefficient because of the high thermal resistance of nanometer-sharp tips and nanometer-sized contact areas, in combination with the other parasitic heat-loss paths present in the system.

In this section, we analyze the various heat-loss paths and mechanisms that can play a role in heated-probe experiments. The analysis in this section is taken from some unpublished results of U. Dürig and B. Gotsmann. The majority of heat-loss paths are undesirable in the sense that they do not contribute to the image contrast in SThM and reduce the efficiency of delivering heat to the sample in other applications. The various heat-loss mechanisms that can contribute to heat loss from a heated tip are illustrated in Figure 4.8 and described in more detail in the corresponding Sections 8.4.1–8.4.6. Which of these mechanisms contribute in a given experiment and the relative magnitudes of their contributions depend on the details of the cantilever design and the experimental conditions. The potentially undesirable heat-loss mechanisms include conductive heat loss through air and the cantilever (Section 4.4.1) and also thermal radiation (Section 4.4.2). Heat conduction through the tip is desirable, but the thermal resistance of the tip (Section 4.4.4) and conduction through a water meniscus (Section 4.4.3) that can form between the tip and sample limit the sensitivity and resolution. The thermal resistance of the tip–surface interface (Section 4.4.6) and the thermal spreading resistance in the sample (Section 4.4.5) are material-specific and determine the image contrast in SThM. The relative magnitudes of these resistances also play an important role in determining the efficiency of heat delivery to the sample. As a quantitative example, we calculate the thermal impedances of the various heat-loss paths for the microfabricated silicon cantilever with integrated silicon heater described at the end of Section 4.2. Finally, in Section 4.4.7, we describe a set of experiments designed to quantify the various thermal resistances.

4.4.1
Heat Transport Through the Cantilever Legs and Air

Microfabricated silicon heater structures integrated into a cantilever structure typically have a minimum size on the order of micrometers. Usually, they are integrated into 'u'-shaped cantilever structures that provide both mechanical support and electrical connections to the heaters. These cantilever structures result in additional heat-loss paths that, however, do not go through the tip but rather into the support structure and from there into the surrounding air. If the cantilever is close to a sample surface, then a fraction of this heat will be conducted through the air into sample, but will not lead to a significant temperature increase in the sample owing to the distributed nature of the heat transfer. The thermal coupling between cantilever and sample is relatively strong if the cantilever–surface distance is comparable to the

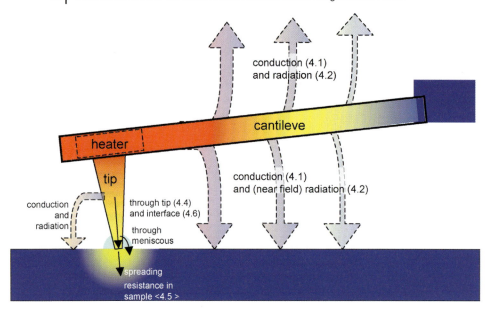
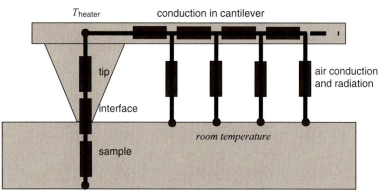

Figure 4.8 (a) Heat paths relevant for experiment using heated probes (numbers refer to text sections in which they are described); (b) Schematic representation as thermal resistances. Note that the distinction between tip, interface and spreading resistances is not possible in every case.

mean free paths of air molecules (~60 nm). For the cantilever design shown in Figure 4.3, the tip height is ~500 nm, resulting in a strong coupling to the substrate. For cantilevers that are long relative to the dimensions of the heater and to the cantilever–surface distance, most of the heat is conducted through the air and into the substrate. For the cantilever in Figure 4.3, the conductivity through cantilever allows the heat to spread along the cantilever by a distance on the order of a few tens of micrometers. Heat conduction along the cantilever and into the air is analogous to a lossy transmission line. Thus, it can be modeled as a series of thermal resistances, describing conduction through the cantilever with a set of parallel resistances at each node that give the conduction through the air to the sample, as illustrated in Figure 4.8b. Earlier studies of the heat transfer through the cantilever–air gap and within the silicon cantilever/heater [14, 16] predicted that for typical dimensions (see Figure 4.3) – that is, a cantilever thickness of ~200 nm – a cantilever/heater–surface distance of 500 nm, a heater size ~5 × 5 μm and a cantilever width of ~5 μm, the thermal resistance of the combined air/cantilever heat loss path is on the order of 10^5 K W^{-1} and the thermal response times are approximately 10 μs. If the cantilever is operated in a vacuum environment, heat loss through the air is eliminated and heat flows directly through the cantilever to the thick silicon cantilever support structure. For the cantilever design shown in Figure 4.3, the thermal resistance of this heat loss path is on the order of 5–10 × 10^5 K W^{-1}.

4.4.2
Heat Transfer Through Radiation

Heat loss due to thermal black-body radiation, which involves the propagation of electromagnetic waves from a hot object, is described by the Stefan–Boltzmann equation

$$S = \frac{\pi^2 k_B^4}{60 \hbar^3 c^2} (T_1^4 - T_2^4).$$

Here, the cooling power per area, S, is expressed in terms of the Boltzmann constant, k_B, the Planck constant, $\hbar$, the speed of light, c, the temperature of the heated body, T_1, and the temperature of the environment, T_2. For a heater temperature 100 K above the environment temperature, this corresponds to a thermal resistance of ~6 × 10^8 K W^{-1} for effective heater dimensions of (9 μm)^2. Compared to the thermal resistance of the cantilever and air heat-loss paths of about 1–10 × 10^5 K W^{-1}, the contribution due to black-body radiation is negligible. In ambient conditions, the overall thermal resistance is dominated by air conduction and in vacuum by conduction through the legs to the support structure.

In heated-probe SPM experiments, the distance between the heater and sample is often less than 1 μm. In this case, the Stefan–Boltzmann equation is only an approximation, and near-field effects must be taken into account. Such effects have a long history of theoretical analysis (see for example Refs [24–28]). Experimentally, however, the effect appears difficult to pin down, and very few reports have been made [29–32]. It has been predicted on a theoretical basis that, compared with Stefan–Boltzmann's law, heat transport by evanescent thermal radiation will depend heavily on the materials involved, with a strong distance dependence ($1/d^2$ for most cases) and a weakened temperature dependence (T^2 for most cases) [25]. The effect is also heavily dependent on the dielectric constants of the heater and the sample material. According to theory, we can expect that the near-field effect for a polymer surface is very small. However, for a silicon surface it can be significantly higher, depending on the doping [26–28], but even in this case the effect is expected to be negligible when compared to the total thermal resistance of the cantilever.

Under ambient conditions, the distant-dependent cooling of the heater/cantilever is dominated by air cooling, and therefore it is not possible to observe near-field cooling effects. Under vacuum conditions, the contribution to cooling due to thermal radiation should become measurable – not so much because of the increased overall thermal resistance without air conduction but rather because we can use the distance dependence to demonstrate the existence of near-field cooling. In air, the distance dependence is dominated by air conduction, whereas in a vacuum the air conduction is of course eliminated and conduction through the cantilever legs does not depend on the heater–sample distance. Thus, on approaching a surface in vacuum, any variation in the thermal resistance that is observed prior to tip–surface contact can likely be attributed to near-field radiation effects.

4.4.3
Thermal Resistance of a Water Meniscus

Under ambient conditions, humidity in the air usually results in the formation of a water meniscus around the tip–sample contact. The size and thermal conductance of the meniscus are a function of both humidity and sample material, and therefore are difficult to control. Thermal conduction through such a water meniscus effectively increases the tip–sample contact area and thereby reduces lateral resolution. On the other hand, the meniscus improves thermal contact, especially on rough surfaces, and may even be necessary to make nanoscale measurements possible in the first place. In a groundbreaking report, Shi and Majumdar concluded that in experiments using a $\sim$100 nm-diameter metal tip on a metal surface, the influence of the water meniscus is of the same order of magnitude as the conduction through the solid–solid tip–sample contact [10]. Moreover, they also concluded that for relatively blunt tips under ambient conditions, thermal conduction through the air gap between the tip sidewalls [33] predominates, whereas for sharper tips, solid–solid and water–meniscus conduction dominate [10]. The effects of conduction through a water meniscus can of course be avoided by operating the heated tip in a low-humidity or vacuum environment.

4.4.4
Heat Transfer Through a Silicon Tip

The thermal resistance of the tip stems from the conductance of phonons in the silicon tip and from the layer of native oxide covering it. In the tip, the resistivity is larger than that of bulk silicon because of enhanced phonon scattering at boundary surfaces [3]. The thermal resistance of the silicon tip can be estimated using predictions for the thermal resistivity of silicon nanowires as a function of diameter [34]. Integrating the expression for the varying diameter of a cone-shaped tip with a typical opening angle of 50° down to the apex with a radius of 5–10 nm yields a thermal resistance on the order of 10^6 to $10^7 \, \mathrm{K \, W^{-1}}$ because of phonon scattering. This is in agreement with finite-element calculations [35].

To develop a 'hands-on' feeling for this rather complex subject, we first derive a simple model for heat conduction in conical structures, in particular with regard to ballistic phonon transport. Let us initially consider a cylindrical rod with a cross-section $A = d^2\pi/4$ (d = rod diameter). Let us further assume that a constant current I_{th} of thermal energy flows through the rod, which is driven by a temperature difference along the rod axis (x-axis). For a temperature gradient $\Delta T/\Delta x$, the heat flow is)

$$I_{th} = \frac{\kappa A}{\Delta x}\Delta T, \qquad (4.1)$$

where κ denotes the thermal conductance of the rod and ΔT is the temperature difference across a cylindrical slice of thickness Δx. In using Equation 4.1, we assume that the phonon mean free path is large compared with the diameter of the rod. For very thin rods, say $d < 100$ nm for crystalline Si, this assumption is no longer valid and one must account for phonon scattering at the boundaries by renormalizing the thermal conductance according to the so-called Mathiessen's rule [36, 37]:

$$\kappa' = \kappa \frac{1}{1+\frac{\lambda}{d}} \approx \kappa \frac{d}{\lambda}, \quad d \perp \lambda, \qquad (4.2)$$

where λ denotes the phonon mean free path.

Let us now consider a rotational symmetric, conical conductor with opening angle Θ (see Figure 4.9). The diameter of a circular slice at distance x from the cone tip is thus $d(x) = 2x|\tan(\Theta/2)|$. If $d \gg \lambda$, we can calculate the temperature profile along the cone axis by approximating the cone by a stack of short cylindrical rods of length $\Delta x \gg \lambda$ and diameter $d(x)$, yielding ($x_1 - x_2 \, \pi \, \Delta x$):

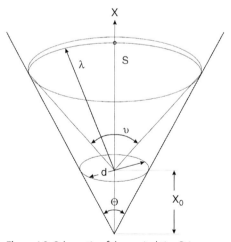

Figure 4.9 Schematic of the conical tip: Θ is cone angle, d the cone diameter at x_0, λ the phonon mean free path, and υ the opening angle for the intersection of the phonon mean free path with the surface of the cone.

$$T_1 - T_2 = I_{\text{th}} \frac{1}{\pi \tan \frac{\Theta}{2}} \int_{x_1}^{x_2} \kappa \frac{dx}{x^2}. \tag{4.3}$$

The cylindrical rod approximation has also been applied for $d \ll \lambda$ by simply replacing the constant thermal conductivity κ by the renormalized value κ' [36]. However, it is not obvious that this approach is applicable because the cone cross-section changes significantly over a distance λ along the tip axis. Therefore, let us examine the problem in more detail.

Consider a point on the cone axis at a distance x_0 from the apex (see Figure 4.9). We assume that $d < \lambda$. Let S be the spherical cap defined by the intersection of the conical tip with a sphere of radius λ centered at x_0. A fraction of the phonons emanating from S arrive at x_0 in a direct path without interference from the tip surface. These unperturbed phonons impinge from a solid angle

$$\tan \frac{\upsilon}{2} = \frac{d(x_0 + \lambda)}{2\lambda} = \frac{d(x_0)}{2\lambda} + \tan \frac{\Theta}{2}. \tag{4.4}$$

The fraction of the total heat transport seen at x_0 due to these direct phonons is

$$\eta^d = \frac{I_{\text{th}}^d(\upsilon)}{I_{\text{th}}^d(\pi)}$$

$$= \frac{2\pi \int_0^\upsilon \frac{T(x_0) - (x_0 + \lambda)}{dT/dx \cdot \lambda} \cos\upsilon/2 \sin\upsilon/2 \, d\upsilon/2}{2\pi \int_0^\pi \cos\upsilon/2 \sin\upsilon 2 \, d\upsilon/2} \tag{4.5}$$

$$= \frac{T(x_0) - T(x_0 + \lambda)}{dT/dx \cdot \lambda} \sin^2\upsilon/2 \approx \frac{d(x_0)}{\lambda} \sin^2\upsilon/2.$$

The factor $(T(x_0) - T(x_0 + \lambda))/(\lambda dT/dx)$ accounts for the reduced thermal energy carried by the impinging phonons with respect to the value calculated from the local thermal gradient at x_0. As the temperature difference $T(x_0) - T(x_0 + \lambda) \cong T(x_0) \propto 1/d^2$ (see below), the factor is equal to d/λ. Similarly, for the fraction of heat transported by the phonons scattered off the wall, one can write

$$\eta^d = \frac{I_{\text{th}}^w(\upsilon)}{I_{\text{th}}(\pi)}$$

$$= \frac{2\pi \int_0^\pi \frac{\Delta T^w(\upsilon')}{\Delta T(\upsilon')} \cos\upsilon'/2/2\sin\upsilon'/2 \, d\upsilon'/2}{2\pi \int_0^\pi \cos\upsilon'/2\sin\upsilon'/2 \, d\upsilon'/2} \tag{4.6}$$

$$\approx \frac{d(x_0)}{\lambda} \cos^2\upsilon/2.$$

4.4 Heat-Transfer Mechanisms

As for the direct phonons, the factor $\Delta T^w/\Delta T$ denotes the fraction of heat carried by a phonon scattered from the wall with respect to a thermal equilibrium phonon. A calculation (A. Dürig, unpublished results) yields

$$\Delta T^w(\upsilon') \approx \Delta T(\upsilon') \begin{cases} \dfrac{d(x_0)}{\lambda} & \dfrac{d(x_0)}{\lambda} < 1 \\ 1 & \dfrac{d(x_0)}{\lambda} \geq 1 \end{cases}, \quad (4.7)$$

where the equality holds for $\upsilon=0$ and deviations for $\upsilon>0$ have been neglected. Hence, we obtain as final result

$$\kappa'(x_0) = \kappa(\eta^d + \eta^w) \approx \kappa \frac{d(x_0)}{\lambda}, \quad (4.8)$$

in exact agreement with Mathiessen's rule.

According to Equation 4.3, the thermal resistance of a conical heat conductor with an apex diameter $d_0 \perp \lambda$ can be written as

$$\begin{aligned} R &= \frac{1}{\kappa} \frac{4\lambda}{\pi} \int_{d_0 \ll \lambda}^{\lambda} \frac{1}{d^3} dx \\ &\approx \frac{1}{\kappa} \frac{2\lambda}{\pi \tan\Theta/2} \frac{1}{d_0^2} \\ &= \frac{3}{8} \frac{1}{\tan\Theta/2} R_s, \end{aligned} \quad (4.9)$$

where

$$R_s = \frac{1}{\kappa} \frac{4\lambda}{3\pi} \frac{1}{(d_0/2)^2} \quad (4.10)$$

is the so-called Sharvin resistance for ballistic transport through a circular aperture (see e.g. Ref. [38]). It is interesting that, despite the short effective mean free path due to boundary scattering, the resistance of a conical conductor still retains the characteristics of ballistic transport expressed by the inverse d_0^2 dependence. Also, the length of the cone does not enter because more than 90% of the temperature change occurs over a distance on the order of three times the apex diameter. Substituting the corresponding values for the thermal conductivity $\kappa = 165$ W K^{-1} m^{-1} and the mean free path $\lambda = 100$ nm, one obtains the following for the thermal resistance of a Si tip:

$$R \approx 3.86 \times 10^8 \text{ KW}^{-1} \text{ nm}^2 \frac{1}{\tan\Theta/2 d_0^2}. \quad (4.11)$$

The cone angle dependence and explicit values of R for a representative set of Θ and d_0 values are tabulated in Table 4.1. Note that the tip resistance increases markedly if the cone angle is less than $\sim 45°$.

Next, we investigate the influence of interface scattering at the apex for a tip in contact with a substrate surface. Specifically, let us assume that the substrate is a

Table 4.1 Normalized thermal resistance of a conical tip as a function of the cone angle and thermal resistance for $\Theta = 90°$ as a function of apex diameter.

$R(\Theta)/R(90°)$	1	1.73	2.41	3.73	7.60
Θ	90	60	45	30	15
$R\,(90°)\,(\text{KW}^{-1})$	3.86×10^8	9.65×10^7	1.54×10^7	3.86×10^6	9.65×10^5
d_0 (nm)	1	2	5	10	20

poorly conducting material, such as a polymer film, which has a correspondingly short phonon mean free path $\perp d_0$. The heat transmitted through the apex is carried away radially, where the temperature drop in the substrate is given by the spreading resistance (see e.g. Ref. [38] and Section 4.4.5):

$$R_{sp} = \frac{1}{2\kappa_s d_0}. \tag{4.12}$$

Here, κ_s denotes the thermal conductance of the substrate.

Consider a heat current impinging on the interface. A fraction η is reflected back into the tip owing to scattering at the apex (see Figure 4.10a). Whether the scattering is elastic or inelastic is not important for the subsequent discussion. Because of the ballistic nature of the propagation, thermalization of the reflected phonon will occur far away from the interface, and therefore will not influence the local thermal equilibrium at the interface. Let T and I_{th} be the interface temperature and the thermal current in the absence of scattering, respectively. The heat transmitted into the substrate is thus $I_{th}^t = (1-\eta)I_{th}$. Correspondingly, the substrate temperature at the interface is $T_s = T - \eta(T - T_0)$, where T_0 is the substrate temperature far from the interface.

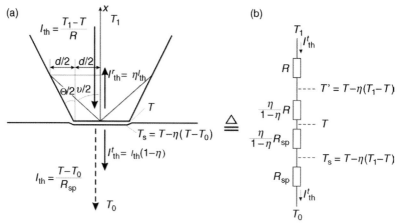

Figure 4.10 Schematic of boundary scattering at the tip apex.

The net heat current flow in the tip is $I_{th}^t = I_{th} - I_{th}^r$, where $I_{th}^r = \eta I_{th}$ denotes the reflected current in the tip. As argued above, the tip temperature at the apex is not altered. Nevertheless, we introduce an effective temperature at a virtual tip interface $T' = T + \eta(T_1 - T)$, where T_1 is the tip temperature far from the interface (see Figure 4.10b). With this definition, the heat balance can be satisfied with regard to the net current, $T' = T_1 - I_{th}^t R$, where the heat conduction in the tip is represented by the single resistive element, R. The virtual tip interface is connected to the substrate surface by means of a resistive element that must satisfy the equation $T' - T_s = I_{th}^t R_b$. Hence, one obtains

$$R_b = \frac{\eta}{1-\eta}(R + R_{sp}). \tag{4.13}$$

The scattering physics is captured in the reflection coefficient

$$\eta = a(1-\cos\upsilon/2) \approx a\left(1 - \frac{1}{\sqrt{4\tan^2\Theta/2 + 1}}\right), \tag{4.14}$$

where $0 \leq a \leq 1$ is a type of accommodation factor for the back-scattering of the phonons into the tip and $(1 - \cos \upsilon/2)$ accounts for the fraction of the solid angle covered by the back-scattered phonons that escape form the tip apex and thermalize in the heat bath. The particular choice for υ is heuristically motivated by the observation that the temperature gradient is roughly one order of magnitude lower at a distance above the apex that corresponds to a cone diameter of twice the aperture diameter.

The substrate temperature is one of the key parameters for studying thermo-mechanical material properties. It can be written as

$$T_s = \frac{R_{sp}}{R + R_b + R_{sp}}(T_1 - T_0) + T_0 = \frac{1-\eta}{1 + R/R_{sp}}(T_1 - T_0) + T_0, \tag{4.15}$$

Note that already for rather weak back scattering the interface resistance accounts for a significant fraction of the total resistance, that is, $a = 0.3$ yields $R_b \lambda 0.5 (R + R_{sp})$ (see Equations 4.13 and 4.14). Figure 4.11 shows the substrate temperature as a function of tip cone angle for $a = 0.5$ and 0.75 and for various ratios of $R + R_{sp} = (\kappa_s \lambda)/(\kappa d_0) < 1$ corresponding to cases in which the spreading resistance is dominating the tip resistance. Under such conditions, the substrate temperature is rather insensitive to the values substituted for the accommodation coefficient a and the cut-off angle υ.

As observed experimentally [8, 14, 35, 39, 40] and also described in Section 4.4.8, the simple model predicts that the temperature rise at the substrate interface is on the order of 0.4 to 0.7 times the total temperature difference between tip and substrate for parameters that correspond to typical experimental conditions. It is clear that the model cannot capture the complexities of phonon scattering in a predictive manner. Instead, a phenomenological parameter a must be introduced to match experimental observations with model predictions. However, the model provides a means for assessing the scaling properties and provides qualitative

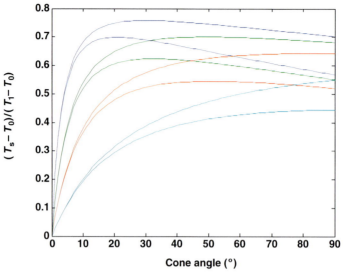

Figure 4.11 Substrate temperature at the tip apex: $a = 0.75$ (solid lines) and 0.5 (dashed lines) and $R/R_{sp} = (\kappa_s \lambda)/(\kappa\, d_0) =$ 0.025 (blue), 0.05 (green), 0.1 (red) and 0.25 (cyan).

insight based on intuitive physical arguments. A deeper discussion of physical mechanisms governing thermal transport of nanometer-scale tip–surface contacts is presented in Sections 4.4.5–4.4.8.

4.4.5
Thermal Spreading Resistance

The spreading resistance in the sample is probably the best understood of all the thermal resistances involved. Commonly, it is well approximated by Equation 4.12, which says that the resistance scales inversely with the contact diameter d_0. The scaling is borne out by the fundamental heat conduction Equation 4.1 by observing that the mean gradient $\Delta T/\Delta x$ scales as $1/d_0$ for diffusive transport in a half-space.

For a thin film on a substrate, one can account for the effect of the substrate by using an approximate solution proposed by Yovanovich et al. [41]:

$$R_{sp} = \frac{1}{2\kappa_s d_0} - \frac{1}{2\pi\kappa_s t}\log\left(\frac{2}{1+\kappa_s/\kappa_{sub}}\right). \tag{4.16}$$

Here, κ_s and t are the thermal conductance and the thickness of the film, respectively, and κ_{sub} denotes the thermal conductivity of the substrate on which the film has been deposited. In all experiments discussed below, the film thickness is at least one order of magnitude larger than the contact diameter, and we can disregard the finite-size correction term.

For polymer films, a value of $\kappa_s \sim 0.2$–0.3 W·K·m^{-1} is typical. The thermal conductance can increase by up to a factor of 2 under a pressure of 1 GPa. As the stress under the tip varies during an experiment and is transient within the tip–surface interaction volume, we must resort to estimating an effective pressure-increased thermal conductivity [42]. For the experiment described below, we use a value of k_{pol} of 0.3–0.6 W K·m^{-1}. For a contact diameter of $d_0 = 10$ nm, we obtain an estimated R_{sp} of approximately $(0.8 - 1.6) \times 10^8$ K W^{-1} for polymers, $\sim 10^6$–10^7 K W^{-1} for oxides, $\sim 3 \times 10^5$ K W^{-1} for silicon, and down to 10^4 K W^{-1} for metals.

4.4.6
Interface Thermal Resistance

As discussed in Section 4.4.4, the thermal resistance of the interface R_{int} defies accurate prediction. The situation is further complicated because the interface resistance usually depends heavily on the quality of the interface and the contact pressure. For most of the cases described in this chapter, we can assume a single-asperity contact characterized by a contact diameter d_0. Contact mechanics models can be invoked to estimate d_0. For a tip with a radius of 10 nm and applied loads of a few tens of nanoNewtons, d_0 is on the order of a few nanometers.

On the other hand, the values of the interface resistance reported in the literature were measured on macroscopically large areas (rather than a tip–surface contact). Typical values for silicon–polymer interfaces are in the range of 10^{-8} to 10^{-7} Km^2 W^{-1} [43, 44]. For a silicon–silicon interface, the corresponding value obtained for phonon scattering is 2.1×10^{-9} Km^2 W^{-1}. The subject of interface scattering has been extensively reviewed by Swartz and Pohl, who discuss the thermal interface (or boundary) resistance between various materials as well as related models [45].

Returning to the nanoscale tip contact, it is not immediately clear how to relate the macroscopic data to the nanoscale interface resistance. One simple approach adopted by King [35] is to treat the interface as a scattering site for phonons in silicon and to assume that the total interface resistance is inversely proportional to the contact area, which yields $R_{int} = 2.1 \times 10^{-9}$ Km^2 W$^{-1} \times 4/(\pi\, d_0^2) \sim 2.7 \times 10^7$ K W^{-1} for $d_0 = 10$ nm. Alternatively, if we substitute the measured value for the polymer–silicon interface resistance, we obtain $R_{int} = 10^{-8}$ to 10^{-7} Km^2 W$^{-1} \times 4/(\pi\, d_0^2) \sim 1.3 \times 10^8$ to 10^9 K W^{-1} for $d_0 = 10$ nm.

Alternatively, it is shown in Section 4.4.4 that the resistance due to boundary scattering at the tip apex is proportional to the sum of the thermal resistances associated with the conduction paths through substrate and tip. Therefore, this has two components: one scaling as $1/d_0$ and corresponding to the spreading resistance in the substrate; and the other scaling as $1/d_0^2$ and corresponding to the tip resistance. The question then arises how this is to be reconciled with the $1/d_0^2$ scaling suggested by extrapolating from the macroscopic scale.

The interface resistance is a somewhat artificial construct, which bridges the gap in the conduction path where the temperature of the phonon gas cannot be defined unequivocally. The temperature is well defined only if the phonons thermalize by means of mutual scattering. Therefore, the gap typically extends over a distance on the order of the phonon mean free path on either side of the interface. For consistency with the concept of a thermal resistance, the interface resistance is defined as the temperature difference divided by the net heat flux across the gap, as measured by an imaginary observer with an apparatus that is in thermal equilibrium with the phonon gas. Note, however, that unlike a regular thermal resistor, the interface resistance cannot be broken up into a string of series resistors to calculate the temperature at any point along the gap. In fact, such temperatures have no meaning and merely serve as a mathematical concept. The interface temperature of the tip, T' (which was introduced in Figure 4.10b), is an example of such a fictitious temperature. Moreover, the ballistic tip resistance, R (see Equation 4.9 in Section 4.4.4) constitutes part of the overall interface resistance. Therefore, we must write the following for the interface resistance:

$$R_{int} = R_b + R, \qquad (4.17)$$

which spans the entire ballistic propagation path of the phonons through the conical tip, including boundary scattering at the tip–substrate interface up to their thermalization in the substrate. It is also clear from the above discussion that we cannot simply extrapolate from macroscopic results to nanoscale thermal contacts without accurately accounting for the conduction path. What one can do, however, is to extract a mean backscattering probability from macroscopic experiments. Using the same type of reasoning as in Section 4.4.4, one obtains the following for the interface resistance for a unit area of a planar contact:

$$r_{int} = \frac{\eta}{1-\eta} \frac{\lambda}{\kappa}. \qquad (4.18)$$

With $\eta \sim a$, and assuming $\lambda \sim 1$ nm and $\kappa \sim 0.3$ W Km^{-1} for the mean free path and the thermal conductance of polymers, respectively, one must substitute $a \sim 0.75$ to 0.97 in order to obtain the experimentally observed values of $r_{int} \sim 10^{-8}$ to 10^{-7} Km^2 W^{-1} [43, 44]. The upper bound for the measured interface resistance yields a somewhat unrealistically high value of 0.97 for the backscattering probability. However, it must be borne in mind that it is difficult to obtain good contact uniformity in a large-scale experiment, and therefore the experimental values must be seen as upper bounds.

4.4.7
Combined Heat Transport Through Tip, Interface and Sample

We define the *heating efficiency c* (similar to – but a simplification of Equation 4.15 and Figure 4.11) as the increase in the sample surface temperature divided by the total temperature difference between heater and substrate:

$$c = \frac{R_{sp}}{R_{tip} + R_{int} + R_{sp}}. \qquad (4.19)$$

Here, R_{tip} denotes the nonballistic, diffusive component of the tip resistance (the ballistic part is captured in R_{int}, as explained in Section 4.4.6). This definition is useful for understanding sensitivity issues when using heated probes. The heating efficiency c is a strong function of tip size – that is, of the lateral resolution. Small values of c imply that also the measured signal will be small, indicating that achieving high lateral resolution becomes increasingly difficult.

As outlined in Section 4.4.4 and inferred experimentally [8, 14, 35, 39, 40], typical values for c range from 0.3 to 0.7 for polymer samples. In the case of better thermal conductors, c can be much lower; for example, on metals we estimate $c \sim 10^{-3}$–10^{-4}, for semiconductors $c \sim 10^{-2}$–10^{-3}, and for oxides $c \sim 10^{-1}$. This points to the challenges expected when extending the SThM method to both the nanoscale (e.g. $a = 5$ nm in the above calculations) and to sample materials having a higher thermal conductivity than the commonly used polymers or oxides.

We note that, although the heating efficiency reflects the temperatures, it does not reflect how efficient a probe is in terms of *heating power*. For the cantilever type shown in Figure 4.3 operating in air, the ratio of power going through the tip to that lost to other heat paths is approximately $10^5/10^8$ (K/W) $= 0.001$. Thus, from a power consumption point of view, the delivery of heat to the sample through the tip is very inefficient. Improving the efficiency requires either a reduction in the interface and the tip thermal resistances or an increase in the air/cantilever thermal resistance. The tip and interface resistances can of course be decreased by using a blunter tip, but at the expense of lateral resolution. Increasing the cantilever thermal resistance requires a decrease in the heater and lead cross-sections and/or an increase in the cantilever length. Among the design issues that restrict the freedom to reduce the heater/lead cross-sections are the mechanical stability, cantilever stiffness, mechanical response time, power consumption of the heater, electrical resistance and noise, thermal response time and fabrication tolerances.

4.4.8
Heat-Transport Experiments Through a Tip–Surface Point Contact

The sheer number of heat-transport paths described above renders an understanding of heat transport in heated probes challenging, let alone the direct measurement of individual components of heat transport. In order to distinguish between different heat paths, we have chosen a threefold approach:

- Bringing the tip into and out of contact with the sample opens and closes heat channels.

- Operation in a vacuum removes the distance-dependent cooling path through the air, which tends to dominate the small change in cooling that occurs when the tip is brought into contact with the sample in ambient conditions and completely eliminates conduction through the water meniscus.

- By varying the contact area, a, only some of the contributions will be affected (interface and spreading part).

The total thermal resistance of the heater, R_{th}, is given by

$$R_{th} = (T_{heater} - \mathrm{RT})/P, \tag{4.20}$$

where T_{heater} is the temperature of the heater, RT is the room temperature, and P the heating power. The thermal resistance due to conduction through the cantilever legs

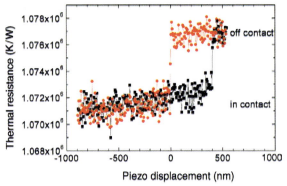

Figure 4.12 Thermal resistance of heated cantilever/tip as a function of displacement and contact with an 80 nm-thick SU8 film on a silicon substrate. (Reproduced from Ref. [14]; © Springer-Verlag.)

and radiation can be determined from the data obtained before the tip contacts the sample. The tip–surface thermal conductance can then be determined by subtracting this value from the data measured with the tip in contact with the sample. Figure 4.12 shows an example of such an experiment performed using a thermal probe similar to that shown in Figure 4.3 and a sample consisting of 80 nm of SU8 (an epoxy-based photoresist) on a silicon substrate. Out of contact, the displacement translates into a distance change between heater and sample. In contact, the displacement translates into a load force as the tip is pressed against the polymer. In this experiment, the average T_{heater} was approximately 315 °C, and the change in T_{heater} resulting from contact was about 1.5 K. From the difference between the thermal resistance out-of-contact ($\sim$1.077 MK W^{-1}) and the thermal resistance in-contact (1.071–1.073 MK W^{-1}), the thermal resistance due to heat transport through the tip–surface contact was calculated to be $2 - 3 \times 10^8$ MK W^{-1} (see Figure 4.14).

The tip–sample resistance is given by

$$R_{ts} = R_{tip} + R_{int} + R_{sp}, \tag{4.21}$$

as illustrated in Figure 4.13, where R_{tip} is the diffusive thermal resistance of tip, R_{int} is the tip–sample interface resistance, and R_{sp} is the spreading resistance in the polymer.

To experimentally quantify the different contributions to R_{ts}, we can vary the individual contributions by varying the sample material and the applied force. As the contributions depend on the contact radius, 0.5 d_0, it is useful to vary this parameter. To study the contact area dependence of the overall thermal resistance of the tip–polymer contact, we vary the force during an approach experiment. The contact area is calculated using the JKR model [46]. For this purpose, the applied force, the pull-off force and the tip radius need to be known. The applied force and pull-off force are determined from the cantilever spring constant and the known motion of the tip holder relative to the sample surface. The tip radius is measured *ex situ* by means of

4.4 Heat-Transfer Mechanisms | 143

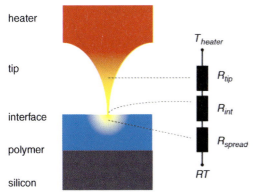

Figure 4.13 Schematic representation of the thermal resistances involved in heat transfer between the heater and a polymer sample. (Reproduced from Ref. [14]; © Springer-Verlag.)

scanning electron microscopy. The data shown in Figure 4.14 were obtained for a tip with $R_{tip} \cong 13.5$ nm corresponding to a variation of the contact diameter from 7 nm just before the contact breaks at a pull-off force of -15 nN to 13 nm at the maximum load of 35 nN.

Clearly, the thermal resistance depends on the tip force and hence on the contact diameter. Motivated by the above discussion, we propose the following ansatz for the thermal resistance:

$$R_{ts} = A_0 + A_1/d_0 + A_2/d_0^2. \tag{4.22}$$

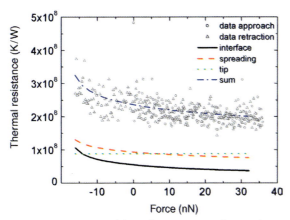

Figure 4.14 Experimental thermal resistance R_{th} of a typical tip–sample contact on a polymer (SU8) film and fit (blue, dash-dotted line) containing the three components R_{tip} or A_0 (green, dotted), R_{int} or A_2 (black, solid) and R_{sp} or A_1 (red, dashed). (Reproduced from Ref. [14]; © Springer-Verlag.)

Indeed, a good fit to the data is obtained with this second-order ansatz. In Figure 4.14, the individual contributions are shown as a green dotted line for the A_0 term, a red dashed line for the A_1/d_0 term, and a solid black line for the A_2/d_0^2 term.

Recalling the results from Sections NaN.2.4–NaN.2.6, it is surprising that there is a significant A_0 term ($\sim 9 \times 10^7$ MKW^{-1}). All components – tip, interface and spreading – should have an explicit dependence on d_0. We attribute the contribution A_0 to the thermal resistance of the tip R_{tip}. This may appear as a rather strong assumption, because we argued in Section 4.4.4 that the thermal resistance of the tip is predominated by ballistic conduction, and therefore the diffusive (nonballistic) thermal resistance of the tip should be vanishingly small in comparison. However, in practical applications the inner structure of the tip may play a role. In particular, the oxide cap covering silicon tips can contribute decisively to the thermal resistance of the tip. The value of this contribution can be estimated on its own as an independent thermal resistance by using approximate dimensions or as a mesoscopic link enabling 'phonon tunneling' between the silicon cone and the sample [47]. The uncertainty remains large, and we estimate 10^7–10^8 MKW^{-1} for the oxide cap. Accordingly, the total value of R_{tip} might be dominated by the value for the oxide cap, and we therefore also expect 10^7–10^8 MKW^{-1} for R_{tip}. This interpretation is supported by control experiments using the same tip on a silicon sample.

The A_2/d_0^2 term is an unequivocal sign for ballistic transport, and therefore it must be assigned to the interface resistance R_{int} (see Section 4.4.6). The A_1/d_0 term stems from the spreading resistance in the polymer sample. As discussed in Sections 4.4.4 and 4.4.6, it is composed of a real diffusive component and an interface contribution. It is difficult to assess the magnitude of the latter without detailed knowledge of the tip structure at the interface, which would allow one to make realistic assumptions on the scattering efficiency at the interface to the polymer. We note, however, that the magnitude is consistent with the diffusive spreading resistance for polymers (see Section 4.4.5). Hence, most likely the interface scattering contribution is rather small. It is interesting to note that all three terms contributing to the overall thermal resistance are of similar magnitude, namely, on the order of 10^8 MKW^{-1}, which results in a heating efficiency of $c \sim 40\%$ in this example (assuming that the A_1/d_0 term corresponds to a purely diffusive spreading resistance).

4.5
Thermomechanical Nanoindentation

Thermomechanical nanoindentation can be viewed as a powerful nanoscale extension to the existing methods of indentation (hardness testing) and dynamic thermomechanical analysis (modulus testing). The process involves pressing a heated tip into a sample using a defined tip temperature, load/heat duration, and load force. The indentation dynamics and the yield of the sample can be used to understand its material properties. Apart from the metrology discussed in this section, the technique has applications in data storage and in nanoscale patterning and lithography.

The indentation process yields considerable insight into the thermomechanical properties of materials – in particular of polymers – on the nanometer scale [48–52]. Traditionally, indentation processes are used to determine the hardness of a material [53], in which experiments an indenter produces a permanent deformation at the surface of the material under investigation. The hardness is determined by the size of the indentation with respect to the loading force. As the geometry of the indenter plays a fundamental role, common hardness definitions are based on individual, defined geometries, such as a ball (Brinell) or a pyramid (Vickers) [53]. In the experiments, typically a specific load is applied for specific time durations to yield comparable results.

Although it has not been widely used in traditional hardness testing, temperature is an additional important parameter in indentation experiments [14, 52]. As the mechanical properties of materials typically depend largely on temperature, the control of this parameter opens up interesting new areas of investigation. Heated probes provide an easy means to vary the temperatures on any given surface. In addition, controlling the probe temperature is relatively straightforward, and the low heat capacity of the probes allows the temperature of the probe to be switched at relatively high rates.

Polymers are one class of materials in which the mechanical and viscoelastic properties change dramatically with temperature, for example, at the polymer's glass transition temperature, T_g [1, 54]. At this temperature, the internal configurational changes within the polymer chain (which are linked to the translational motion of the chain) become slow compared to the typical experimental observation time scale of 1–100 s. As a result, the material drops out of equilibrium into a so-called 'glassy' state, and its materials properties change dramatically. The elastic modulus, for example, increases by orders of magnitude.

Heated probes are ideally suited to investigate the mechanical properties of polymers on a nanometer scale over a wide range of temperatures, from room temperature to several hundred degrees Celsius, and time scales varying over orders of magnitude down to the microsecond regime.

As an example, Figure 4.15 shows the result of an indentation experiment using heated probes. The indentations were written as a function of the load force, F, and the tip temperature, T, for indentation times of 10 μs using a tip having a radius of about 10 nm. The image shows the topography measured in contact mode after writing the indentations using the same tip. Whereas, in the lower left part of the image, no permanent indentations were formed, they appeared very clearly in the upper right part at high temperatures and forces. Clearly, in order to produce permanent indentations a certain minimum force and/or temperature must be applied. The corresponding characteristic line at the onset of indentation formation is called the *writing threshold* (see shaded region in Figure 4.15). For practical purposes, this dividing line can be defined as the load/temperature combination that leads to indentations of 1 nm depth: $T(F)_{d\,=\,1\,\mathrm{nm}}$.

Figure 4.16 exemplifies the threshold behavior of writing indentations with increasing temperature at a given load of 50 nN and a pulse duration of 10 μs. Whereas, below the threshold temperature T_h, no permanent indentations can be formed, above T_h their depth increases linearly with temperature.

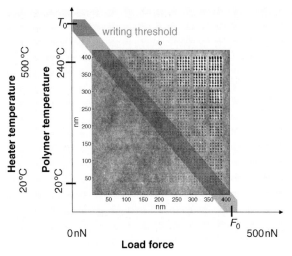

Figure 4.15 Atomic force microscopy image of indentations written into a polymer film at various combinations of load forces and tip temperatures using an indentation time of 10 μs. Blocks of 5 × 5 indentations spaced 36.6 nm apart are written using the same parameters. The blocks are written with increasing force and temperature along the x- and y-axis, respectively. The heater temperature is determined as described in the text. The polymer temperature under the tip is estimated using the results of the heat-transfer experiments described in Section 4.4.

Before turning to the physical interpretation of such temperature–load plots, let us briefly see how they can be used in practice. Of particular interest are the well-defined intersections of the writing threshold curves with the axes – that is, the writing temperature in the limit of no load force applied, T_0, and the load force in the limit of

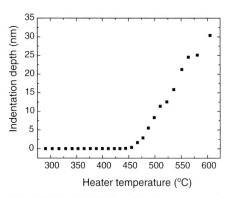

Figure 4.16 Depth of indentations in a polymer film (polymethylmethacrylate, PMMA) as a function of heater temperature for a load of 50 nN and heat- and force-pulse durations of 10 μs. Above a threshold heater temperature T_h of ∼450 °C, the indentation depth increases linearly with the heater temperature. For the parameters applied, T_h is therefore closely related to the glass-transition temperature, T_g, of the polymer.

no heat applied, F_0. The quantities T_0 and F_0 are a function of both indentation time and tip geometry.

T_0 is the writing temperature needed in the limit $F \to 0$, which is also the limit of zero stress. In this limit the T_g is defined and therefore, for a given tip radius and indentation time, T_0 is a measure of the T_g.

For a given heater temperature, the temperature reached in the polymer underneath the tip depends on the geometry of the tip (see Section 4.4). Whereas, the opening angle of the tip cone has relatively little effect, the contact area – and therefore the heat transport from the tip to the polymer – differ for blunt and for sharp tips.

To reach more quantitative statements about the T_g of a polymer, we must normalize the effects of the tip geometry. There are two possible solutions to this:

- First, to obtain comparable results, one can use the same tip for all samples being studied. Assuming that the tip shape stays constant over the range of experiments, this procedure yields comparable results that can be correlated to traditionally measured T_g values of polymers.

- The second approach is to pick one of the polymers as a reference sample and to normalize the results on the other polymers with respect to this reference polymer.

Clearly, the first approach is prone to difficulties relating to the necessary constant geometry of the tip, is restricted to the use of a single tip, and therefore cannot be applied as a general method. In the second method, the T_g values measured are rescaled with respect to a reference tip on the reference sample, which cancels to first order the relative difference in heat-transfer properties resulting from the use of different tips.

By using these two approaches, a correlation to T_g measured by conventional means can be made [6, 14], as shown in Figure 4.17. All experimental data were either

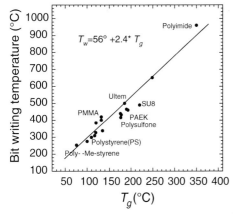

Figure 4.17 Indentation-writing temperature T_0 as a function of the conventionally determined T_g for various polymers. The indentation-writing temperature has been normalized to a reference tip, as described in the text. All data points are within 10% of the linear fit to the data.

obtained with a tip of approximately 10 nm radius or rescaled using reference samples [poly(methylmethacrylate) (PMMA) and polystyrene] to the case of a particular tip with specific opening angle and tip radius ($R_{tip} = 10$ nm), which yielded T_h at 400 °C for 10 μs-long heat pulses. The writing temperatures measured correlate very well with the respective T_g values of the polymers. This holds for uncrosslinked polymers, such as PMMA or Poly-a-Me-styrene, as well as for highly crosslinked polymers, such as the epoxy-resist SU8. In fact, the largest deviation of the data from the linear fit is less than 10%.

The excellent correlation of the indentation-writing and T_g values demonstrates the applicability of the method for determining T_g for unknown samples. This is insofar surprising as the T_g values for the given samples were determined using macroscopic bulk methods of dynamic mechanical analysis (DMA) or differential scanning calorimetry (DSC) and, in particular, also because these methods work on much longer time scales (1–100 s).

The second parameter obtained from the threshold curve, F_0, can be used to determine the hardness of the materials. As a demonstration, measured indentation-writing threshold curves are shown in Figure 4.18 for a class of similar polymers. The samples are thin films (120 nm thick) of polystyrene crosslinked using benzocyclobutene (BCB) as crosslinking groups [55]. This system is an ideal system to study the effects of the crosslink density on the thermomechanical properties of polymers. Increasing the crosslink density by incorporating a larger amount of BCB monomers increases both the hardness and T_g [56].

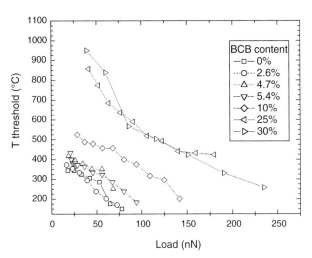

Figure 4.18 Indentation-writing threshold plots determined by writing indentation arrays, as shown in Figure 4.15. Each datum point refers to the temperature needed to write an indentation of 1 nm depth at a given load. Here, thins films of polystyrene that have different crosslink densities are compared. The percentage values in the inset refer to the relative amount of crosslinking benzocyclobutene (BCB) monomers with respect to styrene monomers in the polymer.

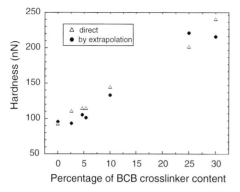

Figure 4.19 Comparison of two methods to determine sample hardness by nanoindentation. In the first method, the value (solid circles) is obtained by extrapolating the writing-threshold data from Figure 4.18 to room temperature. The second method (open triangles) measures the hardness in a more conventional manner: the minimum force required to obtain an indentation of 1 nm is determined, with tip and sample at room temperature. Within experimental uncertainty, the two methods yield identical results.

To a good approximation, the temperature needed to write an indentation decreases linearly with the writing load applied. This linearity can be used to extrapolate T_0 and F_0 from data taken in a limited force/temperature range. For example, in Figure 4.18 the writing-threshold curves reach neither T_0 nor F_0. Nevertheless, both values may be determined by extrapolation to zero force and room temperature, respectively. To verify the validity of this extrapolation, direct measurements of F_0 were performed at room temperature using the same tip. A comparison of the extrapolated and the directly measured hardness values for the polymers is shown in Figure 4.19.

A well-known issue in hardness testing is the wear of the indenter. This poses a problem in particular in the case of nanoindentation because, in order to draw valid conclusions, the indenter geometry must be known. Typically, this issue is circumvented by choosing indenters having relatively blunt apexes and/or using diamond materials. For true nanoscale applications using indentations that are confined in both the normal and the lateral direction, the wear of the probe is an unsolved issue.

The writing-threshold experiment as depicted in Figure 4.15 can be a workable solution, because the forces acting on the tip can be minimized. A grid with limited writing forces and subsequent extrapolation to room temperature can be applied to minimize tip wear. Thus, measuring the temperature dependence of writing permanent indentations is an elegant way to measure the real hardness data by extrapolation.

The indentation experiments shown above demonstrate how sensitively thermomechanical nanoindentation depends on the load and temperature. Underlying, of course, are the material properties of the polymer, such as hardness and the T_g, as well as the tip geometry and heat-transfer properties of the cantilever/tip. All of these govern the indentation formation. As discussed above, the indentation-writing threshold experiment results in a relationship of $T_{\mathrm{thresh}}(F)$ (or $F(T_{\mathrm{thresh}})$) that is linear within the uncertainties given.

One explanation of the existence of a defined threshold for the force needed to write a permanent indentation is to argue that the stress build-up in the polymer has to overcome a critical stress, the yield stress of the polymer. It has been found macroscopically that the yield stress σ_y [57, 58] of polymers is a function of temperature. More precisely, around T_g, σ_y varies linearly with temperature, with $\sigma_y \sim 0$ for $T = T_g$, which is consistent with our observation of the linear shape of the writing-threshold curves. Hence, we can write $F(T_{\mathrm{thresh}}) \propto \sigma_y(T)$. This model of yielding is also supported by the analysis of the indentation shapes as a function of indentation parameters [59] (T. Altebaeumer, unpublished results).

A model that explains the observed indentation behavior simply as a yielding phenomenon, however, is not fully satisfactorily. Yield implies a permanent deformation of the material. In polymers, such a permanent change is linked to a change in the topology of the material, which proceeds via chains sliding with respect to each other. In the case of yielding, this sliding is forced by the external stress, which has to overcome the inherent monomer-sliding friction in the polymer [60].

In macroscopic yield experiments, two types of yield behavior are generally observed, namely *shear yielding* and *crazing*. Shear yielding occurs in partially crystalline and tough polymers (such as polycarbonates) which can extend to a multiple of their initial lengths. Just above a critical yield stress, the polymers often form shear bands on a macroscopic scale. Crazing is observed in brittle polymers such as polystyrenes; these polymers can elongate by only a small percentage before they rupture, and therefore the stress–strain curve is only slightly bent just before fracture. At the same time, one often observes elongated voids in the material called 'crazes'.

Clearly, both macroscopic phenomena encounter difficulties at the nanometer scale. Both typically have a length scale much larger than the length scale of the nanoindentation experiments. Moreover, in many of the materials studied here such mechanisms should be largely constrained by the high crosslink density. Therefore, we note that the macroscopic definition of yielding must be applied with care on the nanometer scale.

In an alternative model, the material is not assumed to undergo yielding but rather a viscoelastic deformation (like rubber) in the heated state at a temperature above T_g. Elastic deformation in this regime still proceeds via the deformation of polymer chains, which implies a relative movement of polymer chains. In contrast to yielding, the chains in rubbery deformation are almost free to slide and are only held in place by entanglement or crosslink sites. The monomer relaxations in the polymer backbone, which couple to the translational motion of the polymer chains, are fast, and the friction between the monomers is reduced to very low values. Polymer motion is mainly limited by the chain-like nature and the network constraints of the material. In viscoelastic deformation, the external force is mainly needed to deform the polymer network because monomer friction is low.

After cooling to temperatures below T_g, monomer relaxations in the backbone become orders of magnitude slower and limit the translational motion of the polymer chains: the indentation is 'frozen in' and the 'loaded rubber springs' are kinetically hindered from relaxing.

This picture of rubbery indentation is more compliant with our nanoscopic length scales because no macroscopic changes in the material are involved. In addition, as has been argued [14, 61], this picture of 'rubbery indentation' also captures some apparent physics better than the yield picture. For example, it works much better at ultrafast indentation times, which are possible experimentally. Moreover, even polymers with extremely high degrees of crosslinking can undergo a rubbery deformation if the deformation is small. On the other hand, rubbery indentation implies temperatures above T_g, in clear contradiction to findings of a linear threshold curve all the way down to F_0. In the threshold curve no apparent transition through the glass transition exists.

In nanoindentation experiments, T_g is not easy to quantify. It can be expected, however, that T_g increases for the high indentation rates typical of these experiments and decreases for the high stresses. Even for the lowest stresses that can be applied in the experiments, significant shear stresses of ~100 MPa will have to be considered. Although at compressive stresses, T_g always increases in macroscopic experiments, it is expected that under shear or tensile stress the underlying alpha-transition is eased. We note that a theory on yielding by Robertson [62] predicts WLF (Williams–Landel–Ferry) kinetics below T_g under shear stress, but this theory was found to be useful only near T_g [63]. All in all, T_g is difficult to predict for such experiments.

As mentioned above, an important aspect of both models is the difference in the indentation dynamics because in the yield picture monomer friction is predominant, whereas in the rubbery picture the chain/network topology of the polymer is the limiting factor.

In the rubbery picture, polymer backbone dynamics above T_g generally follow the so-called 'time–temperature superposition' [1, 64] and their kinetics are well described by the WLF equation:

$$T = \frac{\log(\tau_{ref}/\tau)T_\infty - c_1 T_{ref}}{\log(\tau_{ref}/\tau) - c_1}.$$

Here, τ, T and k_B are the indentation time, indentation temperature of the polymer, and the Boltzmann constant, respectively. The WLF parameters τ_{ref}, T_{inf}, T_{ref} and c_1 are the fit parameters characteristic for individual polymers. Note that usually these parameters are found to be independent of the actual quantity measured, be it shear modulus, viscosity or heat capacity. We therefore expect rubbery indentation to be essentially controlled by backbone kinetics following WLF.

In the yielding model, again the indentation kinetics is controlled by the dynamics of the backbone and is essentially of Arrhenius-type with a single activation energy E_a [65]:

$$\frac{1}{\tau} = \frac{1}{\tau_0} \exp\frac{E_a}{k_B T}.$$

Thus, to distinguish between 'rubbery indentation' and yielding, the indentation kinetics should be investigated.

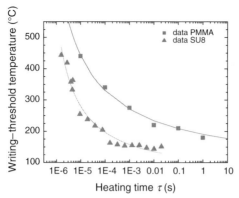

Figure 4.20 Writing-threshold heater temperature (i.e. the temperature required to write an indentation of 1 nm depth at a constant load force) as a function of heating time at a fixed load for a linear polymer PMMA (■) and a highly crosslinked epoxy SU8 (▲). The solid lines are fits using WLF kinetics (for which it was taken into account that the actual polymer temperature is significantly lower than the heater temperature).

First experiments [66] revealed that the indentation kinetics measured using PMMA and SU8 samples between 1 μs and 1 s cannot be fitted by a single activation energy (i.e. Arrhenius kinetics). An overall fit using WLF was satisfactory. An example of such an experiment is shown in Figure 4.20. Writing-threshold curves, defined as the minimum heater temperature needed to achieve an indentation depth of 1 nm at a given indentation time and at constant load force, were measured. Two prototype polymers are used: one is a thin film of PMMA; and the other a highly crosslinked thermoset, the epoxy SU8. A good fit with WLF can be obtained for PMMA. The temperatures needed are clearly above the T_g of about 120 °C, in agreement with our rubbery-indentation picture.

In a highly crosslinked system such as SU8, viscous flow can no longer account for the indentation. In this case, the load force was relatively high (200 nN), so that indentation times down to 1 μs were feasible with limited heater temperatures. Here also, a reasonable WLF fit was obtained despite the fact that, for long indentation times, the writing-threshold temperatures were considerably lower than the T_g of about 200 °C. Note, however, that for these data points the quality of the data does not allow the exclusion of a transition to Arrhenius-type behavior at long indentation times.

More detailed experiments using a crosslinked polystyrene sample were performed to investigate the topic further. These data are shown in Figure 4.21, in the form of an Arrhenius plot. Although the curve can be fitted linearly at long times, there is a clear deviation from the linear fit above a temperature of 180 °C that coincides with the T_g of the material measured using DSC. Above T_g, the data is well-fitted using WLF and a Vogel temperature T_∞ of $T_g - 50$ °C. Below T_g, however, the simple linear Arrhenius model is the best fit. Despite the many uncertainties, the

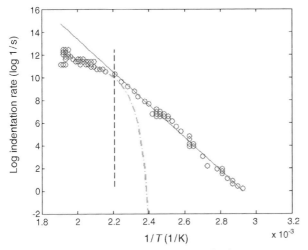

Figure 4.21 Arrhenius plot of thermomechanical indentation kinetics experiment. Below the glass transition temperature (T_g) of 180 °C (dashed line) an Arrhenius fit (solid line) is used. Above T_g, WLF yields a good fit (dash-dotted line). The sample used was a crosslinked polystyrene film; the indentation depth was 4 nm.

activation energy can be quantified and is found to be comparable with the activation energy of macroscopic polystyrene samples (1–2 eV).

It is concluded that WLF kinetics predominates at higher temperatures, and that a smooth transition to Arrhenius kinetics can be observed at lower temperatures/longer times. As expected, the transition occurs close to the T_g of the polymer.

These two physical pictures only manifest themselves in the different indentation dynamics. Only in the rubbery model above T_g does the chain-like nature of the polymers become apparent. On the other hand, it becomes clear that both physical pictures are useful to understand the experiments, and a distinction between them may appear artificial. The reason for this difference to the macroscopic polymer world is twofold: (i) because of the nanometer scale of the experiment and the crosslinked nature of the materials, macroscopic phenomena such as shear bands or crazes are absent; and (ii) the variations of shear stress and indentation rates are rather extreme. This forces a transition between the two conventionally fully separate regimes, which usually are switched only by the temperature and T_g.

In summary, a more unified picture of the indentation process emerges. For better clarity, we would like to propose a schematic, qualitative picture (see Figure 4.22). In this schematic, we capture the mechanics of the material using springs and dashpots. Springs k_s and k_p are connected in series and, correspondingly, in parallel to the dashpot. The elastic part of the medium is symbolized by k_s, whereas k_p is the elastic part linked to conformational changes of the polymer network. The dashpot, γ, is linked to the glass-to-rubber transition in the polymer. Below T_g, the dashpot is locked and can only be deformed by high stress; above T_g, it is open, representing the low-friction sliding of the polymer chains.

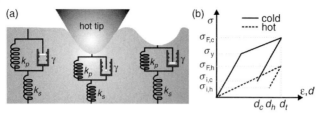

Figure 4.22 (a) Model representation of the polymer and the indentation process. From left to right: Undeformed polymer, polymer heated and deformed by the tip, and relaxed cold indentation; (b) Schematic of the stress–strain curves during an indentation experiment with hot and cold tips. See text for details.

Figure 4.22a describes (from left to right) the events during an indentation using a hot tip. If the hot tip is in contact with the polymer sample, the polymer is above T_g, which means that the dashpot is open and essentially free to move. Upon application of external stress, both springs will be deformed according to their strength, as shown by the dashed line in the stress–strain diagram in Figure 4.22b. Let us assume that the total deformation is d_t. Upon cooling, we lock the dashpot in the deformed state, and by releasing the external stress, $\sigma_{F,h}$, spring k_s relaxes to its uncompressed length. This state is show in the center of Figure 4.22a. We observe a partial loss of the indentation depth (to d_c for the cold case) as we retract the tip which results from elastic recovery in the material.

However, the $k_p - \gamma$ system is still deformed, and a residual stress given by the deformation of the polymer network spring k_p is locked in the deformation. In fact, this residual stress can be used to erase the indention (as will be shown in Section 4.6), and we also found stress-dependent relaxation in retention studies of the indentations. This deformation above T_g implies that the dynamics follows WLF kinetics because the backbone relaxation dynamics also follows this law. And indeed, we did observe this behavior for hot tips, as shown above.

If we deform the polymer below T_g, the situation is rather different, however. Under a cold tip, the dashpot is initially in the locked state. If we apply stress, the entire deformation at low stress values will first be absorbed only by the elastic spring k_s. The dashpot only opens once a critical stress, the yield stress σ_y, has been attained, as indicated by the solid line in Figure 4.22b. At the yield stress, the backbone motion is forced by the external stress and we have reached the writing threshold. As the dashpot opens, k_p is deformed accordingly, producing internal stress and, similarly to the hot case, we obtain elastic stress relaxation upon removal of the external force.

The state of the polymer after indentation is therefore remarkably similar in the two physical pictures discussed. Stress is stored in the deformed polymer network and is frozen in by the glassy state of the cold polymer. Only the amounts of elastic recovery (to d_c and d_h) and of the internally stored stress ($\sigma_{i,c}$ and $\sigma_{i,h}$) differ slightly. Experimentally, there is evidence of a higher remaining stress in cold indentations than in hot written ones because, at elevated temperature, the former relax faster (A. Knoll, unpublished results). The other significant manifestation of the two mechanisms is in the indentation dynamics, where we see a transient crossover from yielding to rubbery deformation with increasing temperature.

It is concluded that nanoindentation is a universal technique to study the deformation physics of polymers at the nanometer scale. By varying load, force, heat and temperature, important material properties – such as glass temperature, hardness, shift factors and yield-activation energies – can be extracted.

4.6
Application in Data Storage: The 'Millipede' Project

The capability of scanning probe techniques to modify and image a surface on the nanometer scale makes these techniques obvious candidates for data-storage applications. In fact, since the invention of the STM, many demonstrations of bit formation and imaging have been reported in the literature using almost every SPM technique and many different storage media and write mechanisms. Perhaps the most impressive of these demonstrations – at least from a density point of view – is the manipulation of individual atoms on a surface [67]. Although the storage densities that could be achieved with such techniques are very impressive, the construction of an actual storage system based on one of these ideas requires that numerous issues be addressed, including automated bit detection, system data rate, error correction, bit retention, power consumption, eraseability/cyclability, servo/tracking, reliability and cost. Many of these requirements are actually in competition with each other. For example, the highest storage densities demonstrated so far – that is, atomic scale – were achieved with very slow read-back speeds and had rather complex system requirements, such as ultra-high-vacuum conditions and low temperatures.

One scanning probe storage technology that achieves a balance between the many competing system requirements is the thermomechanical approach developed by IBM and referred to internally as the 'millipede' project [6, 7, 68]. In order to achieve a data rate comparable to those of conventional storage technologies, IBM has used microelectromechanical systems (MEMS) technology to fabricate large arrays of cantilevers that can be operated in parallel, with each cantilever writing and reading data in its own small storage field. The internal name of the project – 'millipede' – refers to the approximate 1000 cantilevers that were used in one of the first prototype systems.

4.6.1
Writing

The *write mechanism* used is the thermomechanical nanoindentation of polymers, as described in Section 4.5. The basic write process is illustrated in Figure 4.23. Data are written by pulsing the voltages applied to the cantilever to obtain suitable heat and force pulses while the tip is being scanned over the surface. Indentations placed at predefined positions along the data track can be used to encode data, with for

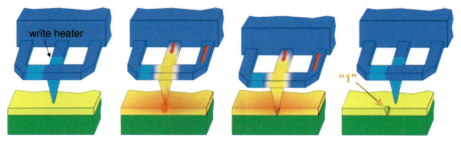

Figure 4.23 The principle of thermomechanical writing. The tip is heated by applying a current pulse to a resistive heater integrated in the silicon cantilever, directly behind the tip.

example, an indentation representing a logical '1' and the absence of an indentation a logical '0'. Storage densities greater than $1\,\text{Tb\,in}^{-2}$ have been demonstrated using this scheme in combination with appropriate coding [69].

4.6.2
Reading

The data are read back by measuring the topography of the polymer surface using same tip that wrote the data. In the IBM approach, this is done using a read-back mechanism based on heat-transport sensing. For this purpose, a second heater has been integrated into the cantilever structure. This second heater is remote from the tip, and can be heated without causing much of an impact on the tip temperature. When operated in ambient air conditions, the thermal resistance of the read heater exhibits a strong dependence on the distance between heater and medium surface, as discussed in Section 4.4.1. This thermal resistance dependence results in turn in a heater temperature dependence and hence also an electrical resistance dependence on the distance between heater and medium surface. (The electrical resistance change with temperature is an intrinsic property of silicon, as discussed in Section 4.2.) This situation can be exploited to sense the topography by applying a constant voltage to the heater and monitoring the changes in the electrical resistance that result as the tip is being scanned over the surface. For example, when the tip moves into an indentation (a '1'), the distance between cantilever and surface is reduced and the heat-transfer rate increased. This leads to a resistance change of the $\sim 1\,\text{k}\Omega$ heater of $\Delta R/R \sim 10^{-4}$ per nanometer (Figure 4.24).

4.6.3
Erasing

Erasing [6] is achieved by exploiting the mechanical stress that is stored in an indentation. The thermomechanical writing process described above results in indentations that are a metastable, deformed state of the polymer with a significant amount of stored elastic energy. If a sufficiently hot tip is pressed against the surface

4.6 Application in Data Storage: The 'Millipede' Project

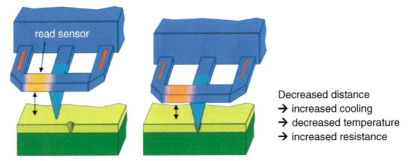

Figure 4.24 The principle of thermomechanical reading. Heat is generated by applying a current to a resistive heater integrated into the silicon cantilever. The heat transfer between heater and medium surface varies as a function of the distance between the cantilever and surface. Decreasing the distance between tip and medium leads to an increase in the cooling, which in turn decreases the temperature and increases the resistance, producing a detectable signal.

in the close vicinity of an indentation – that is, in the region of the rim around the indentation – then the increase in temperature in the indentation will result in a decrease in the viscosity of the polymer, which allows the elastic stress in the indentation to relax, effectively erasing the indentation. This process usually results in the creation of a new indentation, which can then be erased by repeating the procedure. Thus, a previously written data track can be erased by overwriting the data track with a series of closely spaced indentations. With this procedure, each new indentation erases the preceding one such that, at the end of the data track, all indentations will have been erased except for the last one. A demonstration of this principle is shown in Figure 4.25.

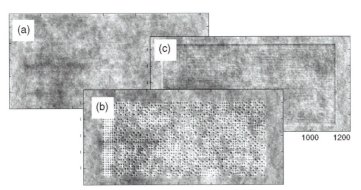

Figure 4.25 Atomic force microscopy topographical images illustrating the principle of thermomechanical writing and erasing. The images show (a) an empty area; (b) an area with indentations written at 1 Tb in^{-2}; and (c) an erased area. The grayscale covers 5 nm in all three images.

4.6.4
Medium Endurance

Medium endurance is a critical issue. The first challenge for polymer media is to be robust against repeated scanning with a sharp tip. In general, polymers tend to quickly roughen (and form ripples) when they are scanned repeatedly with a sharp tip, even at low load forces (see Section 4.6.5). Of the many solutions proposed to overcome medium wear, only a few can be readily applied to the nanoscale. On the nanometer scale, the homogeneity of the medium is crucial for nanoscale data-storage applications, and thus phase separation, filler particles or similar ideas cannot be used.

One elegant way to solve the issue of medium wear is to introduce a high degree of crosslinking into the polymer. This not only solves the roughening issue during sliding (reading) [56] but also facilitates erasing, because it provides the medium with a means of storing elastic energy and results in a type of 'shape memory'. To date, more than 10^4 write/erase cycles have been demonstrated using highly crosslinked polymer media (H. Podzidis *et al.*, unpublished results).

The dramatic improvement in wear endurance that occurs with increasing crosslinking is demonstrated in Figure 4.26, where the wear rate is plotted as a function of crosslink density for a set of polystyrene samples that were repeatedly scanned with a sharp tip. At a critical value of crosslinking, the mobility of the polymer is significantly reduced. This occurs when there is a sufficient number of crosslinks so that each region of cooperative polymer motion (typically ~1–3 nm in size) is affected.

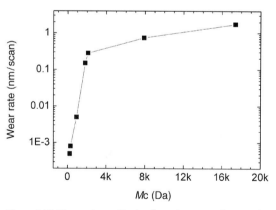

Figure 4.26 The peak amplitude of roughening induced by the wear of crosslinked polystyrene using a sharp tip scanning the surface, as in the reading operation. Here, the wear rate is defined as the cumulative amplitude of surface-topography changes normalized by the number of repeated scans of the same area. The wear rate is a strong function of crosslinkage. The crosslink density is given in units of molecular weight between crosslinks in the polymer chain. (Reprinted from Ref. [56]; © 2006, American Chemical Society.)

4.6.5
Bit Retention

Bit retention and the long-term stability of written data are also governed by the polymer mobility below the T_g value. This mobility is fundamentally provided by the activation energy of a backbone motion – that is, the so-called *alpha-relaxation*. Depending on the polymer, this can be as be as much as several electron volts, and can thus be sufficiently high for typical lifetime requirements. Lifetimes of 10 or more years at operating temperatures of up to $\sim 80\,°C$ have been extrapolated from experimental data.

4.6.6
Tip Endurance

Tip endurance may limit the feasibility of several SPM-based data-storage schemes that involve mechanical contact between probe and surface. The endurance requirements of a tip will, of course vary, depending on the application and the system architecture. In general, however, a single tip will have to scan distances ranging from 10^4 to 10^8 m during the lifetime of the device, without losing its ability to read and write data. In thermomechanical data storage, the polymer medium is relatively soft compared to the hard silicon tips used for reading and writing. However, even for this combination, tip wear still is an important issue, and other – even harder – tip materials are also currently being investigated. Lubrication has proved to be key in improving the endurance of hard-disk drives, and may also prove to be useful for probe-based storage.

The density limits of thermomechanical data storage are predicted to be well above the $1\,\text{Tb in}^{-2}$ mark. Ultimately, the density limits will be determined by the mobility of the polymer that corresponds to finite regions in which cooperative motions of polymer chains or chain segments occur. These regions range in size from 1 to 3 nm. As a small number of such regions must occur in each indented zone, a limit will appear somewhere at or below an indentation spacing of 10 nm.

4.6.7
Data Rate

The data rate is commonly one of the weaker aspects of SPM-related data-storage schemes. In the thermomechanical approach, two factors contribute to data-rate limitations:

- The cantilevered tip must be able to follow the topography mechanically; this translates into the requirement of a high mechanical resonant frequency.
- Temperature-based displacement sensor must be able to respond to these topography-induced height changes, ideally with a low power consumption.

The situation is further complicated by the requirement for low applied forces during the read operation in order to minimize tip and polymer wear. This low-force requirement in turn entails the need for a small spring constant, which tends to reduce the resonance frequency. Finally, during the write operation, the cantilever must also be able to apply and withstand forces on the order of hundreds of nanoNewtons. Thus, in order to achieve a competitive thermomechanical storage technology, all of these competing requirements must be carefully balanced and the cantilever design highly optimized. However, even with optimization, a data rate per cantilever/tip well above 1 MHz appears speculative. Consequently, a high degree of *parallelization* of 10^2 to 10^4 tips operating in parallel is required to achieve a sufficient user data rate, and this is feasible only if the fabrication employs VLSI silicon technology. To date, the fabrication of prototype cantilever arrays with thousands of tips has been demonstrated, as illustrated in Figure 4.27. Moreover, parallel read/write operation at high densities using a small subset of cantilevers has been achieved [70]. Currently, three electrical connections to the array chip are required for each cantilever that is to be operated, and thus the number of cantilevers that can be operated in parallel is limited by the area available for bonding wires. The demonstration of higher degrees of parallelization will require the integration of some of the system electronics behind the cantilevers, and this is an area of current research. The other basic components required to make an actual prototype storage system based on this technology, including a MEMS scanner, a position-sensing and servo-control scheme, a bit-detection scheme and error-correction codes as well as a

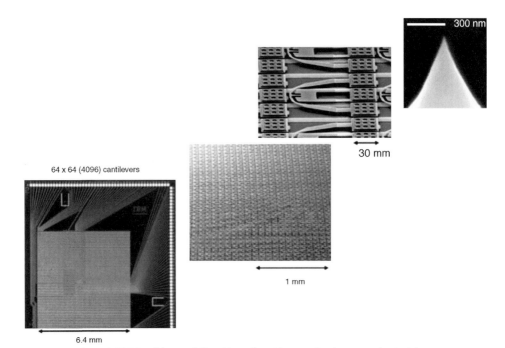

Figure 4.27 Microfabricated 64 × 64 cantilever/tip array for thermomechanical data storage.

system controller, have also been developed. All of this makes the route to highly parallel SPM-based storage appear feasible.

4.7 Nanotribology and Nanolithography Applications

Applications of heated probes going beyond thermal imaging and thermomechanical indentation can more generally be described as exposing surfaces to heated tips. In this section, examples are presented that lead to the modification and patterning of surfaces. First, experiments are described that involve scanning with a hot tip on polymer samples, with the aim of understanding nanoscale wear. Second, the *controlled* removal of material with the application to scanning-probe lithography (SPL) is developed and analyzed. Finally, dip-pen nanolithography using heated tips will be discussed.

4.7.1 Nanowear Testing

Nanowear testing using AFM is commonly performed to understand the nanoscale wear of various materials. Wear in general is a complex phenomenon, and often very different physical mechanisms come simultaneously into play, such as thermally activated bond rupture, adhesion, frictional shear stress, third-body lubrication, and so on [71]. Thus, in wear experiments, certain parameters are varied to elucidate the wear mechanisms. For example the repetition of wear cycles, the load force on the scanning tip, the scanning speed, or environmental conditions, such as humidity, are examined. Temperature is less suitable as a variable in such experiments, as mentioned above in the context of thermal imaging. However, temperature is a decisive parameter controlling wear, in particular of polymers. Let us, for example, consider the wear of polymer surfaces around T_g. The glass-transition region is often very sharp and covers only few degrees; for thin films T_g may vary by several degrees because of finite-size effects. A temperature variation of a sample around T_g therefore involves ramping the temperature in relatively small steps, with each step including settling for minutes, or even hours, to equilibrate the sample. In contrast, the *tip temperature* is easily varied and settles within microseconds. Therefore, heating the tip rather than the sample makes it possible to perform wide temperature variations in a single experiment [8, 14, 72].

As an example, let us consider experiments in which a variably heated tip is raster-scanned over polymer surfaces. While the tip temperature is continuously increased along the slow-scan axis, low loading forces are maintained between the tip and polymer surface. A real-space image of thermal degradation (a 'wear track') is generated, in which each fast scan line corresponds to a certain tip temperature. A wear experiment of a polymer film of polystyrene (PS), a standard linear polymer, is shown in Figure 4.28a. Here, three regimes can easily be identified:

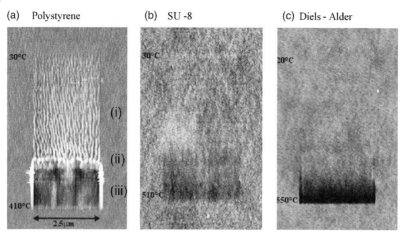

Figure 4.28 Wear tracks on different polymer samples obtained by raster-scanning a heated tip over a surface. The tip temperature is increased along the slow-scan direction (vertical). The wear track consists of 512 lines of 2.5 μm length scanned at 10 Hz. After the wear process, the images were obtained with an unheated tip in AFM imaging mode. (a) A polystyrene surface scanned four times; the grayscale covers an image corrugation of 17 nm. (Reprinted from Ref. [39]; (c) 2004, American Chemical Society.); (b) SU8 thermoset scanned 10 times; the grayscale covers a corrugation of 1.7 nm; (c) Diels–Alder polymer scanned 10 times; grayscale covers a corrugation of 5 nm.

(i) At low temperatures, a *ripple pattern* is generated. The activated kinetics of ripple formation of PS has been studied using both variable sample temperatures [73] and heated tips [39]. Activation energies of the ripple process that exhibit a similarity with the yield process discussed above have been established.

(ii) A second regime is the glass-transition region – that is, at tip temperatures that lead to a heating of the polymer in the interaction region under the tip up to T_g. There rather drastic effects become apparent in a dramatic increase of the ripple amplitude.

(iii) In the third regime, above T_g, the material becomes so ductile that it is swept to the sides of the heated scan.

A very different trend is observed when the same experiment is performed with a thermoset material in which material transfer is more suppressed, for example in a highly crosslinked epoxy, such as a 100 nm-thick film of SU8 (see Figure 4.28b). Because of the high crosslink density of SU8, no ripple pattern can be formed. Overall, the surface remains unchanged by the wear test, and only a marginal depression is observed at the largest temperatures. If any debris is formed it is apparently so volatile that it cannot be traced on the surface with the probing tip after the wear procedure.

The third example (see Figure 4.28c) demonstrates yet another characteristic degradation mode, where a chemical reaction is induced by the heated tip. The material is a highly crosslinked material in which the crosslinks are thermally reversible by virtue of a retro-Diels–Alder reaction [8]. The crosslinks are opened when a temperature of 130 °C is reached, and the material constituents are small molecular fragments that are rather volatile and may either diffuse onto the tip or evaporate. As observed in Figure 4.28c, no change in the surface is seen in areas scanned with heater temperatures of up to 320 °C. On further increasing the temperature, the depth of the wear track increases linearly up to 550 °C, although there is a clear lack of debris. In addressing the mechanism for track formation, compression is ruled out based on the observation that measured wear is cumulative. Repeating the experiment in a given area yields a proportional increase of the wear depth. Hence, it can be concluded that the material is lost by evaporation or diffusion onto the tip. The debris-less removal of material renders this method and the Diels–Alder polymer candidates for lithography applications. However, before we test this idea further let us briefly consider maskless lithography (ML2), in particular SPL.

4.7.2
Nanolithography Applications

In *microelectronics*, the time and expense required to produce a mask set is an issue in the prototyping of integrated circuitry [74]. For this reason, electron-beam lithography (EBL), which is an ML2 technique, is often used to prototype individual devices. The main drawback of ML2 efforts with respect to conventional lithography is the comparatively low exposure throughput. To remedy this, efforts are underway to develop EBL systems to operate a multitude of beams in parallel so as to reduce the overall exposure time [74]. However, further development and research is needed to control the crosstalk due to the high-voltage control signals and source brightness. The need for UHV conditions is also seen as an obstacle.

Currently, numerous SPL-based systems are under development – or have at least been proposed – and, in comparison to EBL tools, these will be more compact and simpler systems. The fabrication of large arrays of probes for massively parallel operation has been realized. One of the most powerful demonstrations of research-scale SPL used electrons extracted from a conducting tip to expose a resist [75]. Structures with resolutions of 30 nm have been transferred, and parallel operation has been demonstrated. The main challenges to this particular technique are the need for a conducting substrate, the high voltage necessary to extract electrons, the reliability of the electron-extraction process, and the lack of a simple overlay strategy.

Heated probes can also be used for SPL. For example, local heating has been used to induce the crosslink reaction of a conventional photoresist locally [76], after which the pattern is developed in the same way as in conventional lithography. Based on the example given in Figure 4.28c, an alternative SPL method can be applied that directly removes material from the exposed region. This approach offers specific advantages:

- It combines exposure and development in a single step, which not only makes the method simpler but also enables direct inspection of the exposure and direct repair in a separate repair step. A prerequisite here is that the same probe can be used for imaging, which is possible because the exposure creates a surface topography that can easily be measured. Thereby, the difficult issue of reliability of SPL is directly tackled.

- It also facilitates the use of a simple overlay strategy based on exposure/development of alignment marks. These marks can be imaged and used to align the new layer to be exposed. Note that the same strategy could be used for stitching error correction or self-calibration of probe positioning.

- In contrast to conventional nanoindentation or nanoscratch-based methods [77], this variant helps to achieve high resolution because the material is not simply displaced but is actually removed from the exposure site [8]. This permits achievements similar to those with other SPL techniques.

A demonstration of a high-resolution exposure of such a material is given in Figure 4.29 [78]. The imaged area was inscribed using pixels spaced 3.0 nm apart in a line, with a spacing of 107 nm between the lines. Each pixel was written using a heat pulse of 20 µs duration and a loading force of 40 nN. The Diels–Alder polymers were shown to permit the production of repeatable lithographic features of 12–15 nm in size (full width at half maximum) by sequential inscription. This makes it possible to overlay line patterns and faithfully reproduce line-crossings, both features which are desirable in lithography.

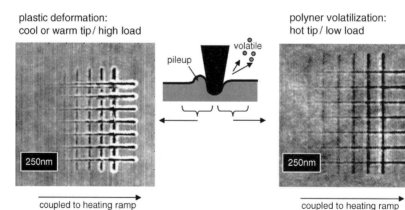

Figure 4.29 Comparison of lithography results using tips to (re)move material in a direct quasi-single exposure and development step. The material used is a reversibly crosslinked polymer described in Ref. [8]. Left panel: Conventional 'scratch'-type lithography using high forces (300 nN). The tip-heater temperature is varied from left to right, from room temperature to 440 °C. Independently of temperature, it is shown that the plastic deformation leads to a *displacement* of material to the sides of the individual lines written, making line-crossing impossible. Right panel: Similar experiment using a low load force (30 nN) and a heating ramp between 300 and 500 °C. In this case, material is *removed* and not just translated, and line-crossings are possible. A sketch of the difference between the two lithography processes is shown in the center panel.

(a)

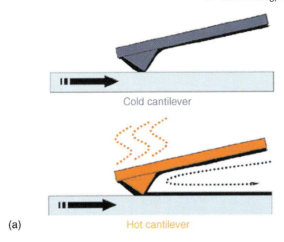

Cold cantilever

Hot cantilever

(b)

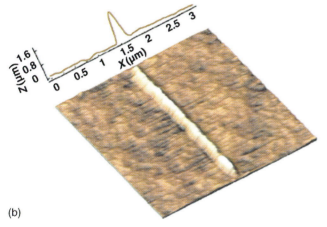

Figure 4.30 (a) Schematic of the operation of thermal dip-pen lithography (tDPN), which uses a heated AFM cantilever with a tip coated with a solid 'ink'. When the tip is hot enough to melt the ink, it flows onto the substrate. No deposition occurs when the tip is cold, thus enabling imaging without unintended deposition; (b) A topographic AFM image of a continuous nanostructure deposited from an indium-coated tip onto a borosilicate glass substrate. (Reprinted from Ref. [80]; (c) 2004, American Institute of Physics.)

The concept of removing material can be amplified by using materials that undergo an exothermic reaction when being volatized; in this regard, explosives have been used by King et al. [72]. The volatilization rate of a thin film of pentaerythritol tetranitrate exhibits a strong dependence on the tip temperature and exposure time. Although the study of King and coworkers emphasized the analytical application of heated probes, the nanostructuring context is evident.

For nanostructuring in a broader sense, it is interesting not only to *transfer* material along a surface (as in 'nanowear' experiments) or to *remove* material (as in the SPL examples above), but also to *deposit* materials. The deposition of material is particularly appealing for biological applications. Although various methods exist to deposit material from liquid or gas phases using a scanned probe tip (such as local oxidation of semiconductor surfaces [79]), the direct deposition of liquid droplets or lines using dip-pen lithography (DPN) has recently attracted considerable attention. In this method, an AFM tip that has been covered with a liquid to be deposited is brought into contact with a surface at locations where the liquid is to be deposited. In numerous experiments performed over the past few years, a range of materials have been successfully deposited, notably those used for biopatterning applications. A major challenge of DPN is the controlled switching on and off of the deposition process. Since in many applications, arrays of probes must be run in parallel in order to achieve sufficient throughput, individual probes cannot easily be brought into and out of contact independently; otherwise, all probes would write the same pattern. One strategy to circumvent this problem relies on heated probes, where the temperature of the tip – the sidewalls of which are the 'ink' reservoir – can be used to turn the deposition on and off. This has been demonstrated by Sheehan et al. [80] (see Figure 4.30), where lines of ink (octadecylphosphonic acid) were written at linewidths down to 100 nm in a controlled manner.

Acknowledgments

The authors would like to thank C. Bolliger for carefully proof reading the manuscript of the chapter. They are also grateful to the 'Millipede' teams at the IBM Research Laboratories in Zurich and Almaden for their continued collaboration and helpful scientific discussions. Previously unpublished results were obtained in collaboration with J. Frommer, C. J. Hawker, J. Hedrick, M. Hinz and R. Pratt.

References

1 Ferry, J.D. (1980) *Viscoelastic Properties of Polymers*, 3rd edn, John Wiley & Sons, New York.

2 Cahill, D.G., Ford, W.K., Goodson, K.E., Mahan, G.D., Majumdar, A., Maris, H.J., Merlin, R. and Phillpot, S.R. (2003) *Journal of Applied Physics*, **93**, 793.

3 Chen, G. (2000) *International Journal of Thermal Sciences*, **39**, 471.

4 Chen, G., Borca-Tasciuc, D. and Yang, R.G. (2004) in *Encyclopedia of Nanoscience and Nanotechnology*, Vol. 7 (ed. H.S. Nalwa), American Scientific Publishers, p. 429.

5 Balandin, A.A. (2005) *Journal of Nanoscience and Nanotechnology*, **5**, 1015.

6 Vettiger, P., Cross, G., Despont, M. et al. (2002) *IEEE Transactions on Nanotechnology*, **1**, 39.

7 Eleftheriou, E., Antonakopoulos, T., Binnig, G. K. et al. (2003) *IEEE Transactionson Magnetics*, **39**, 938.

8 Gotsmann, B., Dürig, U., Frommer, J. and Hawker, C.J. (2006) *Advanced Functional Materials*, **16**, 1499.

9 Reading, M., Price, D.M., Grandy, D.B. et al. (2001) *Macromolecular Symposia*, **167**, 54–62.

10 Shi, L. and Majumdar, A. (2002) *Journal of Heat Transfer*, **124**, 329.

11 Majumdar, A., Lai, J., Chandrachood, M., Nakabeoppu, O., Wu, Y. and Shi, Z.

(1995) *Review of Scientific Instruments*, **66**, 3584.
12 Shi, L., Plyasunov, S., Bachthold, A., McEuen, P.L. and Majumdar, A. (2000) *Applied Physics Letters*, **77**, 4295.
13 Despont, M., Brugger, J., Drechsler, U., Dürig, U., Häberle, W., Lutwyche, M., Rothuizen, H., Stutz, R., Widmer, R., Rohrer, H., Binnig, G. and Vettiger, P. (2000) *Sensors and Actuators A*, **80**, 100.
14 Gotsmann, B. and Dürig, U. (2006) in *Applied Scanning Probe Methods IV: Industrial Applications* (eds B. Bhushan and H. Fuchs), Springer, Berlin, Heidelberg, p. 215.
15 Sze, S.M. (1981) *Physics of Semiconductor Devices*, 2nd edn, John Wiley & Sons, New York.
16 Dürig, U. (2005) *Journal of Applied Physics*, **98**, 044906.
17 Wiesmann, D. and Sebastian, A. in (2006) *Proceedings 19th IEEE International Conference on Micro Electro Mechanical Systems, 2006, Istanbul, Turkey*, IEEE, p. 182.
18 Reading, M., Houston, D.J., Song, M., Pollock, H.M. and Hammiche, A. (1998) *American Laboratory*, **30**, 13.
19 Price, D.M., Reading, M., Hammiche, A. and Pollock, H.M. (1999) *International Journal of Pharmaceutics*, **192**, 85.
20 Majumdar, A. (1999) *Annual Review of Materials Science*, **29**, 505.
21 Shi, L. and Majumdar, A. (2001) *Microscale Thermophysical Engineering*, **5**, 251.
22 (a) Pollock, H.M. and Hammiche, A. (2001) *Journal of Physics D - Applied Physics*, **34**, R23; (b) Shi, L. and Majumbdar, A. (2004) in *Applied Scanning Probe Methods I*, (eds B. Bushan, H. Fuchs and S. Hosaka), Springer, Berlin, Heidelberg, New York, p. 327.
23 Hinz, M., Marti, O., Gotsmann, B., Lantz, M.A. and Dürig, U. (2008) *Applied Physics. Letters*, **92**, 043122.
24 Polder, D. and Van Hove, M. (1971) *Physical Review B - Condensed Matter*, **4**, 3303.
25 Loomis, J.J. and Maris, H.J. (1994) *Physical Review B - Condensed Matter*, **50**, 18517.
26 Volokitin, A.I. and Persson, B.N.J. (2004) *Physical Review B - Condensed Matter*, **69**, 045417.
27 Pendry, J.B. (1999) *Journal of Physics: Condensed Matter*, **11**, 6621.
28 Mulet, J.-P., Joulain, K., Carminati, R. and Greffet, J.-J. (2002) *Microscale Thermophysical Engineering*, **6**, 209.
29 Hargreaves, C.M. (1973) *Philips Research Reports Supplements*, **5**, 1.
30 Xu, J.-B., Laeuger, K., Dransfeld, K. and Wilson, I. H. (1994) *Journal of Applied Physics*, **76**, 7209.
31 Mueller-Hirsch, W., Kraft, A., Hirsch, M.T., Parisi, J. and Kittel, A. (1999) *Journal of Vacuum Science & Technology A - Vacuum Surfaces and Films*, **17**, 1205.
32 DiMatteo, R.S., Greiff, P., Finberg, S.L., Young-Waithe, K.A., Choy, H.K.H., Masaki, M.M. and Fonstad, C.G. (2001) *Applied Physics Letters*, **79**, 26.
33 Chapuis, P.-O., Greffet, J.-J., Joulain, K. and Volz, S. (2006) *Nanotechnology*, **17**, 2978.
34 Volz, S. and Chen, G. (1999) *Applied Physics Letters*, **75**, 2056.
35 King, W.P. (2002) Thermomechanical Formation of Polymer Nanostructures, PhD Dissertation, Stanford University, CA, USA.
36 King, W.P., Santiago, J.G., Kenny, Th.W. and Goodson, K.E. (1999) *Proceedings, American Society of Mechanical Engineering, MEMS*, **1**, 583.
37 Dames, C., Dresselhaus, M.S. and Chen, G. (2004) Phonon thermal conductivity of superlattice nanowires for thermoelectric applications. Materials Research Society Proceedings, **793**, S1.2.1.
38 Jansen, A.G.M., van Gelder, A.P. and Wyder, P. (1980) *Journal of Physics C - Solid State Physics*, **13**, 6073.
39 Gotsmann, B. and Dürig, U. (2004) *Langmuir*, **20**, 1495.
40 Lantz, M., Gotsmann, B., Dürig, U.T., Vettiger, P., Nakayama, Y., Yoshikazu, S.,

Tetsuo, T. and Tokumoto, H. (2003) *Applied Physics Letters*, **83**, 1266.

41 Yovanovich, M.M., Culham, J.R. and Teerstra, P. (1998) *IEEE Transactions on Components, Packaging, and Manufacturing Technology, Part A*, **21**, 168.

42 Ross, R.G., Andersson, P., Sundqvist, B. and Backstrom, G. (1984) *Reports on Progress in Physics*, **47**, 1347.

43 Hu, C., Kiene, M. and Ho, P.S. (2001) *Applied Physics Letters*, **79**, 4121.

44 Govorkov, S., Ruderman, W., Horn, M.W., Goodman, R.B. and Rothschild, M. (1997) *Review of Scientific Instruments*, **68**, 3828.

45 Swartz, E.T. and Pohl, R.O. (1989) *Reviews of Modern Physics*, **61**, 605.

46 Cappella, B. and Dietler, G. (1999) *Surface Science Reports*, **34**, 1.

47 Patton, K.R. and Geller, M.R. (2001) *Physical Review B - Condensed Matter*, **64**, 155320.

48 Briscoe, B.J. (1998) *Journal of Physics D - Applied Physics*, **31**, 2395.

49 VanLandingham, M.R., Villarrubia, J.S., Guthrie, W.F. and Meyers, G.F. (2001) *Macromolecular Symposia*, **167**, 15.

50 Klapperich, C., Komvopoulos, K. and Pruitt, L. (2001) *Transactions of the American Society of Mechanical Engineering*, **123**, 624.

51 Fischer-Cripps, A.C. (2002) *Nanoindentation*, Springer, New York.

52 Hinz, M., Kleiner, A., Hild, S., Marti, O., Dürig, U., Gotsmann, B., Drechsler, U., Albrecht, T.R. and Vettiger, P. (2004) *European Polymer Journal*, **40**, 957.

53 Balta Calleja, F.J. and Fakirov, S. (2000) *Microhardness of Polymers*, Cambridge University Press, Cambridge.

54 van Krevelen, D.W. (1997) *Properties of Polymers*, 3rd edn, Elsevier, Amsterdam.

55 Harth, E., Van Horn, B., Germack, D.S., Gonzales, C.P., Miller, R.D. and Hawker, C.J. (2002) *Journal of the American Chemical Society*, **124**, 8653.

56 Gotsmann, B., Duerig, U.T., Frommer, J. and Hawker, C.J. (2006) *Nano Letters*, **6**, 296.

57 Kody, R.S. and Lesser, A.J. (1997) *Journal of Materials Science*, **32**, 5637.

58 Brooks, N.W.J., Duckett, R.A. and Ward, I.M. (1998) *Journal of Polymer Science Part B - Polymer Physics*, **36**, 2177.

59 Sills, S., Overney, R.M., Gotsmann, B. and Frommer, J. (2005) *Tribology Letters*, **19**, 9.

60 Strobl, G. (1996) *The Physics of Polymers*, 2nd edn, Springer, Berlin, Heidelberg, New York.

61 Binnig, G.K., Cherubini, G., Despont, M., Dürig, U., Eleftheriou, E., Pozidis, H. and Vettiger, P. (2007) *Springer Handbook of Nanotechnology, Part F Industrial Applications* 2nd edn, (ed. B. Bhushan), Springer-Verlag, Berlin, Heidelberg, New York, p. 1457.

62 Robertson, R.E. (1966) *Journal of Chemical Physics*, **44**, 3950.

63 Argon, A.S. and Bessonov, M.I. (1977) *Polymer Engineering and Science*, **17**, 174.

64 Williams, M.L., Landel, R.F. and Ferry, J.D. (1955) *Journal of the American Chemical Society*, 3701.

65 Ree, T. and Eyring, H. (1955) *Journal of Applied Physics*, **26**, 793.

66 Gotsmann, B., Dürig, U., Frommer, J. and Hawker, C.J. (2006) *Advanced Functional Materials*, **16**, 1499.

67 Eigler, D.M. and Schweizer, E.K. (1990) *Nature*, **344**, 524.

68 Pozidis, H., Häberle, W., Wiesmann, D., Drechsler, U., Despont, M., Albrecht, T.R. and Eleftheriou, E. (2004) *IEEE Transactions on Magnetics*, **40**, 2531.

69 Wiesmann, D., Dürig, U., Gotsmann, B., Knoll, A., Pozidis, H., Porro, F. and Vecchione, R. (2007) Innovative Mass Storage Technologies "IMST 2007", Enschede, The Netherlands.

70 Sebastian, A., Pantazi, A., Cherubini, G., Eleftheriou, E., Lantz, M. and Pozidis, H. (2005) Proceedings of the 2005 American Control Conf. "ACC 2005", Portland, OR, June 2005, IEEE, Vol. 6, p. 4181.

71 Bhushan, B.(ed.) (1995) *Handbook of Micro/Nano Tribology* CRC Press, London.

72 King, W.P., Saxena, S., Nelson, B.A., Weeks, B.L. and Pitchimani, R. (2006) *Nano Letters*, **6**, 2145.

73 (a) Schmidt, H.R., Haugstad, G. and Gladfelter, W.L. (2003) *Langmuir*, **19**, 10390; (b) Wang, X.P., Loy, M.M.T. and Xiao, X. (2002) *Nanotechnology*, **13**, 478.

74 Groves, T.R., Pickard, D., Rafferty, B., Crosland, N., Adam, D. and Schubert, G. (2002) *Microelectronic Engineering*, **61–62**, 285.

75 Wilder, K., Quate, C.F., Singh, B. and Kyser, D.F. (1998) *Journal of Vacuum Science and Technology B*, **16**, 3864.

76 Hung, M.-T., Kim, J. and Sungtaek Ju, Y. (2006) *Applied Physics Letters*, **88**, 123110.

77 Kunze, U. and Klehn, B. (1999) *Advanced Materials*, **11**, 1473.

78 Gotsmann, B., Dürig, U., Frommer, J. and Hawker, C.J. (2006) *Advanced Functional Materials*, **16**, 1499.

79 Tello, M., Garcia, F. and Garcia, R. (2006) in *Applied Scanning Probe Methods IV: Industrial Applications* (eds B. Bhushan and H. Fuchs), Springer, Berlin, Heidelberg, p. 215.

80 Sheehana, P.E., Whitman, L.J., King, W.P. and Nelson, B.A. (2004) *Applied Physics Letters*, **85**, 1589.

5
Materials Integration by Dip-Pen Nanolithography
Steven Lenhert, Harald Fuchs, and Chad A. Mirkin

5.1
Introduction

The concept of using a tip coated with an ink – that is, a pen – to write on a surface has been used throughout history and is widely used today for recording or communicating information by hand. Although the most ancient written texts appear to have been carved in surfaces using sharp tools such as a knife or chisel, there are several reasons why the pen has eventually become the hand-writing tool of choice. First, the constructive nature of the pen typically enables a higher contrast than carving, making it possible to distinguish the writing from the surface background without further processing steps. Second, pen writing is relatively independent of the contact force in comparison with carving. And finally, if desired, a variety of different inks can be readily integrated on the same surface.

The same conceptual advantages that make the pen a useful tool on the macroscale also translate to the nanoscale when the tip of an atomic force microscope is used as an ultra-sharp pen to transfer material to a surface with nanometer scale resolution, a method known as dip-pen nanolithography (DPN) [1]. By using this technique, high-resolution chemical patterns can be constructed on surfaces in a single deposition step. Because the ink-transfer is independent of the contact force between the atomic force microscope tip and the substrate in almost all known cases, it is possible to carry out DPN reproducibly and in parallel, without the requirement for feedback from individual tips. By coating different tips with different inks, it then becomes possible to integrate a wide variety of molecules on a surface. As with other scanning probe lithography (SPL) methods (e.g. mechanical modifications, oxidation, local thermal treatments), DPN offers ultra-high lateral resolution, well below 20 nm. As a direct write lithographic method, DPN enables arbitrary patterns to be drawn without the need for a mask, with capabilities comparable to those of electron-beam lithography (EBL). Additionally, it is a tool that is ideally situated to rapidly produce laboratory prototypes and structures that are incompatible with the harsh conditions associated with conventional microfabrication techniques (soft biological structures in particular).

Importantly, DPN makes it possible to integrate different materials on scales (both in size and complexity) that appear impossible to reach by any other direct-write fabrication method. Such a method is clearly desirable for the fabrication of biomolecular arrays, and opens entirely new possibilities in the study and development of nanotechnology. This chapter will introduce the fundamental concepts in DPN technology, with a focus on aspects which enable nanoscale materials integration. In order to gain an understanding of what to expect, theoretical models will be introduced, followed by experimental approaches to controlling ink transport of various ink–substrate combinations, tip-coating methods, driving forces and characterization methods. Examples of unique applications of materials integration by DPN will then be described that cannot be achieved by any other method. Excellent reviews have been produced by Mirkin and others that summarize the DPN literature, and the reader is referred to those works for a more complete description of the vast amount of work already published on DPN to date [2–4].

5.2
Ink Transport

DPN is made possible by the transport of a material (ink) from the tip of an atomic force microscope to a surface at point where the tip contacts the surface (Figure 5.1a). As in the case of a macroscopic pen, the ink must flow from the tip of the pen to the paper, and this transport process is typically driven by an interaction between the ink and the substrate. However, quantitative differences appear when this technique is carried out at the nanoscale tip of an atomic force microscope. Most striking is that DPN is able to produce patterns consisting of a single molecular layer. The most

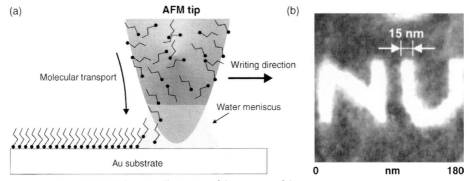

Figure 5.1 (a) Schematic illustration of the concept of dip-pen lithography DPN. (Reprinted from Ref. [1], with permission from the American Association for the Advancement of Science.); (b) Atomic force microscopy (AFM) friction image of patterns of the thiol mercaptohexadecanoic acid patterned on gold ⟨111⟩ with 15 nm line widths. (Reprinted from Ref. [5], with permission from the American Association for the Advancement of Science.)

thoroughly studied (and widely reproduced) ink–substrate combination for DPN is the patterning of alkanethiols on gold surfaces. Alkanethiols spontaneously self-assemble on gold surfaces under the appropriate conditions to form tightly packed self-assembled monolayers (SAMs). Typical thiol inks include octadecanethiol (ODT), which forms a methyl-terminated SAM, and mercaptohexadecanoic acid (MHA), which forms a carboxylic acid-terminated SAM. Figure 5.1b shows an example of a high-resolution DPN pattern of MHA on a single crystalline gold surface, with line widths of 15 nm. As the radius of curvature of the tip used to make that pattern was approximately 10 nm, it has been hypothesized that a sharper tip may enable the fabrication of even smaller features [5]. The ultimate resolution limit of DPN has yet to be determined. It has even been proposed that DPN can be used to generate features consisting of single molecules, and that the practical limit lies in detecting such small features by atomic force microscopy (AFM) [6].

In a typical experiment aimed at characterizing the transport rate of an alkanethiol such as MHA or ODT from the atomic force microscope tip to a gold surface, the coated tip is placed in contact with different areas of the surface for different amounts of time. These contact areas can then be imaged *in situ* by rapidly scanning the patterned area with the same tip in lateral force mode to obtain a friction contrast image such as that shown in Figure 5.2a. Upon plotting the area (or radius, r) of the spots as a function of contact time, it is possible to quantify ink transport rate. It is reproducibly observed that the surface area covered by a single dot is roughly proportional to the contact time, albeit with some exceptions – for example, in the case of very long contact times [7]. Based on this simple assumption, a single parameter can be used to describe the transport rate – namely a spreading constant C expressed in units of $\mu m^2 s^{-1}$ (dashed line in Figure 5.2b). Once this transport rate has been determined, and is considered in a calibration, it then becomes possible to control dot dimensions and fabricate arbitrary patterns in a lithography process [8].

5.2.1
Theoretical Models for Ink Transport

The ability to obtain quantitative data on transport rates of inks, as well as the morphological information obtained by *in situ* AFM imaging, opens the possibility of testing theoretical models for the nanoscale ink transport in DPN. In addition to the perfectly round and sharp spots reproducibly achieved when patterning thiols on gold under optimal conditions, occasionally it can be observed that some spots appear more diffuse, are surrounded by a halo of lower lateral force microscopy (LFM) contrast or consist only of a ring. Furthermore, in some cases an 'anomalous diffusion' is observed where, instead of circles, fractal-like branches appear. Figure 5.3 shows schematics of four models that have been developed to explain and understand the different spreading phenomena observed in DPN experiments. The first three models (see Figure 5.3a–c) focus on the deposition of thiol SAMs on gold, where a strong chemical binding of the ink molecule to the substrate is expected and anomalous diffusion is not observed. The fourth model explains anomalous diffusion in terms of strong intermolecular interactions within the ink.

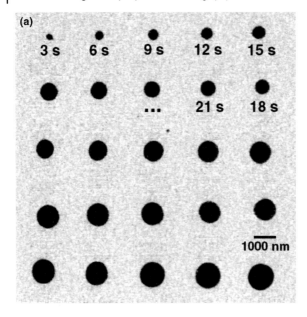
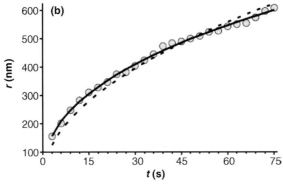

Figure 5.2 (a) Lateral force image of octadecanethiol dots deposited on a gold surface at different contact times; (b) A typical plot of the radius (r) of the spots versus contact time (t), including fits to two theoretical models. (Reprinted with permission from Ref. [9]; © 2002, American Physical Society.)

The first two models (Figure 5.3a and b) are based on diffusion theory, and are similar in that they both assume the tip to be an infinite point source, and use diffusion theory to describe the spreading of a thiol monolayer on a gold surface. The first model (Figure 5.3a) assumes a constant flux of molecules flowing from the tip, with a concentration of zero outside a spread island. That idea is consistent with the experimental observations that the area tends to increase linearly with contact time, and that the monolayer islands have sharp edges in the majority of AFM images. The second model (Figure 5.3b) assumes a constant concentration at the tip and an area of

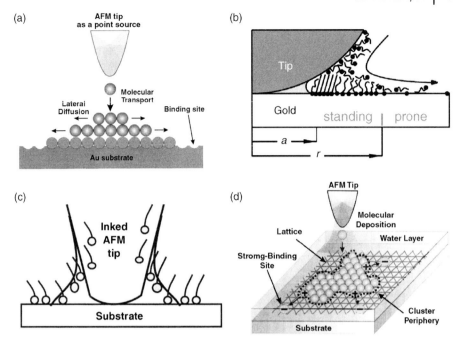

Figure 5.3 Drawings of different models of ink transport. (a) Diffusion model with constant flux from the tip. (Reproduced with permission from Ref. [14]; © 2001, American Institute of Physics.); (b) Diffusion model with a constant concentration at the tip. (Reprinted with permission from Ref. [9]; © 2002, American Physical Society.); (c) Meniscus interface transport model. (Reproduced with permission from Ref. [15]; © 2006, American Institute of Physics.); (d) Model for anomalous diffusion based on collective behavior. Reproduced with permission from Ref. [13]; © 2006, American Institute of Physics.)

a lower density of thiol molecules on the surface diffusing from the tightly packed monolayer. This second model allows for variation in the flux from the source, explains the occasional presence of halos and provides a slightly better fit to the experimental data (solid line in Figure 5.2b). The constant-concentration model also allows the derivation of an absolute diffusion coefficient using three physically relevant fit parameters (tip contact area, ratio of the concentration on the tip to that of a tightly packed monolayer and the diffusion coefficient).

In addition to the general idea of modeling the tip as a point source from which ink molecules diffuse, the effect that a microscopic condensed water meniscus forming at the tip–substrate contact point in the presence of humidity has been considered in order to further unravel the mechanisms by which the ink molecules make their way to the surface [10]. In particular, amphiphilic molecules such as MHA, which are only very slightly soluble in water, might be expected to show an affinity for the air–water interface. The meniscus interface transport model (Figure 5.3c) was therefore developed and can be used to explain the formation of hollow ring patterns. This model can also predict the transport behavior of a variety of amphiphilic molecules, including some that physisorb on the substrate.

Molecular dynamics simulations have suggested that SAM growth on the nanoscale involves a molecular basis that cannot be adequately described by analytical diffusion models [11]. For instance, it was shown that, even in the case of strong binding (e.g. alkanethiols on gold), the monolayer will grow as molecules from the tip displace molecules already bound to the surface, in a mechanism more akin to spreading than to diffusion. Computer simulations have also been able to recreate anomalous diffusion [12] *in silico* by considering the collective behavior of the molecules in a SAM (Figure 5.3d) [13]. There is evidence supporting aspects of each of these models, and there is no consensus as to which is the best. Further innovations from theoreticians, as well as carefully planned experiments designed to test the different models, are necessary to determine the true situation at the DPN tip.

5.2.2
Experimental Parameters Affecting Ink Transport

Several experimental parameters have been observed to influence the ink transport in DPN, including: driving forces (chemical interactions and external fields), ink composition, surface properties (chemistry and roughness), humidity, temperature and tip geometry. It is necessary to understand and control these parameters in order to optimize DPN processes for a particular application. This is especially important if different materials and nanostructures of the desired materials are to be integrated on the same substrate.

5.2.2.1 Driving Forces
In order for the ink to move from the tip to the substrate, a driving force is required, otherwise the ink will simply remain on the tip. Internal driving forces (i.e. in the absence of external fields) are typically based either on a chemical reaction of the ink with the substrate (chemisorption) resulting in a SAM, or on physical adhesion of the ink to the substrate (physisorption). Another approach is to apply an external field, for instance by heating the tip (thermal DPN) or applying a voltage (electrochemical DPN). It is worth noting that all DPN fabrication process require some sort of interaction between the ink and the substrate, even when the driving force is provided externally.

5.2.2.2 Covalent Reaction with the Substrate
The covalent reaction of the ink molecules with the substrate to form a SAM is the most straightforward and widely used driving force in DPN. The most reproduced and well-studied system is the formation of thiol SAMs, as described in Section 5.2.1. However, covalent self-assembly has also been used as a driving force for other ink molecules. As an example, considerable effort has been made to develop reproducible methods for DPN on semiconducting or insulating surfaces such as silicon and glass, as thiols are limited to self-assembly on metallic surfaces.

Functional silane molecules are widely used for the fabrication of SAMs on semiconductor surfaces, and are the therefore the first choice. However, a drawback of patterning silanes by DPN in air is that they polymerize in the presence of water, and tend to be liquid in their monomeric state (thiol inks in contrast are typically solid at room temperature). Despite these challenges, it has been shown that through careful optimization of the experimental conditions – for example, selection of the molecule, control of humidity and functionalizing the AFM tip – it is possible to pattern SAMs of functional silanes [16–18]. Another approach is to choose a functional group for self-assembly that is not as sensitive to water; for example, silazanes have been shown to be suitable inks on semiconductor surfaces by covalent reaction with OH groups on the surface [19].

Perhaps the most generally applicable strategy for depositing and integrating arbitrary molecules on an arbitrary surface by DPN is to prefunctionalize the desired surface with a bulk SAM, which can then be covalently linked to the ink molecule of choice. This approach has been particularly useful for the patterning of biofunctional molecules (this will be discussed in more detail in Section 5.6). Briefly, it has been used successfully for direct DPN of synthetic macromolecules [20], peptides [21] and DNA [22].

5.2.2.3 Noncovalent Driving Forces

While chemisorption of the DPN ink results in highly stable patterns, covalent reactions tend to be highly specific – and therefore in some cases it is desirable to be able to deposit an ink noncovalently. Examples include patterning on inert substrates, the integration of different materials on the same substrate, and/or the fabrication of multilayer structures. Such noncovalent deposition is, for example, the method used for patterning with macroscopic pens.

Numerous examples of noncovalent patterning have been reported in the literature. For example, the first instance of controlled deposition of organic materials from an AFM tip was the deposition of thiols on mica [23]. Electrostatic interactions have been used as a driving force to pattern charged conducting polymers [24], as well as polyelectrolytes which could be used as templates for layer-by-layer assembly [25] on silicon substrates. Inorganic nanostructures were fabricated by depositing inorganic precursors dispersed in a copolymer surfactant or dissolved in an ethylene glycol solvent which wets the substrate [26, 27]. Luminescent polymer nanowires were patterned on glass using only adhesion as a driving force [6, 28]. Nanoparticles (Fe_3O_2 and gold) have been picked up by an AFM tip and deposited noncovalently in controlled fashion onto mica surfaces in air by DPN [29, 30]. Semiconductor precursors, which are expected to react with each other and precipitate CdS only in the water meniscus, were used as inks to fabricate semiconductor nanostructures on mica [31]. Another approach is to mix a functional molecule with a well-characterized ink, as has been demonstrated by the DPN patterning of binary ink mixtures [32]. Finally, surfactants can be added to the ink in order to tune the 'wettability' of the ink on the substrate, providing another parameter that can be used to control ink transport [33].

5.2.2.4 Tip Geometry and Substrate Roughness

In addition to chemical interactions between the ink and substrate, it is also apparent that the topography of the substrate and geometry of the tip play a role in ink transport. The effect that these parameters have on the minimum feature size was systematically investigated in the case of alklythiol patterning on gold surfaces, where the smallest line widths (14 nm) could be achieved with the sharpest tips, and on the smoothest gold available [8]. In another study on the effect of tip-geometry, the AFM tips were deliberately made blunt using laser ablation [34]. It was then found that not only did the minimum feature size depend on the tip radius, but also the rate of ink transport – an idea consistent with several of the ink transport models described above.

5.2.2.5 Humidity and Meniscus Formation

A significant amount of evidence is available which suggests that the transport of ink molecules with polar groups (such as the thiol MHA) are heavily dependent on humidity, with higher humidity showing higher diffusion constants. Although there seems to be only a slight (if any) humidity dependence for the nonpolar molecule ODT [9, 10], the effect cannot be ignored in the patterning of just about all other molecules. *Humidity* is therefore an important parameter that must be controlled in order to optimize DPN conditions. Ideally, this is achieved by encasing the DPN apparatus in an environmental chamber, or locally by placing a water-containing capillary tube near the atomic force microscope tip. Although the possibility that the humidity might affect the ink properties or substrate reactivity has not been excluded, it is generally thought that the mechanism for humidity dependence depends on the presence of a meniscus that condenses at the tip of an AFM when it contacts a surface in the presence of humidity [35].

Theoretical studies of meniscus formation at an atomic force microscope tip based on Monte Carlo simulations have predicted that the meniscus should depend not only on humidity, but also on the tip geometry and surface chemistry [36]. Striking images confirming the presence of such meniscus formation (and indeed showing that the meniscus can grow larger than expected) have been made possible by using environmental scanning electron microscopy (ESEM), as shown in Figure 5.4 [37]. Interestingly, studies of the kinetics of meniscus formation between an atomic force microscope tip and gold surfaces showed similar trends as the early patterning rates of thiols on gold. It has therefore been hypothesized that growth of the water meniscus may in some cases be the rate-limiting step in DPN ink transport [15, 38].

Figure 5.4 Environmental scanning electron microscopy (ESEM) series of meniscus formation on a cantilever tip at three different relative humidities (left, 40%; center, 60%; right, 99%). (Reproduced with permission from Ref. [37].)

It should be noted that, although the humidity clearly influences ink transport in DPN, it does not appear to be a prerequisite, as patterning has been achieved at 0% humidity and even in ultra-high vacuum [9, 10].

5.2.2.6 External Driving Forces

Another approach to controlling the transport of materials from an atomic force microscope tip to the surface is to apply an external driving force between the tip and the substrate. Although this poses an engineering challenge in fabricating externally addressable tips, the ability to switch a particular pen on and off greatly increases the versatility, especially when one considers parallel arrays of tips where each may be addressable for large scale integration. Two examples are the use of heatable cantilevers (thermal DPN) and the application of a voltage between the tip and sample (electrochemical DPN).

5.2.2.7 Thermal DPN

The idea of thermal DPN is similar to that of a soldering iron; that is, the material on the tip of the microscope should be heated above its melting temperature in order to facilitate transport to the surface. The concept is shown schematically in Figure 5.5. Although heating the ink can be a disadvantage for biological inks, which may be sensitive to high temperature and dehydration, it provides a useful means of patterning other materials. For example, octadecylphosphonic acid (OPA) was the first compound to be patterned on silicon by using thermal DPN [39]. It was observed that OPA only began to write when the tip was heated above a critical temperature.

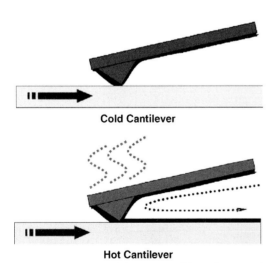

Figure 5.5 Schematic of the concept of thermal DPN. At low temperature, the ink does not transfer from the tip (top), while upon heating the tip the ink flow can be controlled (bottom). (Reproduced with permission from Ref. [39]; © 2004, American Institute of Physics.)

The method has since been applied to the deposition of conducting polymers [40]. Although it was initially suggested that thermal DPN is necessary for organic compounds with high melting points, it has since been shown that such compounds can also be patterned at room temperature (well below their melting temperatures) by humidity-controlled DPN. For instance, OPA as well as other compounds with melting points up to 230 °C have been patterned at room temperature under the appropriate humidity [41]. The mechanism of transport in thermal DPN of organic inks therefore remains unclear, although it appears to be possible to control the ink transport by controlling the tip temperature. Most striking is that indium metal nanostructures could be directly written by thermal DPN [42]. Furthermore, by filling a single carbon nanotube with molten copper and then dispensing it under observation with a transmission electron microscope, it has been suggested that using such a carbon nanotube-based spotwelder in thermal DPN might enable the ultra-high resolution of molten metals by thermal DPN [43]. Such direct, nanoscale writing of water-insoluble metals has not been shown to be possible below the melting point of the metal.

5.2.2.8 Electrochemical DPN

Another approach to controlling the transport of ink from the atomic force microscope tip is to use the water meniscus as a nanoscale electrochemical cell, where metal salts can be dissolved, reduced and precipitated to form metal nanostructures on the surface; this method, which is referred to as electrochemical DPN (E-DPN), is illustrated schematically in Figure 5.6 [44]. This method was further applied to the controllable transport of his-tagged proteins, which have an affinity to certain metal ions such as Ni^{2+}. By carrying out E-DPN on nickel-coated surfaces, the surface could be locally ionized, thereby allowing proteins on the microscope tip to transport the surface and bind. The same approach of combining local surface oxidation with material transport from the atomic force microscope tip was used to locally oxidize a pre-existing SAM and to simultaneously deposit organic inks to those same areas [45].

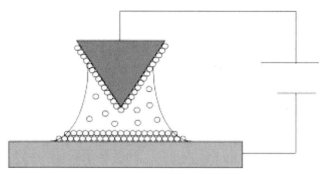

Figure 5.6 Schematic of electrochemical DPN (E-DPN). The water meniscus that condenses in air between the AFM tip and sample is used as a nanometer-sized electrochemical cell. (Reproduced with permission from Ref. [44].)

5.3
Parallel DPN

5.3.1
Passive Arrays

The constructive and chemically driven nature of DPN makes it uniquely amenable to being carried out in parallel, using arrays of tips, and without the need for accessing each tip electronically for force feedback. A rough alignment of the tip-array with the surface is sufficient, since if the tips are touching the surface then the ink will be transported at a constant rate which is determined primarily by the ink–substrate combination. The first demonstration that DPN could be readily carried out in parallel, employed micromachining processes to fabricate one-dimensional arrays with 32 silicon nitride tips or eight boron-doped silicon tips, the latter having sharper tips at the expense of pen densities [46]. The number of tips in a single linear array was then scaled up to the centimeter scale, using arrays of up to 250 tips, all of which wrote simultaneously [47]. Parallel DPN was then scaled up again to a two-dimensional arrays of tips that covered a square centimeter and consisted of 55 000 probes writing simultaneously; an example is shown in Figure 5.7 [48]. The parallel and constructive capabilities of DPN are what give it the potential to integrate materials on unprecedented scales.

5.3.2
Active Arrays

One factor which limits the complexity of patterns that can be generated by parallel DPN as described above, is that every tip necessarily writes the same

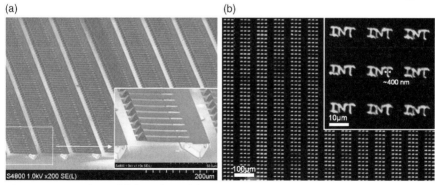

Figure 5.7 Massively parallel dip-pen nanolithography (DPN) with two-dimensional (2-D) tip arrays. (a) Scanning electron microscopy image of a small section of a 55 000 tip array covering an area of 1 cm^2. (Image courtesy of NanoInk.); (b) Fluorescence image of phospholipid patterns generated with the 2-D arrays at a throughput of 5 cm^2 min^{-1} [49].

pattern, provided that it is coated with ink. That is, it would be impossible to get each tip in an array to write a different pattern using passive tips. In addition to the possibility of controlling the driving force by external fields (e.g. by thermal DPN or E-DPN, as described earlier), another innovative step in the development of DPN probes for parallel materials integration was taken in which the cantilevers could be externally actuated. The first approach of this type used thermal bimorph cantilevers, where heating one side of the cantilever caused it to bend down so that the tip contacted the substrate and the ink could flow [50]. Again, by optimizing the tip fabrication the resolution of patterns generated by active pen arrays could be brought down below 100 nm [51]. Another approach would be to use electrostatically actuated cantilevers, where the cantilever bends towards the surface based on an applied electric field in order to avoid the possibility of unwanted heating of the tip or thermal crosstalk between neighboring cantilevers [51]. Actuated probes not only increase the available complexity of patterns that can be generated by a single ink, but also open the door to the integration of different materials from different tips in an array on an area smaller than the dimensions of the tip array itself. This can be done by writing with one tip, and then moving the array such that a neighboring tip writes on or near the same area that has already been patterned.

5.4
Tip Coating

5.4.1
Methods for Inking Multiple Tips with the Same Ink

An essential part of any DPN process is to bring the ink onto the atomic force microscope tips. This is typically achieved either by thermally evaporating the ink onto the tips, or by immersing the tips in the ink in a type of dip-coating process. Thermal evaporation is rather straightforward and typically results in a homogeneous ink coating, for instance of ODT. However, the majority of ink molecules – and especially the more polar ones – do not seem suited to evaporation, and tend to function better when the tip is coated from solution. For solution coating, the entire cantilever chip can be dipped by hand into an ink solution and, upon removing the tip and allowing the solvent to dry (e.g. under a stream of inert gas), a typically homogeneous coating results. However, the distribution of the ink on the tip when coated in this way will inevitably depend on exactly how the ink wets (or de-wets) the tip, and also on how the ink solutes concentrate on the tip as it is dried. Functionalization of the atomic force microscope tip before coating is therefore sometimes beneficial. For instance, in order to reproducibly pattern proteins it was found useful to precoat the tips with a SAM of a thiolated polyethylene glycol [11-mercapto-undecylpenta(ethylene glycol)disulfide(PEG)], which makes the tip hydrophilic but prevents the denaturing of adsorbed proteins [52].

5.4.2
Ink Wells

Inking the tips in the ways described above is well suited for single tips, or in the case that the requirement is to coat all tips with the same ink. However, in order to take advantage of the potential for DPN to integrate different materials on the same substrate in parallel, it is necessary to deliver different inks selectively to different tips in an array. For this purpose, microfluidic ink-delivery systems have been developed [53, 54]. An example of tips being dipped in wells that coat only every second tip in an array is shown in Figure 5.8a. An array where only every second tip is coated is useful for many experiments that require uncoated tips as negative controls, or in cases where there is a need to have the patterns spaced further apart than the spacing of the tips in the array. Today, ink wells are available commercially (from the company NanoInk) that allow the integration of up to 24 different inks on a one-dimensional array. Figure 5.8b shows an example of two different fluorescently labeled phospholipids integrated on a single cantilever array [49]. In similar fashion, a chip has also been developed that allows local vapor coating onto tip arrays [55].

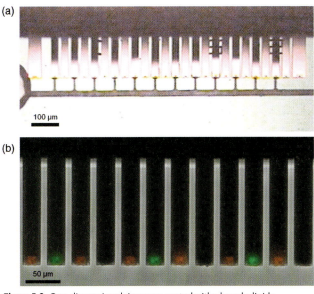

Figure 5.8 One-dimensional tip arrays coated with phospholipids using ink wells. (a) Optical micrograph of tips in contact with the ink wells. Every second tip in the array is being dipped with one ink; (b) Multi-channel fluorescence and bright-field micrograph of phospholipids doped with different inks. Every second tip is coated with a red dye, every fourth tip with a green dye, and the remaining tips function as negative controls.

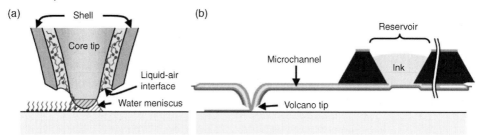
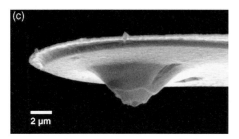
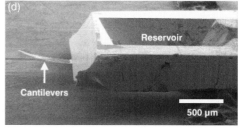

Figure 5.9 Fountain pens. (a) A cross-sectional schematic of the volcano-like tip in the process of writing; (b) A schematic of the entire chip including the reservoir; (c, d) SEM images of the tip and entire chip, respectively. (Reproduced with permission from Ref. [56]; © Wiley-VCH Verlag GmbH & Co. KGaA.)

5.4.3
Fountain Pens

One particularly elegant approach to delivering ink to the atomic force microscope tip is through the integration of microfluidic channels directly onto the tip itself, in order to generate a nanofountain probe (NFP), such as that shown in Figure 5.9 [56]. As standard microfabrication techniques were used, it has been possible to generate parallel arrays of NFPs integrated on a single chip, with different ink reservoirs leading to different tips in the same array, thus enabling the parallel integration of different inks [57]. As in the case of macroscopic pens, NFPs can be expected to be particularly useful for the patterning of inks where the solvent must remain in the ink until after patterning, as tips coated by dipping in solution are subject to drying.

5.4.4
Nanopipettes

Although micropipettes and nanopipettes differ technically from DPN (in that they do not necessarily utilize an atomic force microscope tip), they are conceptually similar to DPN and NFPs in several ways, and are therefore worthy of brief mention at this point. Cantilevered micropipettes similar to those used for scanning near-field optical microscopy (SNOM) have been used for the local delivery of an etchant to a chrome film, with a resolution of 1 μm [58]; indeed, subsequent studies led to the fabrication of spots of 280 nm diameter [59]. In similar manner, an electrochemical fountain pen was used to fabricate freestanding platinum nanowires with a diameter of 150 nm. By using a voltage-controlled feedback circuit derived from scanning ion conductance microscopy (SICM), submicron and multicomponent features consisting of biomolecules such as DNA and protein have also been fabricated [60, 61]. Although the practical resolution of micropipettes and nanopipettes is much lower than for DPN, and it is difficult to imagine them being used in parallel, they do have a significant advantage for the patterning of biomolecules in that they are able to function under water [62].

5.5 Characterization

In addition to tip inking and writing, a third indispensable part of the DPN process is the characterization and quality control of the resultant patterns. A convenient capability of DPN, which is also shared by most scanning probe-based lithography processes, is that the same tip can be used for both patterning and imaging, in the case of DPN by AFM. In particular, lateral force imaging is typically used for the characterization of chemical contrast in covalently bound inks such as thiols on gold (as described earlier and shown in Figure 5.2). There are two practical issues to be aware of when characterizing DPN patterns generated by the same tip that has been used for imaging:

- An inked tip will typically continue to write during imaging, and therefore high scan speeds must be used to minimize this effect.
- The vast majority of DPN is carried out in contact mode, using cantilevers with a too-low spring constant for use in intermittent contact or tapping mode imaging in air. This is especially the case for parallel DPN, where it is impractical to have a separate tapping feedback mechanism for each tip. As the contact mode typically provides inaccurate heights in air, it is often necessary to change the tip and realign it to find the patterned area in order to obtain quantitative height information.

Although these two issues represent disadvantages in a high-throughput lithography process, they can actually serve as significant advantages when characterizing the ink transport. For instance, in an early study using DPN, monolayer growth could be observed *in situ* by scanning the same area repeatedly at high resolution; this allowed observation of the monolayer growth dynamics as alkanethiols were transferred from the tip to a gold substrate [63]. It is also possible to carry out DPN in tapping mode by using a single tip for the simultaneous deposition and imaging of soft materials, as well as obtaining accurate height information [64]. When this method was applied to the deposition of poly-D,L-lysine hydrobromide onto mica surfaces, it was possible to observe the nucleation and growth dynamics of polymer crystals at a submicron scale that is inaccessible to other methods (Figure 5.10) [65].

In the DPN patterning of a new ink or substrate, it is essential to determine that the patterns generated are indeed composed of the intended ink. If a particular

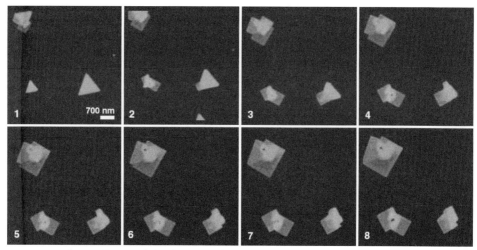

Figure 5.10 Topographical AFM image sequence (1 to 8) of epitaxial crystal nucleation and morphology changes from threefold to fourfold symmetry during growth of poly-D,L-lysine crystals on a mica surface as molecules are transferred from the microscope tip during each scan. (Reproduced from Ref. [65] with permission from the American Association for the Advancement of Science.)

topographical morphology is known, then this information can be obtained from AFM measurements. For instance, the AFM images of DPN-patterned collagen fibrils showed a helical repeat of 65 nm, which was consistent with scanning electron microscopy (SEM) observations of collagen fibrils. Another example is the formation of anisotropic structures formed during the DPN patterning of peptide amphiphiles [66]. Another way of distinguishing DPN patterns from artifacts that might result from mechanical contact of the tip, condensed water or residual solvent, would be to carry out the appropriate negative controls using either uncoated tips or tips dipped only in the solvent, in the absence of the desired ink molecules.

More often than not, it is necessary to confirm the chemical identity of the ink molecules on the surface by other analytical methods. For example, X-ray photoelectron spectroscopy (XPS) is often used to identify elements present in DPN patterns [17, 21, 25, 29, 31, 67–69]. Infrared spectroscopy [68] and mass spectrometry [67] are also powerful characterization tools that have the added advantage of providing structural information. Furthermore, AFM is a rather slow characterization method and, in the case of high-throughput, parallel DPN characterization can easily become the rate limiting step in the fabrication process. In that case, optical characterization is ideal, and this can be achieved by using DPN-generated thiol monolayers as a resist against chemical etching so that the patterns become visible under optical microscopy [48]. Another approach would be to use a fluorescently labeled ink, which not only enables rapid characterization but also provides some chemical information about the patterns [6, 28, 49, 70–72]. Optical characterization using scanning near-field optical microscopy (SNOM) detection of single molecules deposited by DPN suggests that the practical resolution limits of DPN may not be in the patterning, but rather in the ability to detect patterns with dimensions smaller than the AFM tip used to fabricate them [6]. Finally, it is often useful to characterize the AFM tips as well as the presence and distribution of the ink on the tips, which can typically be achieved using SEM or optical microscopy.

5.6
Applications Based on Materials Integration by DPN

Functional chemical patterns fabricated by DPN have been used for a wide variety of scientific applications, and it can be expected that industrial applications will follow. Even in the many published applications, including for example etch resists [73] or templates for selective deposition [74], when only a single ink molecule is patterned onto a single surface, DPN has several advantages over conventional direct-write lithographic techniques such as EBL. While the latter is able to provide competitive lateral resolution, it is severely limited in its throughput as well as its cost. Furthermore, in contrast to DPN, EBL involves removing material from the substrate, which requires an extra development step. One advantage of placing resists and template molecules directly onto the surface is that the remainder of the surface is left free of contaminants. That being said, the ability to generate multicomponent nanostructures opens entirely new possibilities that are inaccessible by any other method. Some of the more striking examples will be briefly described in the following sections, in order to provide an overview of the types of unique application made possible by DPN. Whilst the examples are categorized based on selective adsorption, combinatorial chemistry and biological arrays, these categories are by no means complete and a significant amount of overlap is apparent between them.

5.6.1
Selective Deposition

Although the selective deposition of materials onto patterned surfaces is not limited to DPN patterns, the rapid prototyping capabilities and ability for DPN to generate multicomponent nanostructures on a small scale adds a new dimension to the field. A few of the strategies that can be used to immobilize different particles from solution by selective adsorption or templating are shown in Figure 5.11. In addition to the adsorption strategies shown, covalent binding, nonpolar adhesion forces and entropic effects can also be used to direct binding towards desired areas of the substrate. The surface passivation of the background is often a crucial step in fabricating templates for selective adsorption. Most often, successful selective adsorption will involve combinations of more than one of these strategies.

As a first example of electrostatic templating on DPN patterns, positively charged colloidal particles were immobilized electrostatically onto negatively charged MHA patterns. By fabricating a variety of MHA dot array patterns with different dot sizes and spacings, it was possible to screen the pattern dimensions for those capable of

Figure 5.11 Schematic representation of different strategies for selective adsorption or templating. Left to right: charge-based recognition; macromolecular encoding (i.e. DNA); and specific binding of ligands. (Image courtesy of NanoInk.)

organizing the particles such that each dot had exactly one particle bound, in a combinatorial fashion [74]. Such an approach was later applied to the fabrication of arrays of individual bacterial cells [75]. In another approach, the positive and negatively charged polyelectrolytes poly(diallyldimethylammonium) chloride (PDDA) and poly(styrenesulfonate) (PSS) have been directly patterned on silicon surfaces by DPN. Upon the addition of a complementary polyelectrolyte, selective adsorption was observed which suggested a compatibility of the method with layer-by-layer assembly [25]. Furthermore, such electrostatic templates have been used in combination with *molecular combing* to organize *aligned DNA strands*, a biological molecule which also falls into the polyelectrolyte category and can readily be adsorbed electrostatically [76].

DNA-directed self-assembly on DPN patterns has been used to organize two different-sized nanoparticles into nanoarrays, with the spacing between different-sized particles in an array being as low as 500 nm [74]. This method of self-sorting was improved by using noncomplementary DNA as a *passivation layer*, enabling larger particles to be immobilized [87]. Protein nanoarrays selectively bound to MHA patterns [77], or patterned by direct write methods [52], have been used to subsequently immobilize other protein molecules by molecular recognition. Covalent linking of the biofunctional group biotin was carried out selectively on DPN patterns, enabling the binding of streptavidin protein by molecular recognition and subsequent binding of biotinylated materials [72]. Similarly, the covalent coupling of proteins to DPN templates was carried out using succinimide chemistry [78]. Finally, nonpolar carbon nanotubes were selectively adsorbed to DPN templates based on differences in surface energy between the patterned and passivated regions [79].

Although not often emphasized in the literature, *passivation* is a crucial step in selective adsorption, such that the materials to be integrated on the surface do not simply bind everywhere. In a DPN experiment, the background is typically blocked by a low-surface energy (methyl-terminated) compound such as ODT [74] or, in the case of protein or cells, the adsorption of PEG-terminated passivation layers is preferred due to their nondenaturing character [52]. Nonspecific adsorption is in fact a major problem that limits the applicability of templating-based fabrication methods, although it can be overcome to some degree through chemical modification of the surface and optimization of the solution conditions (e.g. pH and ionic strength). Nonetheless, the fact that nonspecific adsorption cannot be completely eliminated is a qualitative advantage that constructive, parallel DPN has over serial methods based on sequential selective adsorption and patterning steps.

5.6.2
Combinatorial Chemistry

The idea of placing different chemical compounds on the same surface, for exposure to identical solution conditions, lends itself well to combinatorial chemistry. In addition to the possibility of ultra-high-density chemical arrays, the nanoscale resolution of DPN also enables studies of collective molecular interactions, as well as how the properties of nanoscale aggregates might differ from bulk behavior. For instance, the ability for nanopatterned SAMs to function as resists against the chemical etching of metallic films has been investigated combinatorially as a function of pattern dimension in order to minimize feature sizes [73]. Solid-state nanostructures with features as low as 15 nm have been fabricated by the direct deposition of etchant [67], while nanostructures of various metals such as gold, silver and palladium have been generated with 35 nm dot diameters and 53 nm line widths [80, 81].

As a first demonstration of the ability to screen the chemical behavior of different compounds on the same surface, four different thiol molecules were patterned within an area of 5 µm^2 on a single gold surface to form combinatorial libraries, as shown in Figure 5.12. The libraries were then used to study molecular

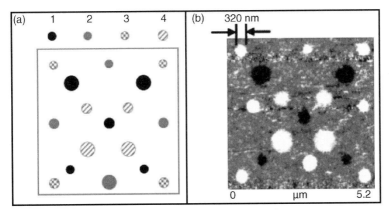

Figure 5.12 (a) Schematic representation of the combinatorial library design consisting of the four different molecular inks; (b) Lateral force microscopy image of the library described in panel (a). (Reproduced with permission from Ref. [82].)

displacement from the surface by repeatedly scanning the same area with an ink-coated tip and observing the order in which the spots disappeared as they were exchanged by new molecules from the tip [82]. In another creative approach to combinatorial chemistry, DPN was used to pattern molecules on cantilevers which could, in principle, be used as force sensors for determining interactions between molecular libraries [83].

Another unique application for DPN is in the study of microscale and nanoscale phase separation. For instance, phase separation and pattern formation in conjugated polymers spin-coated onto combinatorially nanostructured DPN templates was studied as a function of polymer concentration and MHA dot diameter [84]. The information resulting from those screenings of pattern dimensions enables one to control pattern formation in thin organic films, with potential applications in the fabrication of organic electronic and optical devices. Furthermore, by coating the tip with a mixture of two (or more) inks it becomes possible to observe if, and how, the molecules de-mix during the DPN writing process by using *in situ* AFM measurements [85]. The use of such separated phases for selective adsorption provided the ability to reduce line widths down to 10 nm.

5.6.3
Biological Arrays

The complex integration of biological molecules such as DNA, protein and phospholipids on hierarchical scales, ranging from molecular dimensions up to the size of entire organisms, provides the basis for the molecular machinery that makes life possible. The ability to understand, control and even mimic such interactions has long been a dream, not only for biologists but also for nanoscientists in a variety of disciplines. Robotic spotting methods have already proven valuable in the biotechnology industry for the fabrication of biological arrays, even with spot sizes on the order of hundreds of micrometers. Smaller spots not only have evolutionary advantages of higher spot density, sensitivity and lower requirements for sample volumes, but also open entirely new possibilities. For instance, the ability to integrate more than one biomolecule or biological entity (e.g. a virus) under the surface area covered by a single adherent cell in a spatially defined manner is especially exciting for unraveling the roles of intermolecular interactions in biological systems.

The patterning of DNA by DPN was first achieved by the selective deposition of oligonucleotides onto thiol-patterned surfaces [86]. Later, reproducible protocols for direct deposition from the AFM tip were developed [22, 87]. Figure 5.13 shows an example of a binary DNA array that was used to reversibly hybridize both fluorescently labeled oligonucleotides as well as gold nanoparticles of two different diameters [22]. This illustrates the ability for DNA arrays to organize nanomaterials in a highly specific manner. Furthermore, whilst fluorescence is an invaluable tool for characterizing biomolecular arrays, many biological materials are not easily obtainable in functional form with fluorescent labels. Therefore, label-free detection methods such as AFM topography show great promise [71]. The physical–chemical

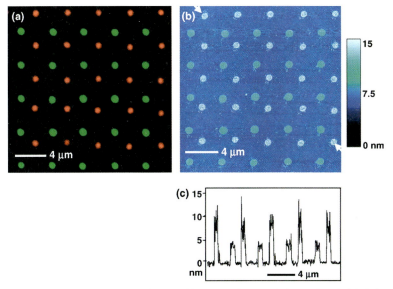

Figure 5.13 A binary DNA array fabricated by direct-write DPN. (a) Fluorescence image showing two different fluorescent probes selectively hybridized to the different spots; (b) AFM topography of the same area after dehybridization of the fluorescent probes and attachment of oligonucleotide-labeled nanoparticles of two different diameters (5 nm and 13 nm). (Reproduced from Ref. [22] with permission from the American Association for the Advancement of Science.)

similarity between different DNA strands of the same sequence renders them promising for parallel arraying, as DNA strands of different sequence can be patterned under the same environmental conditions.

A variety of different proteins have also been patterned by DPN, both by selective adsorption and by direct-write processes. In the majority of cases, selective adsorption using various coupling strategies appears to be most successful in fabricating functional protein nanoarrays to date, as selective adsorption can be carried out without dehydration of fragile proteins [70, 72, 77, 78, 88, 89]. For example, an antibody nanoarray-based detection assay for HIV in patients' plasma exceeded the limit of detection of conventional enzyme-linked immunosorbent assay (ELISA)-based immunoassays (5 pg ml^{-1} plasma) by more than 1000-fold [90]. Direct-write approaches hold the promise of larger-scale integration, and have also proven their ability to successfully generate functional protein nanoarrays. For example, collagen was the first protein to be directly deposited onto gold by first reducing disulfide bridges in the biopolymer, and then allowing the collagen fibrils to reassemble on the gold surfaces. Functional his-tagged proteins have been deposited directly onto nickel oxide surfaces via metal affinity [91]. Unmodified antibodies were directly patterned onto gold surfaces by nonspecific adsorption [52].

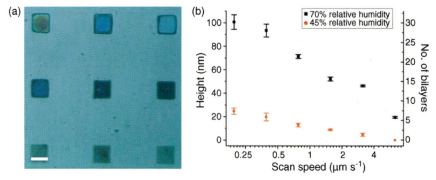

Figure 5.14 Phospholipid DPN patterns with control of multibilayer stacking. (a) Reflection-mode optical micrograph of phospholipid squares patterned on plasma-oxidized silicon at various speeds (scale bar = 5 μm); (b) The height of phospholipid multilayers (and corresponding number of bilayer stacks) measured by AFM is plotted as a function of scan speed (on a logarithmic scale) at two different relative humidities. (Reproduced with permission from Ref. [49]; © Wiley-VCH Verlag GmbH & Co. KGaA.)

In addition to DNA and protein, phospholipids represent another ubiquitous biomolecule that can be patterned noncovalently by DPN on a variety of surfaces, with linewidths as small as 100 nm [49]. In contrast to the transport behavior of most other inks – where the dot and linewidth can be controlled by the tip-contact time and scan speed, as well as humidity – phospholipid inks tend to stack into multilayer structures where the thickness of the film can be controlled by those same parameters (Figure 5.14). Upon immersion into aqueous solution, the multilayers can be spread on the surface to form a single monolayer, a lipid bilayer membrane, or remain as stable multilayers, depending on the substrate. A unique property of phospholipids is that they amphiphilic and are lyotropic liquid crystals that tend to organize into different supramolecular structures, where their fluidity depends on the hydration. Their transport rate from ink wells onto the tips, as well as from the tip to the substrate, can therefore be precisely controlled.

If one views viruses as nanoparticles, the methods described earlier for templating can be applied directly to the immobilization of viruses. For example, arrays of virus particles have been generated using genetically modified cow pea mosaic virus that covalently couple to maleimides patterned by DPN [32, 92]. Another approach based on several steps of selective adsorption to the DPN pattern of MHA has resulted in arrays of influenza virus particles [70] and of single tobacco mosaic virus particles [93]. In this process, Zn^{2+} are first bound to the MHA nanopatterns, after which antibodies specific for the virus in question are selectively adsorbed; finally, the virus particles are adsorbed on top of those two layers. Remarkably, it was shown that this method could be used to position functional viruses capable of infecting cells cultured on the arrays, as illustrated in Figure 5.15. The use of DPN-based methods to organize biological molecules and particles on a subcellular level, and to interface them with living systems, opens numerous possibilities in the emerging field of nanobiology.

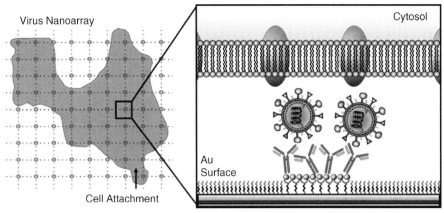

Figure 5.15 Schematic illustration of experiments where cells were cultured on and infected by DPN-patterned virus arrays. (Reproduced with permission from Ref. [94]; © Wiley-VCH Verlag GmbH & Co. KGaA.)

5.7
Conclusions

In conclusion, DPN provides the ability to simultaneously integrate and nanostructure a diverse range of materials on a variety of surfaces, at unprecedented levels of both spatial resolution and complexity. In principle, several thousand different materials could be integrated in thousands of different combinations with nanoscale resolution over square-centimeter (or larger) surface areas. Although much remains to be done before this dream is achieved, the barriers appear to be surmountable. Clearly, a basic understanding of the mechanisms behind the nanoscale transport of ink will be necessary in order to reproducibly carry out and extend the capabilities of DPN. In addition, innovative probe designs and inking strategies will be essential in order to expand the capability of DPN for large-scale materials integration.

References

1 Piner, R.D., Zhu, J., Xu, F., Hong, S.H. and Mirkin, C.A. (1999) *Science*, **283**, 661.
2 Ginger, D.S., Zhang, H. and Mirkin, C.A. (2004) *Angewandte Chemie - International Edition*, **43**, 30.
3 Salaita, K., Wang, Y.H. and Mirkin, C.A. (2007) *Nature Nanotechnology*, **2**, 145.
4 Haaheim, J. and Nafday, O.A. (2008) *Scanning*, **30**, 137.
5 Hong, S.H., Zhu, J. and Mirkin, C.A. (1999) *Science*, **286**, 523.
6 Noy, A., Miller, A.E., Klare, J.E., Weeks, B.L., Woods, B.W. and DeYoreo, J.J. (2002) *Nano Letters*, **2**, 109.
7 Hampton, J.R., Dameron, A.A. and Weiss, P.S. (2005) *The Journal of Physical Chemistry B*, **109**, 23118.

8 Haaheim, J., Eby, R., Nelson, M., Fragala, J., Rosner, B., Zhang, H. and Athas, G. (2005) *Ultramicroscopy*, **103**, 117.

9 Sheehan, P.E. and Whitman, L.J. (2002) *Physical Review Letters*, **88**, 156104.

10 Rozhok, S., Piner, R. and Mirkin, C.A. (2003) *The Journal of Physical Chemistry B*, **107**, 751.

11 Ahn, Y., Hong, S. and Jang, J. (2006) *The Journal of Physical Chemistry B*, **110**, 4270.

12 Manandhar, P., Jang, J., Schatz, G.C., Ratner, M.A. and Hong, S. (2003) *Physical Review Letters*, **90**, 4115505

13 Lee, N.K. and Hong, S.H. (2006) *Journal of Chemical Physics*, **124**, 11471

14 Jang, J.Y., Hong, S.H., Schatz, G.C. and Ratner, M.A. (2001) *Journal of Chemical Physics*, **115**, 2721.

15 Nafday, O.A., Vaughn, M.W. and Weeks, B.L. (2006) *Journal of Chemical Physics*, **125**, 144703

16 Kooi, S.E., Baker, L.A., Sheehan, P.E. and Whitman, L.J. (2004) *Advanced Materials*, **16**, 1013.

17 Sheu, J.T., Wu, C.H. and Chao, T.S. (2006) *Japanese Journal of Applied Physics Part 1 - Regular Papers Short Notes and Review Papers*, **45**, 3693.

18 Jung, H., Kulkarni, R. and Collier, C.P. (2003) *Journal of the American Chemical Society*, **125**, 12096.

19 Ivanisevic, A. and Mirkin, C.A. (2001) *Journal of the American Chemical Society*, **123**, 7887.

20 Salazar, R.B., Shovsky, A., Schonherr, H. and Vancso, G.J. (2006) *Small*, **2**, 1274.

21 Cho, Y. and Ivanisevic, A. (2004) *The Journal of Physical Chemistry B*, **108**, 15223.

22 Demers, L.M., Ginger, D.S., Park, S.J., Li, Z., Chung, S.W. and Mirkin, C.A. (2002) *Science*, **296**, 1836.

23 Jaschke, M. and Butt, H.J. (1995) *Langmuir*, **11**, 1061.

24 Lim, J.H. and Mirkin, C.A. (2002) *Advanced Materials*, **14**, 1474.

25 Yu, M., Nyamjav, D. and Ivanisevic, A. (2005) *Journal of Materials Chemistry*, **15**, 649.

26 Su, M., Liu, X.G., Li, S.Y., Dravid, V.P. and Mirkin, C.A. (2002) *Journal of the American Chemical Society*, **124**, 1560.

27 Fu, L., Liu, X.G., Zhang, Y., Dravid, V.P. and Mirkin, C.A. (2003) *Nano Letters*, **3**, 757.

28 Su, M. and Dravid, V.P. (2002) *Applied Physics Letters*, **80**, 4434.

29 Gundiah, G., John, N.S., Thomas, P.J., Kulkarni, G.U., Rao, C.N.R. and Heun, S. (2004) *Applied Physics Letters*, **84**, 5341.

30 Wang, Y., Zhang, Y., Li, B., Lu, J.H. and Hu, J. (2007) *Applied Physics Letters*, **90**, 133102.

31 Ding, L., Li, Y., Chu, H.B., Li, X.M. and Liu, J. (2005) *The Journal of Physical Chemistry B*, **109**, 22337.

32 Smith, J.C., Lee, K.B., Wang, Q., Finn, M.G., Johnson, J.E., Mrksich, M. and Mirkin, C.A. (2003) *Nano Letters*, **3**, 883.

33 Jung, H., Dalal, C.K., Kuntz, S., Shah, R. and Collier, C.P. (2004) *Nano Letters*, **4**, 2171.

34 John, N.S. and Kulkarni, G.U. (2007) *Journal of Nanoscience and Nanotechnology*, **7**, 977.

35 Su, M., Pan, Z.X., Dravid, V.P. and Thundat, T. (2005) *Langmuir*, **21**, 10902.

36 Jang, J.Y., Schatz, G.C. and Ratner, M.A. (2002) *Journal of Chemical Physics*, **116**, 3875.

37 Weeks, B.L., Vaughn, M.W. and DeYoreo, J.J. (2005) *Langmuir*, **21**, 8096.

38 Weeks, B.L. and DeYoreo, J.J. (2006) *The Journal of Physical Chemistry B*, **110**, 10231.

39 Sheehan, P.E., Whitman, L.J., King, W.P. and Nelson, B.A. (2004) *Applied Physics Letters*, **85**, 1589.

40 Yang, M., Sheehan, P.E., King, W.P. and Whitman, L.J. (2006) *Journal of the American Chemical Society*, **128**, 6774.

41 Huang, L., Chang, Y.H., Kakkassery, J.J. and Mirkin, C.A. (2006) *The Journal of Physical Chemistry B*, **110**, 20756.

42 Nelson, B.A., King, W.P., Laracuente, A.R., Sheehan, P.E. and Whitman, L.J. (2006) *Applied Physics Letters*, **88**, 033104.

43 Dong, L.X., Tao, X.Y., Zhang, L., Zhang, X.B. and Nelson, B.J. (2007) *Nano Letters*, **7**, 58.

44 Li, Y., Maynor, B.W. and Liu, J. (2001) *Journal of the American Chemical Society*, **123**, 2105.
45 Cai, Y.G. and Ocko, B.M. (2005) *Journal of the American Chemical Society*, **127**, 16287.
46 Zhang, M., Bullen, D., Chung, S.W., Hong, S., Ryu, K.S., Fan, Z.F., Mirkin, C.A. and Liu, C. (2002) *Nanotechnology*, **13**, 212.
47 Salaita, K., Lee, S.W., Wang, X.F., Huang, L., Dellinger, T.M., Liu, C. and Mirkin, C.A. (2005) *Small*, **1**, 940.
48 Salaita, K., Wang, Y.H., Fragala, J., Vega, R.A., Liu, C. and Mirkin, C.A. (2006) *Angewandte Chemie - International Edition*, **45**, 7220.
49 Lenhert, S., Sun, P., Wang, Y.H., Fuchs, H. and Mirkin, C.A. (2007) *Small*, **3**, 71.
50 Bullen, D., Chung, S.W., Wang, X.F., Zou, J., Mirkin, C.A. and Liu, C. (2004) *Applied Physics Letters*, **84**, 789.
51 Bullen, D. and Liu, C. (2006) *Sensors and Actuators A - Physical*, **125**, 504.
52 Lee, K.B., Lim, J.H. and Mirkin, C.A. (2003) *Journal of the American Chemical Society*, **125**, 5588.
53 Ryu, K.S., Wang, X.F., Shaikh, K., Bullen, D., Goluch, E., Zou, J., Liu, C. and Mirkin, C.A. (2004) *Applied Physics Letters*, **85**, 136.
54 Banerjee, D., Amro, N.A., Disawal, S. and Fragala, J. (2005) *Journal of Microlithography, Microfabrication and Microsystems*, **4**, 230.
55 Li, S.F., Shaikh, K.A., Szegedi, S., Goluch, E. and Liu, C. (2006) *Applied Physics Letters*, **89**, 173125.
56 Kim, K.H., Moldovan, N. and Espinosa, H.D. (2005) *Small*, **1**, 632.
57 Moldovan, N., Kim, K.H. and Espinosa, H.D. (2006) *Journal of Micromechanics and Microengineering*, **16**, 1935.
58 Lewis, A., Kheifetz, Y., Shambrodt, E., Radko, A., Khatchatryan, E. and Sukenik, C. (1999) *Applied Physics Letters*, **75**, 2689.
59 Taha, H., Marks, R.S., Gheber, L.A., Rousso, I., Newman, J., Sukenik, C. and Lewis, A. (2003) *Applied Physics Letters*, **83**, 1041.
60 Bruckbauer, A., Ying, L.M., Rothery, A.M., Zhou, D.J., Shevchuk, A.I., Abell, C., Korchev, Y.E. and Klenerman, D. (2002) *Journal of the American Chemical Society*, **124**, 8810.
61 Bruckbauer, A., Zhou, D.J., Ying, L.M., Korchev, Y.E., Abell, C. and Klenerman, D. (2003) *Journal of the American Chemical Society*, **125**, 9834.
62 Suryavanshi, A.P. and Yu, M.F. (2007) *Nanotechnology*, **18**, 105305.
63 Hong, S.H., Zhu, J. and Mirkin, C.A. (1999) *Langmuir*, **15**, 7897.
64 Agarwal, G., Sowards, L.A., Naik, R.R. and Stone, M.O. (2003) *Journal of the American Chemical Society*, **125**, 580.
65 Liu, X.G., Zhang, Y., Goswami, D.K., Okasinski, J.S., Salaita, K., Sun, P., Bedzyk, M.J. and Mirkin, C.A. (2005) *Science*, **307**, 1763.
66 Jiang, H.Z. and Stupp, S.I. (2005) *Langmuir*, **21**, 5242.
67 Zheng, Z.K., Yang, M.L., Liu, Y.Q. and Zhang, B.L. (2006) *Nanotechnology*, **17**, 5378.
68 Cho, Y. and Ivanisevic, A. (2006) *Langmuir*, **22**, 8670.
69 Cho, Y. and Ivanisevic, A. (2005) *The Journal of Physical Chemistry B*, **109**, 6225.
70 Vega, R.A., Maspoch, D., Shen, C.K.F., Kakkassery, J.J., Chen, B.J., Lamb, R.A. and Mirkin, C.A. (2006) *Chembiochem: A European Journal of Chemical Biology*, **7**, 1653.
71 Lynch, M., Mosher, C., Huff, J., Nettikadan, S., Johnson, J. and Henderson, E. (2004) *Proteomics*, **4**, 1695.
72 Hyun, J., Ahn, S.J., Lee, W.K., Chilkoti, A. and Zauscher, S. (2002) *Nano Letters*, **2**, 1203.
73 Weinberger, D.A., Hong, S.G., Mirkin, C.A., Wessels, B.W. and Higgins, T.B. (2000) *Advanced Materials*, **12**, 1600.
74 Demers, L.M. and Mirkin, C.A. (2001) *Angewandte Chemie - International Edition*, **40**, 3069.
75 Rozhok, S., Shen, C.K.F., Littler, P.L.H., Fan, Z.F., Liu, C., Mirkin, C.A. and Holz, R.C. (2005) *Small*, **1**, 445.

76 Nyamjav, D. and Ivanisevic, A. (2003) *Advanced Materials*, **15**, 1805.

77 Lee, K.B., Park, S.J., Mirkin, C.A., Smith, J.C. and Mrksich, M. (2002) *Science*, **295**, 1702.

78 Lee, S.W., Oh, B.K., Sanedrin, R.G., Salaita, K., Fujigaya, T. and Mirkin, C.A. (2006) *Advanced Materials*, **18**, 1133.

79 Wang, Y.H., Maspoch, D., Zou, S.L., Schatz, G.C., Smalley, R.E. and Mirkin, C.A. (2006) *Proceedings of the National Academy of Sciences of the United States of America*, **103**, 2026.

80 Zhang, H., Amro, N.A., Disawal, S., Elghanian, R., Shile, R. and Fragala, J. (2007) *Small*, **3**, 81.

81 Zhang, H. and Mirkin, C.A. (2004) *Chemistry of Materials*, **16**, 1480.

82 Ivanisevic, A., McCumber, K.V. and Mirkin, C.A. (2002) *Journal of the American Chemical Society*, **124**, 11997.

83 Wu, S.Y., Berkenbosch, R., Lui, A. and Green, J.B.D. (2006) *Analyst*, **131**, 1213.

84 Coffey, D.C. and Ginger, D.S. (2005) *Journal of the American Chemical Society*, **127**, 4564.

85 Salaita, K., Amarnath, A., Maspoch, D., Higgins, T.B. and Mirkin, C.A. (2005) *Journal of the American Chemical Society*, **127**, 11283.

86 Demers, L.M., Park, S.J., Taton, T.A., Li, Z. and Mirkin, C.A. (2001) *Angewandte Chemie - International Edition*, **40**, 3071.

87 Plutowski, U., Jester, S.S., Lenhert, S., Kappes, M.M. and Richert, C. (2007) *Advanced Materials*, **19**, 1951.

88 Kwak, S.K., Lee, G.S., Ahn, D.J. and Choi, J.W. (2004) *Materials Science and Engineering C - Biomimetic Materials Sensors and Systems*, **24**, 151.

89 Valiokas, R., Vaitekonis, A., Klenkar, G., Trinkunas, G. and Liedberg, B. (2006) *Langmuir*, **22**, 3456.

90 Lee, K.B., Kim, E.Y., Mirkin, C.A. and Wolinsky, S.M. (2004) *Nano Letters*, **4**, 1869.

91 Nam, J.M., Han, S.W., Lee, K.B., Liu, X.G., Ratner, M.A. and Mirkin, C.A. (2004) *Angewandte Chemie - International Edition*, **43**, 1246.

92 Cheung, C.L., Camarero, J.A., Woods, B.W., Lin, T.W., Johnson, J.E. and De Yoreo, J.J. (2003) *Journal of the American Chemical Society*, **125**, 6848.

93 Vega, R.A., Maspoch, D., Salaita, K. and Mirkin, C.A. (2005) *Angewandte Chemie - International Edition*, **44**, 6013.

94 Vega, R.A., Shen, C.K.F., Maspoch, D., Robach, J.G., Lamb, R.A. and Mirkin, C.A. (2007) *Small*, **3**, 1482.

6
Scanning Ion Conductance Microscopy of Cellular and Artificial Membranes

Matthias Böcker, Harald Fuchs, and Tilman E. Schäffer

6.1
Introduction

Eukaryotic cells are enclosed by a plasma membrane, which creates an internal environment that is separated from the outside. The membrane defines a physical border and is impermeable to macromolecules. Integral proteins in the membrane play an essential role for inter- and transcellular processes [1, 2]. In order to understand the properties of membranes and membrane proteins, it is important that cell biology, medicine and pharmacology gain insight into the complex barrier-crossing transport mechanisms. In particular, knowledge concerning the permeability of barriers for substances such as drugs is of great relevance.

Special electrochemical and microscopic methods are used to study the ion-permeability of barrier-forming cell structures. For example, transepithelial electrical resistance (TER)-spectroscopy provides information about the barrier properties of cell layers [3–5]. The development of artificial membranes was an important step in the characterization of membranes [6, 7]. One advantage of artificial membranes is the possibility of inserting selected proteins into the membrane, which in turn creates the possibility of performing single-channel measurements on these selected proteins [8].

For the microscopic characterization of local sample properties, scanning probe microscopes have been developed in various forms. Scanning probe microscopes are based on a small, locally confined probe that is sensitive to various types of physical quantity. To date, several instruments have been developed for the characterization of different sample properties, although only a few are suitable for application in an aqueous environment – an essential requirement when analyzing biological samples under native conditions. Perhaps the most prominent member of the group is the atomic force microscope [9], which allows the creation of high-resolution topographical images of biological samples in buffer solutions [10]. The atomic force microscope employs the mechanical interaction between a sharp tip and the sample surface under investigation on the nanometer scale. In addition to topography, it is possible to measure mechanical sample properties such as elasticity [11]. Many analyses have already been conducted on cellular membranes [12, 13], artificial membranes such as solid supported membranes [14, 15] and membrane proteins [16–19].

Unfortunately, the investigation of soft and fragile samples often proves to be problematic due to mechanical interactions between the tip and the sample. In the case of cells, a force-induced deformation reduces the resolution of atomic force microscopy (AFM) imaging, and native conditions are therefore not always reproduced [20]. Additionally, the sample can be damaged irreversibly or be compressed, leading to errors, for example, in the measurement of sample height [21]. In particular, the investigation of free-standing, pore-suspending artificial membranes with AFM has proved difficult, with very few successful measurements on such membranes having been reported [22–25]. All such reports refer to the problem of interaction forces between the atomic force microscope tip and the suspended membrane, which often leads to rupture of the membrane, even at minimal imaging forces.

6.1.1
Scanning Ion Conductance Microscopy

Scanning ion conductance microscopy (SICM), as invented by Hansma et al. in 1989 [26, 27], is based on the measurement of an ion current through a small aperture, which is usually formed by a nanopipette. In order to provide a medium for ion conduction, the nanopipette is filled with an electrolyte. A silver/silver chloride (Ag/AgCl) electrode is placed inside the pipette (the pipette electrode), while the sample is placed in a dish that is filled typically with the same electrolyte. A second Ag/AgCl electrode is placed inside the electrolyte in the dish (the bath electrode). By applying a voltage between both electrodes and recording the ion current through the pipette (typically on the nanoampère scale), locally resolved images of sample topography can be generated when scanning the sample. The advantage of the SICM is that no mechanical forces are necessary for imaging a sample surface.

In SICM there is a heavy dependence of the ion current on the distance between the pipette tip and sample surface for generating the feedback signal for scanning. This distance dependence is based on a 'current squeezing effect', which allows the presence of the sample surface to be sensed at a distance where no mechanical interactions between tip and sample occur. In this way, soft and delicate samples such as living cells can be imaged in a 'noncontact' configuration [28, 29]. In addition, single ion channels on living cells can be recorded at specified positions with this technique [30]. Improved imaging techniques have been developed based on the principle of an ac measurement. For example, the tip–sample distance can be modulated, thereby modulating the measured ion current; the amplitude of the modulated current can then be used for feedback control [31–33]. The technique of SICM has proved useful in obtaining high-resolution images of fine surface structures such as microvilli [34, 35] or membrane proteins [36].

For many applications, however, it is important to measure the ion current independently of the sample topography, and for this purpose several different extensions to SICM have been developed. In one revision, the scanning ion conductance microscope was combined with an atomic force microscope [37, 38]. For this, a bent nanopipette [39, 40], coated with a reflective metal layer, was used as the atomic force microscope 'tip' and scanned over the sample surface. A laser beam was then focused onto the bent pipette and the reflected light projected onto a split photo-diode. The mechanical deflection of the pipette provided the feedback signal for topography imaging, while the ion current was recorded simultaneously. As an option, the bent pipette was driven in the tapping-mode [41, 42], which simplified the imaging of soft samples in liquid solutions.

Another modification involved the combination of SICM with scanning near-field optical microscopy (SNOM). This combined microscope allowed living cells to be investigated and topography images with additional optical information to be recorded [43–45]. Yet another extension of SICM was the combination with shear force microscopy [46, 47], a technique that is also used in conjunction with the scanning near-field optical microscope [48, 49]. In SICM with complementary shear force distance control, the pipette is transversally oscillated at a mechanical resonant frequency. The sample topography is then detected by shear forces between the pipette tip and sample surface, causing a reduction in the oscillation amplitude.

6.2
Methods

6.2.1
The Basic Set-Up

In a basic SICM set-up a nanopipette with a small tip opening diameter is positioned close to a sample surface (Figure 6.1a). The nanopipette is filled with an electrolyte and the sample is placed in an electrolyte-filled dish. Typically, the electrolytes in the nanopipette and in the dish are identical, so that no osmotic flow in or out of the pipette occurs. For the current measurement a voltage is applied between two silver/silver chloride (Ag/AgCl) electrodes. Ag/AgCl electrodes have electrochemical properties that make them well suited to applications in SICM. For example, they have a very small equilibrium constant at room temperature so that only a small amount of Ag^+-ions exists in the electrolyte. Additionally, they are easily fabricated, for example by the electrolytic deposition of silver chloride on a silver wire. One of the electrodes is placed inside the pipette (the pipette electrode), while the other electrode is place inside the electrolyte in the dish (the bath electrode). The ion current from the pipette electrode through the pipette, and through its small tip opening to the bath electrode, is measured with a high-impedance current amplifier. The applied voltage is in the range of some hundred millivolts, and the measured ion current is in the nano- and picoampère range. The closer the tip is to the sample surface, the more the

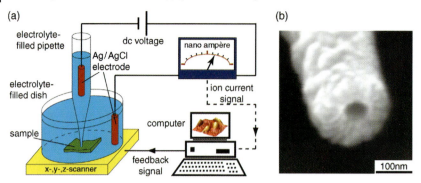

Figure 6.1 (a) Schematic of the scanning ion conductance microscope. A dc voltage between two Ag/AgCl electrodes induces an ion current through the pipette. The ion current signal is passed to a computer, which generates a feedback signal that controls the position of a z-scanner. Images are generated by scanning the sample in x,y-direction; (b) Scanning electron microscopy (SEM) image of the tip of a nanopipette. The inner opening diameter is 60 nm. For SEM imaging the pipette was sputter-coated with a 10 nm-thick aluminum layer.

ion current is restricted ('squeezed') by a narrow gap formed between the pipette tip and the sample surface [26, 50]. The measured ion current is therefore indicative of the tip–sample distance. The ion current signal is passed to a computer, where a feedback signal is generated and used to drive an x-, y-, z-scanner, consisting of piezoelectric actuators (abbreviated as piezos). Depending on the pipette geometry and on the imaging parameters used, topographical imaging without mechanical interaction between tip and sample is possible. Hence, noncontact images of sample topography can be recorded, which is especially valuable for the imaging of soft biological samples such as living cells and artificial membranes.

6.2.2
Nanopipettes

In order to deliver an ion current through a small opening, pipettes proved to be a practical solution. Pipette fabrication is a well-known technique, and pipettes have a wide range of applications in extracellular physiology, for example, in patch–clamp recording [51, 52]. Usually, pipettes are drawn (using a 'pipette puller') from glass capillaries with an initial diameter of 1–2 mm. The principle of a pipette puller is based on local heating of the capillary with a heated coil or a laser beam. While heating, a pulling force is applied to both ends of the capillary such that, when the melting point of the glass is reached, the pulling force leads to a separation of both ends of the capillary, thereby forming fine pipettes. The most commonly used glass is borosilicate; this softens at 825 °C and, as it is pulled, maintains its ratio of inner to outer diameter over the total drawn-out length [53]. A wide variety of opening diameters can be generated with borosilicate glass pipettes, down to some tens of nanometers (Figure 6.1b). Finer pipettes for higher-resolution imaging can be produced from quartz capillaries, with inner tip opening diameters of approximately 12 nm [36].

6.3
Description of Current–Distance Behavior

When placing the electrolyte-filled nanopipette with the pipette electrode into an electrolyte-filled dish with the bath electrode and applying a voltage U between the electrodes, an ion current is measured. When the tip of the pipette is far away from the sample surface the current is maximal (saturation current I_{sat}):

$$I_{sat} = \frac{U}{R_p}. \qquad (6.1)$$

For calculating the pipette resistance R_p a simple analytical model can be used [46]. The geometry of the pipette is assumed to be conical (Figure 6.2a). The resistance can be approximated through sectioning the cone into successive disks of infinitesimal thicknesses, leading to

$$R_p = \frac{1}{\kappa} \cdot \frac{L_p}{\pi r_p r_i} \qquad (6.2)$$

where κ is the specific conductivity of the electrolyte, L_p is the length of the drawn-out end of the pipette, r_p is the inner radius of the capillary, and r_i is the inner diameter of the tip opening. For typical borosilicate glass pipettes and electrolytes, R_p is in the range of 50–200 MΩ. The resistance of the electrolyte bath inside the dish is on the order of kilo-ohms, so that the total resistance of the circuit can be approximated by R_p. With this, the inner tip opening diameter can be estimated as

$$r_i \approx \frac{I_{sat}}{U} \cdot \frac{L_p}{\pi \kappa r_p}. \qquad (6.3)$$

For standard 1 mm outer diameter borosilicate glass pipettes, for example, $r_p = 0.29$ mm. In many cases, a phosphate-buffered saline (PBS) solution with 137 mM NaCl ($\kappa \approx 1.3$ Ωm) is used as electrolyte. L_p is typically in the range of 5 mm, and $U = 200$ mV. For example, from a measured current $I_{sat} = 1$ nA, an inner tip opening radius of $r_i \approx 21$ nm is deduced.

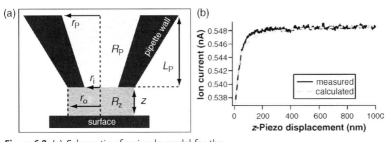

Figure 6.2 (a) Schematic of a simple model for the current–distance behavior of a cone-shaped pipette; (b) Measured and calculated ion current versus z-piezo displacement. Measurement were made with a borosilicate glass pipette over a mica surface, applying a voltage of $U = 200$ mV.

When the tip is at a distance to the sample surface that is comparable to the tip opening diameter, the measured ion current shows a heavy dependence on the tip–sample distance. This is because the current then has to 'squeeze' through the narrow gap between the pipette wall and the sample surface, which leads to an increase in the resistance. Nitz et al. [46] constructed an analytical model for the distance-dependent resistance R_z, obtaining

$$R_z(z) \approx \frac{3}{2\pi} \cdot \frac{\ln(r_o/r_i)}{\kappa z}, \tag{6.4}$$

where r_o is the outer tip diameter of the pipette and z is the tip–sample distance. This leads to a distance-dependent ion current of

$$I(z) \approx I_{sat}\left(1 + \frac{z_0}{z}\right)^{-1}, \tag{6.5}$$

with

$$z_0 = \frac{3}{2} \cdot \frac{r_p r_i \ln(r_o/r_i)}{L_P}. \tag{6.6}$$

The ratio of r_i to r_o usually stays constant while pulling the pipette from the borosilicate capillary [53]. In Figure 6.2b, a measurement of ion current versus z-piezo extension (the tip–sample distance with an unknown offset) is displayed (black trace). The range of the z-piezo extension is 1 μm. At a large tip–sample distance (right-hand side in Figure 6.2b), the ion current is independent of distance, yielding a saturation current of $I_{sat} = 0.549$ nA. By using Equation 6.3, r_i can be approximated as 12 nm. The calculated ion current $I(z)$ obtained from the analytical model (Equation 6.5) is also displayed (gray, dashed trace); a horizontal offset was applied here to optimize the match. The analytical model matches the measured data well. From the offsetting procedure, a tip–sample distance of 18 nm is found at the leftmost position in the graph (at a z-piezo displacement of 0 nm).

6.4
Imaging with SICM

6.4.1
Modulated Scan Technique

The dependence of the measured ion current on tip–sample distance can be utilized in a feedback loop to keep the tip–surface distance constant while scanning, thus allowing topographical images of a sample surface to be recorded. In practice, however, this imaging mode has proved vulnerable to current drift. One method of reducing the influence of current drift is to apply short voltage or current pulses [54, 55] instead of a constant voltage between the pipette and bath electrode.

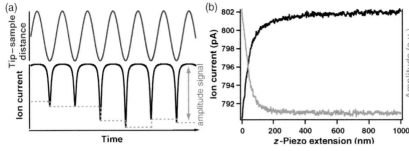

Figure 6.3 (a) Schematic diagram to illustrate the principle of the modulated scan technique. Upper trace: sinusoidal modulation of the tip–sample distance. Lower trace: the resulting ion current. The current decreases periodically at the points of closest approach between pipette tip and sample. The value of the maximal current drop provides the amplitude signal which is used for feedback; (b) Measured ion current (black trace) and ion current amplitude (gray trace) versus z-piezo extension while modulating the tip–sample distance with an amplitude of 120 nm at a frequency of 800 Hz.

Another method is based on modulating the tip–sample distance [31–33]. This distance is modulated sinusoidally with an amplitude of up to some hundred nanometers and with a frequency of a several hundred Hertz (Figure 6.3a, upper graph). For large tip–sample distances, this modulation does not have any influence on the ion current. For small tip–sample distances, the behavior of the ion current is displayed in Figure 6.3a (lower graph). At distances furthest from the surface, the ion current reaches the saturation current (plateaus in the graph). When approaching the sample surface during the modulation cycle, the influence of the squeezing effect becomes greater so that the current decreases. At the point of closest approach the current is smallest, but increases again when the tip retracts from the surface. An amplitude signal can be generated either by using a lock-in amplifier or a trigger-based method [47]. The amplitude of the current signal is used as input to a feedback loop, which regulates the mean tip–sample distance to keep the amplitude signal constant.

Measurement of the ion current (black, left axis) and of the amplitude signal (gray, right axis) versus z-piezo extension is shown in Figure 6.3b. The ion current shows the distance-dependent behavior (as discussed above), while the amplitude signal is constant for large tip–sample distances (right-hand side in Figure 6.3b) and increases for decreasing tip–sample distances. The amplitude signal is less sensitive to current drift than the ion current, and may also help minimizing the risk of lateral forces being applied to the sample.

6.4.2
Cellular Membranes

Use of the modulated scan technique allows the gentle imaging of delicate biological samples such as living cells, and also allows the recording of well-resolved images of fine surface structures such as microvilli [34] or membrane proteins [36]. Living,

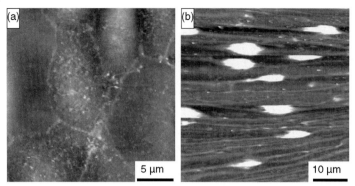

Figure 6.4 (a) Scanning ion conductance microscopy (SCIM) topographic image of living MDCK-II cells in phosphate-buffered saline (PBS) solution. The gray scale ranges from black to white over 1.8 μm; (b) SICM topographic image of living rat Schwann cells in PBS solution. The grayscale ranges from black to white over 5 μm.

confluent MDCK-II cells, for example, exhibit a large cell body and locally raised cell–cell contacts (Figure 6.4a); substructures on the cell membrane are also visible. Another example is the imaging of living Schwann cells (Figure 6.4b); these occur in the peripheral nervous system and are essential for generating the myelin sheath and maintaining the metabolism and integrity of the nerve. The SICM image shows the typical spindle-shaped body of the Schwann cell. On the basis of the excellent long-term stability provided by modulated scanning techniques it is possible to image specific surface areas, continuously, for several hours and thus to observe dynamic processes occurring on the cell membranes [35, 56]. As an example, Gorelik *et al.* were able to study the mechanism by which aldosterone activates sodium reabsorption via the epithelial sodium channel [57].

6.4.3
Artificial Membranes

The study of artificial membranes represents a new application area for SICM. The investigation of pore-suspending membranes with other scanning probe techniques such as AFM has proved difficult due to mechanical interactions that can cause damage to the membrane. In contrast, the noncontact imaging characteristic of SICM allows the mapping of such pore-suspending membranes without mechanical interaction between the pipette tip and the sample surface.

In a study of black lipid membranes (BLMs), highly ordered porous silicon proved to be a useful substrate for spanning the membrane over the pores, although for this purpose the substrate must first be functionalized. In the Müller–Rudin technique [6, 7], the membrane is applied in a solvent locally to the substrate; this causes it to spread over the surface, delivering a membrane monolayer (the spreading process is shown in Figure 6.5a). The scan was started directly after applying the membrane to the surface (scan direction: downwards). Initially, all of the pores were

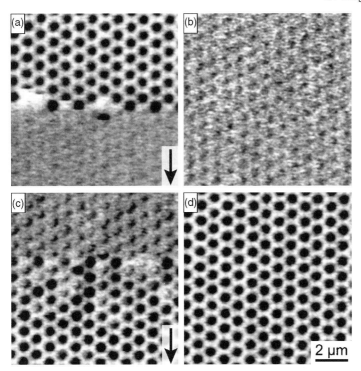

Figure 6.5 Scanning ion conductance microscopy topographic images of a highly ordered porous silicon substrate at different stages of coverage with a lipid membrane. (a) Topography imaged directly after applying the membrane solution to the silicon substrate (scan direction: downwards). In the lower section of the image, the membrane is already suspended over the pores; (b) After spreading over the surface the membrane is suspended over all pores and was stable for hours; (c) Applying a small amount of detergent (Tween 20) to the PBS solution destroyed the membrane (scan direction: downwards). In the upper section of the image, the membrane was still present, while in the lower section it was already destroyed; (d) Finally, the porous silicon substrate was totally free of membrane again. The scan rate was 5 s per line with a resolution of 256 × 256 data points. The grayscale ranges from black to white over 120 nm.

open (Figure 6.5a, upper half), but when reaching the vertical center of the image the first suspended pores became visible, with only suspended pores being observed in the lower section of the image. This effect was due to the solvent reaching the scan area while scanning, thus suspending the pores with membrane. The same area, after re-scanning, is shown in Figure 6.5b. Here, all the pores are suspended with membrane, although the porous structure of the substrate can still be recognized by the small depressions in the membrane surface. This state remained stable over several hours, after which time a droplet of a membrane-dissolving detergent was applied to the electrolyte (Figure 6.5c). Initially, the membrane remained unaffected (upper section), but after some minutes some pores began to re-open (middle section), and a few minutes later only open pores remained (lower section). A re-scan of the area showed only open pores (Figure 6.5d). Taken together, these measurements demonstrated the gentle imaging character of SICM.

6.4.4
SICM with Shear Force Distance Control

The fact that SICM can be used for imaging biological samples, with a resolution in the nanometer range, makes it an interesting candidate for combination with other scanning microscopy techniques. Of particular interest is the measurement of ion current independent of sample topography, and for this purpose the nanopipette can be used as a shear force sensor. Shear forces are well known from SNOM set-ups, where they serve to keep the optical fiber at a constant distance from the sample during scanning [48, 49]. In a shear force configuration, the probe is vertically oriented with respect to the sample surface, while a dither piezo excites a transverse mechanical vibration of the probe. The amplitude of the vibration depends heavily on the tip–sample distance: at small distances, the shear forces between the tip and sample reduce the vibration amplitude, which therefore can be used as a measure of tip–sample distance. To date, several methods have been established for detecting the vibration amplitude, including optical readout [48] and the use of a piezoelectric tuning fork sensor [58]. Although the latter method faced problems when adapted for use in liquids (due to electrical short circuits), several solutions were described, including a custom piezoelectric detection design [59], an electrically insulating layer [60] and a diving bell concept [61].

In a combined shear force and scanning ion conductance microscope, the nanopipette acts both as probe for the ion current and as probe for the shear force measurement. For the detection of vibration amplitude an optical readout was used [46], as this system functions equally well in liquid and in air [62]. An improved optical detection design was based on the use of a pair of periscopes (Figure 6.6) [47]. Here, a collimated laser beam from a laser diode passes down through one periscope tube. Inside the tube, the beam is reflected by a mirror and passes through a lens that provides the interface to the liquid and focuses the beam onto the thin end of the pipette. In the second periscope tube, a two-segment photo-diode detects the light scattered by the pipette, resulting in a vibration amplitude signal. With this periscope-based detection system the vibration amplitude can be detected close to the pipette tip, where the sensitivity is highest. Furthermore, the second tube can be positioned to detect either the transmitted or the reflected light from the pipette. The optical reflectivity of the pipette can be increased by coating with a thin metal layer. In order to induce pipette vibrations a dither piezo is used. By using the vibration amplitude as input to the feedback loop controlling the tip–sample distance, the sample topography can be imaged. The simultaneously recorded ion current then yields a complementary image of ion current at a constant tip–sample distance.

The combined scanning ion conductance and shear force microscope can be used for the investigation of local variations in ion conductance of biological specimens. This is of special interest for research in the field of the barrier-forming structures such as endothelial or epithelial cell layers. In multicellular organisms, these structures form the interface between different fluid compartments and play an important role in inter- and transcellular processes [1, 2]. Gaining insight into the complex barrier-crossing transport mechanisms is a common interest of cell biology,

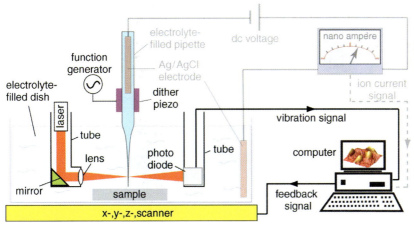

Figure 6.6 (a) Schematic of the combined ion current and shear force measurement set-up. The components for the ion current measurement are grayed out. The pipette is vibrated by a dither piezo parallel to the surface. A periscope design based on two tubes is used to measure the vibration amplitude optically. The first tube is used to focus a laser beam onto the pipette tip; the second tube contains a split photo-diode, which detects the modulated beam and thereby the vibration amplitude of the pipette. The vibration signal provides a complementary input to the computer.

medicine and pharmacology, as malfunctioning of these barriers may have pathological implications.

Special electrochemical and microscopic methods are required to study the ionpermeability of barrier-forming cell structures. For example, experimental techniques such as the measurement of TER may provide valuable information about the barrier properties of cell layers [63, 64]. In addition to such integrating measurements of the total cell layer impedance, the combined scanning ion conductance and shear force microscope can provide further insight into cellular transport mechanisms, as it allows the simultaneous recording of topographic data and local ion conductance with high lateral resolution. One such example is the investigation of the functionality of tight junctions between living MDCK-II cells (Figure 6.7). Here, the shear force image shows the topography of the cells, while the ion current image reveals lines of increased conductance at the position of the cell–cell contacts.

6.5
Outlook

The gentle character of SICM provides for a broad range of possible applications. Examples include the possible study of proteins in pore-suspending membranes, as well as combinations with other imaging techniques. By using the ion current signal of SICM to maintain a constant tip–sample distance, complementary signals such as

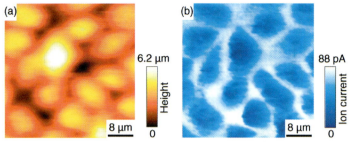

Figure 6.7 (a) Shear force topographic image of living MDCK-II cells; (b) Simultaneously recorded ion current image. The color shows the variation of the measured ion current while scanning.

local optical intensity can be recorded simultaneously. It has been shown that the end of a tapered nanopipette can serve as a near-field light source for SNOM [32, 43, 65], this being achieved by coupling a laser light into the nanopipette via an optical fiber. Coating the outside of the nanopipette with a reflective metal layer then helped to confine the laser light to the aperture (i.e. the tapered end of the nanopipette). Provided that the sample and substrate were transparent, the SNOM signal could then be collected through an objective and detected using a photomultiplier located beneath the SICM set-up. In this way, living cells have been successfully investigated using a combined SICM/SNOM set-up. An alternative use of the SICM probe as a confined light source for SNOM was suggested by Bruckbauer *et al.* [44, 45]. This method was based on the fluorescence that occurred when a calcium indicator (with which the nanopipette is filled) binds with calcium in the sample solution and is illuminated with laser light. The mixing zone where the fluorescent complex forms serves as a localized light source. Another viable combination was that of scanning confocal microscopy (SCM) [66], where the set-up comprised an inverted light microscope fully configured for SCM, on which the SICM set-up was placed. During lateral scanning the vertical position of the sample was controlled by SICM and, as a result, the optical confocal volume, which was located just beneath the end of the nanopipette, followed the topography of the sample. This allowed fluorescence images of a surface to be recorded simultaneously with topographic data.

Acknowledgments

The authors thank Boris Anczykowski, Yuri Korchev, Andrew Shevchuk, Roger Proksch, Eva Schmitt and Claudia Steinem for their stimulating discussions and support. The Schwann cells were a gift from Ilka Kleffner and Peter Young of the University Clinic Münster, while the MDCK-II cells were kindly provided by Joachim Wegener of the Institute of Biochemistry at the University of Münster. The authors are grateful to Asylum Research for support. In addition, the DFG is gratefully acknowledged for financial support (SCHA 1264/1 and STE 884/5). T.E.S. thanks the Gemeinnützige Hertie-Stiftung/Stifterverband für die Deutsche Wissenschaft for support.

References

1. Powell, D.W. (1981) *The American Journal of Physiology*, **241**, G275.
2. Simionescu, M. and Simionescu, N. (1986) *Annual Review of Physiology*, **48**, 279.
3. Cereijido, M., Gonzalez-Mariscal, L., Contreras, R.G., Gallardo, J.M., Garcia-Villegas, R. and Valdes, J. (1993) *Journal of Cell Science - Supplement*, **17**, 127.
4. Diamond, J.M. (1977) *The Physiologist*, **20**, 10.
5. Wegener, J., Sieber, M. and Galla, H.J. (1996) *Journal of Biochemical and Biophysical Methods*, **32**, 151.
6. Müller, P., Rudin, H.T., Tien, H.T. and Wescott, W.C. (1963) *The Journal of Physical Chemistry B*, **67**, 534.
7. Montal, M. and Müller, P. (1972) *Proceedings of the National Academy of Sciences of the United States of America*, **69**, 3561.
8. Schmitt, E.K., Vrouenraets, M. and Steinem, C. (2006) *Biophysical Journal*, **91**, 2163.
9. Binnig, G., Quate, C.F. and Gerber, C. (1986) *Physical Review Letters*, **56**, 930.
10. Drake, B., Prater, C.B., Weisenhorn, A.L., Gould, S.A., Albrecht, T.R., Quate, C.F., Cannell, D.S., Hansma, H.G. and Hansma, P.K. (1989) *Science*, **243**, 1586.
11. Radmacher, M., Fritz, M., Kacher, C.M., Cleveland, J.P. and Hansma, P.K. (1996) *Biophysical Journal*, **70**, 556.
12. Butt, H.J., Wolff, E.K., Gould, S.A., Dixon Northern, B., Peterson, C.M. and Hansma, P.K. (1990) *Journal of Structural Biology*, **105**, 54.
13. Yamashina, S. and Katsumata, O. (2000) *Journal of Electron Microscopy*, **49**, 445.
14. Mou, J., Yang, J. and Shao, Z. (1994) *Biochemistry*, **33**, 4439.
15. Hui, S.W., Viswanathan, R., Zasadzinski, J.A. and Israelachvili, J.N. (1995) *Biophysical Journal*, **68**, 171.
16. Butt, H.J., Downing, K.H. and Hansma, P.K. (1990) *Biophysical Journal*, **58**, 1473.
17. Hoh, J.H., Sosinsky, G.E., Revel, J.P. and Hansma, P.K. (1993) *Biophysical Journal*, **65**, 149.
18. Janshoff, A., Ross, M., Gerke, V. and Steinem, C. (2001) *ChemBioChem*, **2**, 587.
19. Mueller, H., Butt, H.-J. and Bamberg, E. (2000) *The Journal of Physical Chemistry B*, **104**, 4552.
20. Hansma, H.G. and Hoh, J.H. (1994) *Annual Review of Biophysics and Biomolecular Structure*, **23**, 115.
21. Jiao, Y. and Schäffer, T.E. (2004) *Langmuir*, **20**, 10038.
22. Hennesthal, C. and Steinem, C. (2000) *Journal of the American Chemical Society*, **122**, 8085.
23. Hennesthal, C., Drexler, J. and Steinem, C. (2002) *ChemPhysChem*, **3**, 885.
24. Goncalves, R.P., Agnus, G., Sens, P., Houssin, C., Bartenlian, B. and Scheuring, S. (2006) *Nature Methods*, **3**, 1007.
25. Ovalle-Garcia, E. and Ortega-Blake, I. (2007) *Applied Physics Letters*, **91**, 093901.
26. Hansma, P.K., Drake, B., Marti, O., Gould, S.A. and Prater, C.B. (1989) *Science*, **243**, 641.
27. Prater, C.B., Drake, B., Gould, S.A.C., Hansma, H.G. and Hansma, P.K. (1990) *Scanning*, **12**, 50.
28. Korchev, Y.E., Bashford, C.L., Milovanovic, M., Vodyanoy, I. and Lab, M.J. (1997) *Biophysical Journal*, **73**, 653.
29. Korchev, Y.E., Milovanovic, M., Bashford, C.L., Bennett, D.C., Sviderskaya, E.V., Vodyanoy, I. and Lab, M.J. (1997) *Journal of Microscopy*, **188**, 17.
30. Korchev, Y.E., Negulyaev, Y.A., Edwards, C.R., Vodyanoy, I. and Lab, M.J. (2000) *Nature Cell Biology*, **2**, 616.
31. Pastre, D., Iwamoto, H., Liu, J., Szabo, G. and Shao, Z. (2001) *Ultramicroscopy*, **90**, 13.
32. Mannelquist, A., Iwamoto, H., Szabo, G. and Shao, Z.F. (2001) *Applied Physics Letters*, **78**, 2076.
33. Shevchuk, A.I., Gorelik, J., Harding, S.E., Lab, M.J., Klenerman, D. and Korchev, Y.E. (2001) *Biophysical Journal*, **81**, 1759.
34. Gorelik, J., Gu, Y., Spohr, H.A., Shevchuk, A.I., Lab, M.J., Harding, S.E., Edwards, C.R., Whitaker, M., Moss, G.W., Benton,

D.C., Sanchez, D., Darszon, A., Vodyanoy, I., Klenerman, D. and Korchev, Y.E. (2002) *Biophysical Journal*, **83**, 3296.

35 Gorelik, J., Shevchuk, A.I., Frolenkov, G.I., Diakonov, I.A., Lab, M.J., Kros, C.J., Richardson, G.P., Vodyanoy, I., Edwards, C.R., Klenerman, D. and Korchev, Y.E. (2003) *Proceedings of the National Academy of Sciences of the United States of America*, **100**, 5819.

36 Shevchuk, A.I., Frolenkov, G.I., Sanchez, D., James, P.S., Freedman, N., Lab, M.J., Jones, R., Klenerman, D. and Korchev, Y.E. (2006) *Angewandte Chemie - International Edition*, **45**, 2212.

37 Proksch, R., Lal, R., Hansma, P.K., Morse, D. and Stucky, G. (1996) *Biophysical Journal*, **71**, 2155.

38 Schäffer, T.E., IonescuZanetti, C., Proksch, R., Fritz, M., Walters, D.A., Almqvist, N., Zaremba, C.M., Belcher, A.M., Smith, B.L., Stucky, G.D., Morse, D.E. and Hansma, P.K. (1997) *Chemistry of Materials*, **9**, 1731.

39 Shalom, S., Lieberman, K., Lewis, A. and Cohen, S.R. (1992) *Review of Scientific Instruments*, **63**, 4061.

40 Lewis, A., Taha, H., Strinkovski, A., Manevitch, A., Khatchatouriants, A., Dekhter, R. and Ammann, E. (2003) *Nature Biotechnology*, **21**, 1377.

41 Hansma, P.K., Cleveland, J.P., Radmacher, M., Walters, D.A., Hillner, P.E., Bezanilla, M., Fritz, M., Vie, D., Hansma, H.G., Prater, C.B., Massie, J., Fukunaga, L., Gurley, J. and Elings, V. (1994) *Applied Physics Letters*, **64**, 1738.

42 Putman, C.A.J., Werf, K.O.V.d., Grooth, B.G.D., Hulst, N.F.V. and Greve, J. (1994) *Applied Physics Letters*, **64**, 2454.

43 Korchev, Y.E., Raval, M., Lab, M.J., Gorelik, J., Edwards, C.R., Rayment, T. and Klenerman, D. (2000) *Biophysical Journal*, **78**, 2675.

44 Bruckbauer, A., Ying, L.M., Rothery, A.M., Korchev, Y.E. and Klenerman, D. (2002) *Analytical Chemistry*, **74**, 2612.

45 Rothery, A.M., Gorelik, J., Bruckbauer, A., Yu, W., Korchev, Y.E. and Klenerman, D. (2003) *Journal of Microscopy*, **209**, 94.

46 Nitz, H., Kamp, J. and Fuchs, H. (1998) *Probe Microscopy*, **1**, 187.

47 Böcker, M., Anczykowski, B., Wegener, J. and Schäffer, T.E. (2007) *Nanotechnology*, **18**, 145505.

48 Betzig, E., Finn, P.L. and Weiner, J.S. (1992) *Applied Physics Letters*, **60**, 2484–2486.

49 Toledo-Crow, R., Yang, P.C., Chen, Y. and Vaez-Iravani, M. (1992) *Applied Physics Letters*, **60**, 2957.

50 Bard, A.J., Denuault, G., Lee, C., Mandler, D. and Wipf, D.O. (1990) *Accounts of Chemical Research*, **23**, 357.

51 Hille, B. (1992) *Ionic Channels of Excitable Membranes*, 2nd edn, Sinauer Associates, Sunderland, Mass.

52 Sakmann, B. and Neher, E. (1995) *Single-Channel Recording*, 2nd edn, Springer, Heidelberg.

53 Brown, K.T. and Flaming, D.G. (1986) *Advanced Micropipette Techniques for Cell Physiology*, John Wiley & Sons, New York.

54 Happel, P., Hoffmann, G., Mann, S.A. and Dietzel, I.D. (2003) *Journal of Microscopy*, **212**, 144.

55 Mann, S.A., Hoffmann, G., Hengstenberg, A., Schuhmann, W. and Dietzel, I.D. (2002) *Journal of Neuroscience Methods*, **116**, 113.

56 Gorelik, J., Zhang, A., Shevchuk, A., Frolenkov, G.I., Sanchez, D., Lab, M.J., Vodyanoy, I., W, E.C.R., Klenerman, D. and Korchev, Y.E. (2002) *Molecular and Cellular Endocrinology*, **217**, 101.

57 Gorelik, J., Zhang, Y., Sanchez, D., Shevchuk, A., Frolenkov, G., Lab, M., Klenerman, D., Edwards, C. and Korchev, Y. (2005) *Proceedings of the National Academy of Sciences of the United States of America*, **102**, 15000.

58 Karrai, K. and Grober, R.D. (1995) *Applied Physics Letters*, **66**, 1842.

59 Brunner, R., Hering, O., Marti, O. and Hollricher, O. (1997) *Applied Physics Letters*, **71**, 3628.

60 Rensen, W.H.J., van Hulst, N.F. and Kammer, S.B. (2000) *Applied Physics Letters*, **77**, 1557.

61 Koopman, M., de Bakker, B.I., Garcia-Parajo, M.F. and van Hulst, N.F. (2003) *Applied Physics Letters*, **83**, 5083.

62 Lambelet, P., Pfeffer, M., Sayah, A. and Marquis-Weible, F. (1998) *Ultramicroscopy*, **71**, 117.

63 Wegener, J., Abrams, D., Willenbrink, W., Galla, H.J. and Janshoff, A. (2004) *Biotechniques*, **37**, 590.

64 Wegener, J., Zink, S., Rösen, P. and Galla, H.J. (1999) *European Journal of Physiology*, **437**, 925.

65 Mannelquist, A., Iwamoto, H., Szabo, G. and Shao, Z. (2002) *Journal of Microscopy*, **205**, 53.

66 Gorelik, J., Shevchuk, A., Ramalho, M., Elliott, M., Lei, C., Higgins, C.F., Lab, M.J., Klenerman, D., Krauzewicz, N. and Korchev, Y. (2002) *Proceedings of the National Academy of Sciences of the United States of America*, **99**, 16018.

7
Nanoanalysis by Atom Probe Tomography
Guido Schmitz

7.1
Introduction

The emergence of nanotechnology is closely related to progress in microscopy. Certainly, the existence of atoms as indivisible units of matter may be postulated, as did Greek philosophers several centuries BC (Demokritos 460-371). Also the important properties of small clusters of these elementary units may be deduced by a series of clever physical experiments and suitable reasoning. But it is almost impossible to master the technology of nanostructured devices and to establish their mass production without being able to monitor – to 'see' – the real structure of the fabricated devices. Thus, any important step in reducing the size scale of a technology will require the development of a suitable means of microscopy. Since the structural width of modern devices scales down to only a few nanometers, much interest persists in imaging techniques of atomic resolution. Furthermore, to go beyond mere imaging, chemical analysis of the structure with atom-by-atom accuracy and, perhaps, in three-dimensional (3-D) spatial resolution in desired.

With transmission electron microscopy (TEM), scanning probe microscopy (SPM) and field ion microscopy (FIM), three major branches of microscopy are currently able to achieve atomic resolution in everyday practical studies. Interestingly, the first demonstration of imaging individual atoms was achieved in 1951 [1], using FIM. Since that time, whilst both TEM and SPM have found widespread applications, FIM techniques have remained in a tiny niche, with very few laboratories being able to master and develop the related methods. This is all the more surprising as the fundamental process of FIM – the field evaporation of atoms – offers the exciting possibility to perform a chemical analysis simply by counting the individual atoms.

Instruments which use this approach are referred to as atom probes (APs), and have been used for about 40 years. Unfortunately, their detector and computing possibilities remained rather limited until the 1980s, with the technique being reserved to specialized laboratories. However, based on the recent substantial progress in instrumentation this situation has changed significantly and, indeed,

Nanotechnology. Volume 6: Nanoprobes. Edited by Harald Fuchs
Copyright © 2009 WILEY-VCH Verlag GmbH & Co. KGaA, Weinheim
ISBN: 978-3-527-31733-2

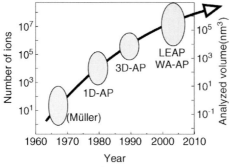

Figure 7.1 Evolution of the atom probe method in terms of analyzed number of atoms or total volume per measurement. For an explanation of the abbreviations, see Section 7.2.

is about to undergo further dramatic development. Depending on their standpoint, some research groups have claimed this to be a 'revolution' [2], while even those less euphoric scientists have admitted 'quite remarkable progress' over the past 20 years. By exploiting the achievements of fast electronics and modern computing, a real 3-D atomic reconstruction of the analyzed volume has become possible, with the maximum number of identified atoms being pushed beyond the hundred million benchmark (see Figure 7.1). Today, atom probe tomography (APT) is capable of providing the chemical analysis of a functional nanodevice as a complete unit, with atomic accuracy.

A second branch of innovation has emerged with the development of efficient pulsed laser sources. By assisting in the process of field evaporation by short laser pulses (of down to a few hundred femtoseconds duration), the practical analysis of nonconductive materials has become a realistic perspective. Recent successful measurements of former difficult classes of materials, such as semi-conductors and ceramics, are encouraging and have initiated intensive methodical research. The concept of laser assistance dates back almost 30 years when the technical limitations of the lasers available prevented their widespread use. As a consequence, the idea was buried among many other interests and, only very recently, has experienced a renaissance.

In the following sections we will examine atom probe tomography in greater detail which, with its increasingly widespread application, can no longer be neglected as an important branch of analytical microscopy. As the typical reader will most likely have only a basic knowledge of the subject, we will first describe the functional principles of the method, the basic algorithms of the 3-D reconstruction, and its accuracy. The instrumental technique and practical specimen preparation will then be considered. Applications of the method will be illustrated by some case studies which address actual problems of thin-film nanoscience. The final section is devoted to the most recent trends in atom probe tomography, including the use of pulsed laser sources to extend the method to complex materials, semiconductors, silicides and oxides as are found in microelectronic devices. It should be noted that the selected examples were biased by the author's personal interests, and are not aimed to provide a complete overview of the extensive studies conducted in recent years. Those readers seeking additional information should consult recent reviews [2, 3] and textbooks and other reports on the atom probe method [4, 5] and FIM [6].

7.2
Historical Development

Atom probe tomography applies the principle of FIM, and represents the latest progress in this area. The method dates back to the pioneering studies of E. W. Müller, who invented the field ion microscope in 1951 [1] after years of experimentation with electron emission microscopy for which the resolution was limited by the finite thermal energy of the electrons. With FIM, Müller was able to demonstrate atomic resolution images of tungsten surfaces as long ago as 1957 [7]. Although capable of achieving image magnifications in the range of one million – and thereby of atomic resolution – a field ion microscope is a surprisingly simple instrument, when compared to the complex electron optics of electron microscopes. Usually, no imaging lens is needed here. Owing to its simple projective geometry, the instrument does not suffer from the problems of stability associated with electron microscopy.

In 1965, Müller and colleagues were the first to combine FIM with time-of-flight (ToF) mass spectrometry, thus creating the so-called one-dimensional atom probe (1D-AP), the first tool to be used for quantitative chemical analysis in the nanometer range [8]. Rapid progress in detector technology during the 1980s led to the creation of single-ion detectors with sufficient spatial resolution and high detection rates. Important milestones here were the introduction of microchannel plates, of CCD cameras, and of rapid charge-to-digital converters which allowed picosecond time measurements. With this equipment, the early atom probe of Müller experienced a remarkable improvement and, by combining chemical identification by ToF mass spectrometry with the spatial information of the atom position, the numerical 3-D reconstruction of the spatial arrangement of the atomic species became possible. The atom probe had truly advanced to become a modern tool of real 3-D analysis!

Meanwhile, a variety of instrumental concepts of three-dimensional atom probes (3D-APs) were designed and put into operation. The first effectively functioning instrument was the 'position-sensitive atom probe' (PoSAP), which was described in 1986 by Cerezo and Smith [9]. A second, improved, instrument which could handle data from multiple atoms in parallel, the tomographic atom probe (TAP), was presented later (in 1993) by Blavette and coworkers [10]. In fact, as this instrument became more popular and was used in many laboratories worldwide, it lent its name to the general term for the method, namely atom probe tomography (APT). Today, the technical development of APT is in a continual state of flux, with new instruments being introduced on a regular basis. The field of view, and in turn the size of the investigated volume, was significantly improved essentially by reducing the specimen–detector separation, which led to the wide- angle tomographic atom probe (WATAP). Likewise, the addition of a laser beam line led to the method of laser-assisted field evaporation (LATAP) while, more recently, different energy focusing devices have been developed in order to improve the mass resolution [11, 12].

One remarkable achievement here has been that of the local electrode atom probe (LEAP) [13]. By placing a micrometer-sized electrode in front of an array of microtips, this instrument moderates the serious restrictions in specimen geometry, namely that a needle of high aspect ratio is needed. At the same time, the total number of atoms analyzed per measurement – and thus the size of the reconstructed volume – is increased by one to two orders of magnitude.

7.3
The Physical Principles of the Method

7.3.1
Field Ionization and Evaporation

All FIM techniques utilize the fact that electrical fields are concentrated at tips of sharp curvature. By supplying moderate voltages, enormous field strengths in the range of some $10\,\mathrm{V\,nm^{-1}}$ are easily obtained at the apex of nanometer-sized tips, whereas fields of such magnitude could be never obtained with macroscopic geometries. Thus, a typical field ion microscope consists of an ultra-high-vacuum (UHV) chamber with a specimen stage which holds the sample tip, a high-voltage supply and a viewing screen with the capability of imaging the ion impacts (Figure 7.2). A positive potential is supplied to the metallic specimen, while the entrance face of the screen is kept at ground. In order to reduce thermal energies, a

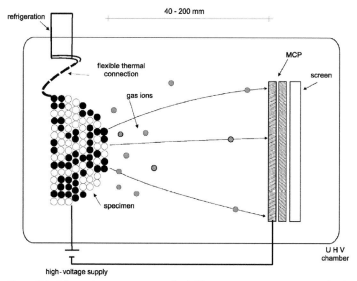

Figure 7.2 Schematic representation of a field ion microscope. MCP = multi-channel plate.

cryostat is required to cool the tip to 20–50 K. The field at the tip surface is controlled by the supplied voltage according to the relationship

$$F = \frac{V}{\beta \cdot R}, \tag{7.1}$$

in which V denotes the voltage supplied to the tip, and β denotes a dimensionless factor that varies with the exact geometry of the tip but is found in the range of 5 to 10. With increasing distance from the tip surface the field decays logarithmically, which means that the dominant drop of the field appears on the first millimeter from the tip surface.

In order to produce a field ion micrograph, an imaging gas (usually He, Ne, or a mixture of both) is introduced into the vacuum chamber. Close to the apex of the sample, the gas atoms are polarized and drawn towards the surface by the inhomogeneous field around the tip. Provided that there is sufficient field strength and a suitable distance between the gas atom and surface, a finite probability exists that an electron will tunnel from the gas atom into the band structure of the specimen. The potential well for electron transfer by tunneling is sensitively controlled by the local field strength. As a consequence, the ionization rate of the gas atoms is a function of: (i) the tip voltage; and (ii) the surface topography. After being ionized, the positively charged particle is accelerated towards the imaging screen where, by means of a multichannel plate and a phosphorus anode, the ion impact produces a visible light flash.

The trajectory of the ionized gas atom is determined by the shape of the electric field. As their thermal energy is negligible compared to the energy gain within the field, the ions are practically starting at rest. As a consequence, there is a one-to-one correspondence between the location of ionization at the tip surface and the impact position on the screen. Because of the discrete atomic structure of the sample, the local field at the surface varies in correlation to the surface corrugation. In particular, the edges of atomic terraces are protruding features and thus, are regions of elevated field strength and pronounced ionization rate. Therefore, the protruding edges of atomic terraces in crystalline structures are imaged as bright concentric rings that surround low-indexed pole directions of the crystal. This is illustrated in Figure 7.3, which shows a comparison of a field ion micrograph and the corresponding ball model of the imaged structure.

In order to achieve a clear field ion micrograph, the so-called 'best imaging field' must be established at the tip surface. To a good approximation, this field strength is only a function of the imaging gas (e.g. 35 V nm^{-1} and 44 V nm^{-1} for Ne and He, respectively). By increasing the field strength beyond this point, an alternative process of 'field evaporation' is observed. As soon as the field reaches a threshold which is characteristic of the sample material, atoms of the tip themselves are ionized, desorbed from the surface, and accelerated towards the screen thereby following very similar trajectories as the former gas atoms.

It is important to understand that the electrical field does not simply tear away the atoms, causing significant damage to the surface structure. Rather, the process remains controlled by a finite energy barrier, which must be overcome by thermal

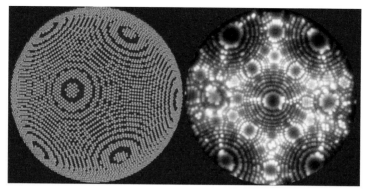

$V=6$ kV
$R=10$ nm $R=15$ nm

Figure 7.3 Ball model of a hemi-spherical apex (left) in comparison to an experimental field ion micrograph. Protruding atoms are represented by bright dots. The structure of the experimental FIM image at the right, comprising of concentric rings, is a natural consequence of the crystalline periodicity and the apex geometry. (Illustration courtesy of V. Vovk, University of Münster.)

activation or by quantum mechanic tunneling at very low temperatures. In other words, there is a clear justification in using the term 'evaporation', as it resembles the situation of vaporization in thermal equilibrium. This can be explained by the most simple model suitable for understanding field evaporation.

In Figure 7.4, the one-dimensional potential curves of a surface atom in its neutral and its n-fold charged ionic state, are plotted against the distance to the sample surface. Transferring an atom into the n-fold charged state requires an ionization energy, I_n, while placing the free electrons back into the band structure of the metallic

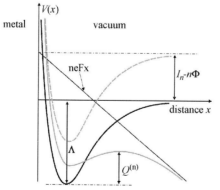

Figure 7.4 Potential curves of neutral atom (black) and ion (gray) close to the surface of the tip. The ionic curve is shown with (solid) and without (dashed) affecting field. For details of further variables, see the text.

sample delivers a payback of n times the work function, Φ. Thus, without the application of a field, the ionic potential experiences a constant shift of

$$\Delta V^{(n)} = I_n - n\Phi, \tag{7.2}$$

with respect to the potential of the neutral atom. If in addition the electrical field F is supplied, the ionic potential steadily decreases with increasing distance to the specimen surface, so that both potential curves necessarily intersect. It is also assumed in the so-called 'image hump model' that the ionic potential develops an intermediate maximum, which represents the important activation barrier. If short-ranged repulsive interactions are neglected, then the ionic potential can be approximated by

$$V(x) = -\frac{n^2 e^2}{16\pi\varepsilon_0 x} - neFx + \Delta V^{(n)}. \tag{7.3}$$

The first term on the right-hand side of the equation is caused by the image force of the metallic surface, while the second term is due to the influence of the field. Straightforward calculation yields the maximum at the hump as

$$V_{max} = -\sqrt{\frac{n^3 e^3 F}{4\pi\varepsilon_0}} + \Delta V^{(n)}. \tag{7.4}$$

so that the activation barrier reads

$$Q^{(n)} = \Lambda + I_n - n\Phi - \sqrt{\frac{n^3 e^3 F}{4\pi\varepsilon_0}} =: Q_0^{(n)} - \alpha \cdot F^{1/2}. \tag{7.5}$$

In Equation 7.5, Λ denotes the sublimation energy of the sample material. If this activation barrier is overcome by thermal excitation, the temperature dependence of the evaporation rate is expected to follow an Arrhenius relationship:

$$v_{evap} = v_0 \exp\left(-\frac{Q_0^{(n)} - \alpha \cdot F^{1/2}}{k_B T}\right). \tag{7.6}$$

Although Equation 7.6 is derived under rather simplifying assumptions, it describes the evaporation behavior at least in a qualitative sense correctly. Some aspects – for example, the characteristic evaporation thresholds of different materials – are even in surprisingly good quantitative agreement with experimental observations. For metals, this evaporation threshold is found to lie in the range of 20–60 V nm^{-1}, a field strength which is easily achieved with specimens of approximately 30 nm curvature radius. Several modifications have been suggested to describe the evaporation, including the so-called 'charge exchange model' and also quantum mechanical concepts. These deliver partly different exponents of the field dependence in the numerator of the argument of the exponential in Equation 7.6. For the range of practical interest, the logarithm of evaporation rate varies almost linearly with the applied field [14], so that Equation 7.5 is frequently replaced by the empirical relationship

$$Q^{(n)} \approx Q_0^{(n)} \left(1 - \frac{E}{E_0^{(n)}}\right), \tag{7.7}$$

in which $E_0^{(n)}$ denotes the field strength of vanishing barrier.

However, with all of these variations the obvious interpretation of Equation 7.6 is preserved: The desorption rate of the sample atoms is sensitively controlled by adjusting the field strength and, to a lesser extent, by the variation of temperature. In particular, the rate can be maintained at such a low level that the evaporation process can be studied in an atom-by-atom manner.

In order to perform a chemical analysis, a time- and position-sensitive detector system is placed opposite the sample, instead of a viewing screen (see Section 7.4.1). A positive dc voltage is then supplied to the tip, this being slightly too low to affect measurable field evaporation. Short high-voltage pulses of only a few nanoseconds duration are superposed to the base voltage to trigger the field evaporation. Measuring the time between the triggering pulse and the detection of a species allows identification of the evaporated particles by means of ToF spectroscopy. Typically, the pulse height is adjusted so that an event is detected after only about 1% of the pulses. Under these circumstances, the probability of evaporating multiple events comprising several atoms (which the detector system may not be able to split correctly into individual species) becomes negligibly small. In this way the sample atoms can be identified and counted, one-by-one.

For practical measurements, the choice of correct evaporation parameters, base and pulse voltages, pulse frequency and sample temperature, is an art which can only be mastered with profound experience. By continuously desorbing atoms, the samples become increasingly blunted. In order to maintain constant evaporation conditions, the tip voltage must be increased steadily during the measurement. In most cases, this is achieved under computer control, so that a constant evaporation rate in terms of the number of atoms per pulse is preserved. In order to avoid early specimen fracture, a rather low pulse amplitude and not too-low temperatures would be preferred. A low pulse amplitude would also improve the resolution of the mass spectra. However, on the other hand too-low pulses and high temperatures corroborate the accuracy of analysis, as alloy components with a low evaporation threshold may desorb in between the pulses and so become lost in the composition statistics. Suitable compromises are typically found at specimen temperatures between 30 and 50 K, and with a pulse fraction of $V_{pulse}/V_{d.c.} = 20\%$.

7.3.2
Ion Trajectories and Image Magnification

In order to understand the properties of field ion micrographs and the quality of the volume reconstruction, the ion trajectories must be discussed in more detail. An idealized specimen may be represented by the geometric model of a truncated cone closed by a hemispherical cap, as sketched in Figure 7.5 (the field lines and model

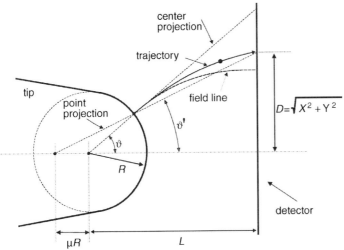

Figure 7.5 Electrical field and trajectory of an evaporated atom. The impact position and initial location at the tip surface are related by a point projection. The center of projection is shifted relative to the center of the spherical cap by μR. As an alternative, the ratio between polar angle ϑ and projection angle ϑ' may be used for evaluation.

trajectories are shown here for clarity). Owing to axial symmetry, the potential and the field can be considered in a two-dimensional space with r and z, the coordinates perpendicular and parallel to the rotational axis of symmetry, respectively. For convenience, the expression for the electrical potential Φ is split into the absolute tip voltage V and a spatial distribution function φ:

$$\Phi(r,z) = V \cdot \varphi(r,z) \tag{7.8}$$

while the equations of motion can be written in accordance to classical mechanics as:

$$\frac{d^2 r}{dt^2} = -\frac{neV}{m}\frac{\partial \varphi}{\partial r}; \quad \frac{d^2 z}{dt^2} = -\frac{neV}{m}\frac{\partial \varphi}{\partial z}, \tag{7.9}$$

where n and m denote the charge state and mass of the ion, respectively. Without any further calculation, it is seen that the acceleration in both coordinate directions depends on mass, charge state and voltage, in the same manner. Thus, when the ion is initially at rest, the shape of the trajectory becomes independent of all these variables. At given tip geometry, different species will follow the same path, and only the required flight time will vary and be characteristic for the given charge state and mass. This has the important consequence that the fundamental imaging relationship between the tip surface and the detector is universal for all species and that, besides minor modifications due to a slight difference in initial position, the imaging gas and specimen atoms will follow the same trajectory.

Since the ions start at rest, their motion follows initially the field lines, and they leave the spherical apex in radial direction. Later, after having gained considerable kinetic energy, the trajectory becomes almost straight and deviates from the field lines only because of the forces of inertia. Let us define (as in Figure 7.5) the initial position of the ion by the polar angle ϑ, and the imaged position at the detector by the smaller angle ϑ'. In order to determine the original position at the tip's surface, only the function between both angles must be known. This function may vary from tip to tip, although from a practical point of view a simple proportionality holds, which is conveniently described by an imaging compression factor:

$$\kappa := \vartheta'/\vartheta. \tag{7.10}$$

This factor can be calibrated by means of field ion micrographs of single crystalline specimens, such as depicted in Figure 7.3. The angle between the different pole directions, and thus the polar angle ϑ, is known from crystallography, while the detection angle ϑ' is determined from the position of the pole in the FIM micrograph and the flight distance L between tip and screen. Typical data determined for an electropolished tungsten tip are shown in Figure 7.6. Clearly, the linear relationship of Equation 7.10 is well fulfilled; a compression factor of $\kappa = 0.54$ is determined for this exemplary specimen.

It is straightforward enough to quantify the magnification of the analytical microscope on the basis of the geometric model of Figure 7.5. The polar distance in the image $D = L \cdot \sin\vartheta' \approx L \cdot \vartheta'$ must be compared to the distance at the hemispherical cap $d \approx R \cdot \vartheta$, where R denotes the current radius of the apex. Thus, the magnification is given by

$$M := \frac{D}{d} = \frac{L \cdot \kappa}{R}. \tag{7.11}$$

Recalling that a typical tip radius amounts to about 30 nm, and that the distance between the detector and tip may reach 50 cm, a magnification of 10^7 is easily obtained.

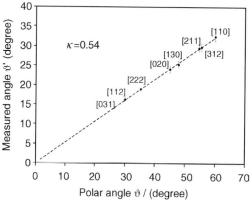

Figure 7.6 The relationship between measured angle ϑ' and polar angle ϑ, as determined for electropolished tungsten tips. The polar angle has been determined from crystallography. Tip axis aligned parallel to [011]. (Illustration courtesy of P. Stender, University of Münster.)

In Equation 7.11 an interesting detail is noteworthy. The magnification of the microscope depends on the tip radius of the specimen or, in other words, the specimen itself represents the essential lens of the microscope. Therefore, APT will only function in a reliable manner, if the specimens are prepared carefully and are notably reproducible in shape. Furthermore, the tip radius increases during the measurement, as the specimen field-evaporates continuously and, consequently, the magnification will decrease during the measurement. In order to reconstruct the spatial arrangement of the atoms after the measurement, the evolution of the radius must be recorded or estimated in a suitable manner.

The introduced geometric model of the tip neglects the roughness of the surface on the atomic scale. In reality, the edges of the atomic terraces and faceting of the spherical surface (low-indexed surface orientations become more pronounced in size due to anisotropy of the evaporation probability) will lead to slight modifications of the trajectories (this point is discussed further in Section 7.3.4).

7.3.3
Tomographic Reconstruction

During each measurement, several million events, the respective flight times and impact positions are recorded. From these raw data the original spatial arrangement of the atomic species is reconstructed by efficient, yet surprisingly simple, algorithms. In order to reduce the mathematical effort, we assume the specimen axis to be aligned perpendicular to the detector plane. The outlined scheme follows the studies of Bas *et al.* [15], and the general case of taking into account a relative rotation between tip and detector can be treated in an analogous manner (appropriate formulas are available in Refs [5, 16]).

The evaluation of data is conveniently subdivided into three steps:

- The specific mass m/n is calculated from the ToF.
- The lateral position at the tip surface is calculated from the impact position at the detector.
- The depth scale along the symmetry axis of the specimen is determined from the data sequence.

In the following section we use the geometric parameters as defined in Figure 7.5.

As the field lines are concentrated at the tip apex, the ions gain the major fraction of their kinetic energy during only the first millimeter of their trajectory. Later, the motion is almost straight and uniform, so that from conservation of energy the specific mass is calculated to sufficient approximation by:

$$\frac{m}{n} = \frac{2 t_{\text{ToF}}^2 e (V_{\text{tip}} + V_{\text{pulse}})}{L^2 + X^2 + Y^2}. \tag{7.12}$$

From the geometric detection angle

$$\tan \vartheta' = \sqrt{X^2 + Y^2}/L, \tag{7.13}$$

the Cartesian coordinates of the position at the tip's surface can be determined by

$$x = \frac{X}{D} R \sin \vartheta = \frac{X}{D} R \sin(\vartheta'/\kappa)$$
$$y = \frac{Y}{D} R \sin(\vartheta'/\kappa) \quad (7.14)$$
$$\tilde{z} = R(1-\cos(\vartheta'/\kappa)).$$

if the image compression factor κ has been calibrated before. In Equation 7.14 the axial coordinate $\tilde{z}$ has been marked by a tilde to express that this axial position is only preliminary, as it is still given relative to the position of the tip front z_0. As this reference point shifts during the measurement, we must correct the depth position in a final evaluation step. With each evaporated atom, the specimen is eroded by one atomic volume; thus, the number of detected atoms represents a natural depth scale. To establish this scale, the actual image magnification, which relates the sensitive area of the detector to the investigated area at the apex, must be taken into account. By expressing all of this in a differential equation, we obtain

$$dz_0 = \frac{\Omega}{\rho A_{\text{measured}}} dN = \frac{\Omega \cdot M^2}{\rho A_{\text{detector}}} dN, \quad (7.15)$$

where Ω and N denote the average volume per atom and the number of detected atoms, respectively. The factor ρ takes into account the limited detection probability of the detector ($\rho \approx 0.5$) and the magnification M is calculated by means of Equation 7.11. By applying Equation 7.15, the total shift of the tip front relative to its initial position at the start of the measurement is found by integration, and the final z-coordinate results from summing both contributions: $z = z_0 + \tilde{z}$.

In order to evaluate Equation 7.14 or 7.15, the instantaneous tip radius R must be known. If the evaporation properties of the investigated material are reasonably homogeneous, this radius is concluded from the total voltage (dc plus pulse voltage). The preset evaporation rate since is known to be obtained at the critical field strength E_{evap} of the investigated material, we can use by inversion of Equation 7.1.

$$R = \frac{U_{\text{tot}}}{\beta \cdot E_{\text{evap}}} \quad (7.16)$$

to determine the actual radius. However, this scheme is only feasible if the evaporation properties of the sample are reasonably constant. In heterogeneous specimens, for example the thin-film layer type, the radius must be concluded from geometric considerations (see e.g. Ref. [17]).

A typical reconstruction, calculated by the outlined formulas, is shown in Figure 7.7. In this case, the analyzed volume stems from a Cu/Co multilayer specimen. The position and chemical identity of each detected atom is marked by a color-coded dot. With modern wide-angle instruments, the lateral width of the reconstructed volumes reaches about 50 nm, but with blunted tips even 100 nm can be achieved. In view of the rather simple algorithms used, it comes as a surprise that even lattice planes of the

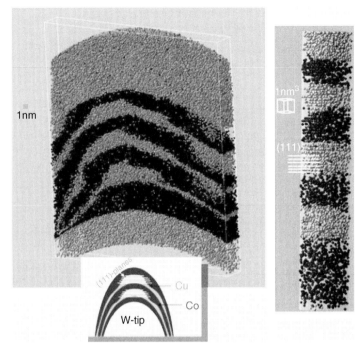

Figure 7.7 Atom probe tomography. Positions of individual atoms are represented by color-coded dots in a perspective representation. This is part of a larger data set of a Cu/Co multilayer after 60 min annealing at 450 °C. The detail on the right-hand side documents lattice plane resolution. (Measurements performed with the WATAP at University of Münster by V. Vovk [60].)

crystalline structure are resolved, as shown in the detail of Figure 7.7. This very welcome feature is explained by the physics of the evaporation process. As atoms at the edges of the atomic terraces have the highest evaporation probability, low-indexed lattice planes that are aligned parallel to the specimen surface have the tendency to desorb in a layer-by-layer mode. Resolved lattice planes demonstrate the outstanding spatial resolution of the method and allow the calibration of important parameters of the reconstruction algorithm. If the critical evaporation field strength, the field and image compression factors β, κ, and the detection probability are chosen correctly, indeed the correct lattice spacing expected for the material is reproduced.

After having reconstructed the spatial arrangement of the atoms, various averages, composition profiles, 2-D compositional maps and iso-concentration surfaces may be derived by sorting and counting the species in suitable subvolumes. Different algorithms were proposed, and are indeed in use, to detect precipitates and the shape of interfaces automatically [18]. The further analysis by Fourier methods [19, 20] or the calculation of pair correlation functions [21] similar to image analysis in the 2-D world of electron microscopy, has also been demonstrated.

7.3.4
Accuracy of the Tomographic Reconstruction

As discussed earlier (see Section 7.3.2), the sample itself represents the critical 'lens' in the projective geometry of the atom probe. Thus, it is not surprising, that a well-prepared shape of the specimen, usually produced by continuous field evaporation prior to the measurement, is the most critical issue in practical work. If the process of 'field development' is conducted too rapidly or insufficiently, then a variety of artifacts may be induced. Of particular danger here is the partial fracture of a specimen before or during measurement, as this usually produces surface topologies that are unsuitable for a reliable spatial reconstruction. In addition, the depth scale will be erroneously calibrated due to the intermediate loss of material. Yet, even if the experimentalist obeys all rules of good experimental practice, the positioning of the atoms cannot be perfect – at least as long as the evaluation scheme outlined in Section 7.3.3 is used. Apart from large angle corrections, this scheme is currently the 'state of the art'.

As illustrated in Figure 7.7 (an even clearer example is shown in Figure 7.20b), it is quite common to reproduce a set of lattice planes in volume reconstructions of pure metals. Usually, these planes are aligned almost parallel to the tip surface. In exceptional cases, several different lattice sets, inclined to each other, could be detected at the same time. By careful Fourier analysis of such examples, the accuracy of the atomic positions in the reconstruction was quantified [19, 20]. By deriving static Debye–Waller factors from the intensity of higher order Fourier components, the standard deviation of the individual atoms from their ideal lattice positions was determined at the example of a pure iron sample to $\sigma_\perp = 0.03$ nm in the direction perpendicular to the local tip surface, and to $\sigma_{II} = 0.1\ldots0.15$ nm in lateral direction [20]. The strong anisotropy in resolution is a characteristic of the atom probe method. It should be noted here that this outstanding high accuracy has been observed under 'best-case' conditions – that is, the investigation of a pure, coarse-grained metal at rather low temperature. Frankly, a microscopic analysis of such a specimen is useless. In relevant cases, of heterogeneous samples measurements are significantly less accurate. In particular, it is impossible to exploit the impressive depth resolution, way better than 1 Å, in order to determine the shift of interstitial defects out of the host lattice planes. Owing to the principle of depth scaling, these defects will be assigned to either of the neighboring lattice planes. Even worse, the localization of these defects is determined by their relative evaporation threshold rather than their original physical position in between the host lattice.

It is instructive to consider the origin of these inaccuracies. The reconstruction scheme outlined in Section 7.3.3 is based on two important assumptions: (i) the tip apex may be represented by a perfect sphere; and (ii) atoms evaporate individually in a predictable layer-by-layer sequence. The former is the prerequisite for accurate lateral positioning, while the latter is critical for exact depth scaling. It is clear that already with pure metals neither assumption is perfectly met, as the atomic scale roughness and surface facets (which are unavoidable due to anisotropy of the evaporation threshold) are neglected. With alloys, the situation becomes even worse.

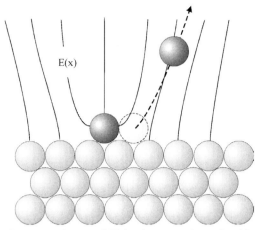

Figure 7.8 Schematic field distribution at a low indexed lattice plane just before being completely evaporated. The last but one atom is accelerated by a field which is significantly deformed by the last atom.

As the various components evaporate at quite different field levels, the sequence of evaporation can be severely disturbed in this case.

Furthermore, in heterogeneous microstructures, so-called 'local magnification effects' prevent the simple mapping of the projective geometry. For example, the embedded particles of a material of the higher evaporation threshold tend to protrude from the tip surface, which leads locally to an increased curvature of the surface and thus to an inhomogeneous magnification by the microscope. The undesired consequence is that the particles appear artificially broadened and that their atomic density is underestimated in the reconstructions. Worse still, due to overlapping of the trajectories artificial mixing of the materials from both sides of the particle interface is erroneously measured.

In order to understand why positioning in lateral direction is less accurate, we consider the situation sketched in Figure 7.8, when only very few atoms are left shortly before evaporating a further low-indexed lattice plane completely. In this case, the electrical field that controls the trajectory of the atom next in evaporation line will definitely deviate from the idealized field surrounding a spherical surface, due to local disturbance of the remaining protruding atoms. Thus, in order to calculate the trajectories exactly, the structure must be known in advance. We are faced with an implicit problem *when* the atoms should be localized exactly on the atomic scale. The solution of this problem remains a goal for the future. Meanwhile, in order to achieve a sound evaluation of atom probe data, the only promising strategy is to simulate the evaporation sequence of hypothetical specimen structures and to compare the simulated reconstructions with those of real experiments. Such a procedure is quite analogous to the normal practice in high-resolution electron microscopy, where experimental images are compared to those simulated from hypothetic structures until a good match is found.

For that task, Vurpillot and coworkers [22, 23] derived a simulation scheme that allowed an investigation of the spatial accuracy of tomography and the influence of heterogeneous evaporation on theoretical grounds. As the simplest geometric model, which still reflects microscopic features on the atomic scale, these authors suggested constructing the apex from a simple cubic arrangement of Wigner–Seitz cells. In a first step, the electrical field surrounding the model tip is calculated by solving numerically Poisson's equation for the electrical potential by means of a finite element method. The electrical field strength at the locations of the surface atoms is determined, and the position of the highest field strength identified. In a second step, the atom at this specific position (meaning the corresponding Wigner–Seitz cell) is removed from the apex model and the field is recalculated for the new configuration. In a final step, the trajectory of the removed atom is calculated between the tip surface and detector in accordance with classical mechanics which considers the acceleration of the ion within the electrical field. In the case of alloys or heterogeneous systems, the different evaporation probabilities of the atomic species must be taken into account. Thus, before the position of highest field is selected in the first step, the electrical field is scaled artificially by a factor varying from atom to atom in order to reflect the respective evaporation probability.

By repeating this scheme recursively to predict the impact positions of several thousand atoms, a part of a measurement can be simulated quite realistically. The general features of the experimental data are well reproduced, although an artificially short flight distance between the tip and the detector, and rather small specimens of approximately 20 nm radius, had to be used to limit the computational effort. The statistical nature of thermal activation has also been neglected. In Figure 7.9, the

Figure 7.9 Simulated impact positions on the detector area, during evaporation of a few lattice planes of a simple cubic model alloy. The tip axis is aligned along [001]. (Reproduced with permission from Ref. [22].)

impact positions of several thousands of atoms on the detector are shown, as calculated from the simulated evaporation of a few lattice planes. In contrast to the behavior of an ideal microscope, the spatial density of the events is by no means homogeneous. Rather, lines of significant redistribution of the atoms are seen, which are related to low-indexed zones of the crystal structure. Clearly, this redistribution is related to the situation sketched in Figure 7.8. The last few atoms sticking on a flat, low-indexed surface are affected by severe field distortions and therefore, their trajectory is significantly disturbed in comparison to a simple point projection.

Deviations of the trajectories induced by the local surface topology on the atomic scale are the main limiting factor for the instrument's lateral resolution along the tip surface. Although various ideas have been proposed to correct for the pronounced deviations at zone lines in pure samples, a practicable improved reconstruction algorithm has not yet been presented. In the case of statistical alloys, the situation is particularly difficult, as the chemical neighborhood of the evaporated atom is not known. One may imagine iterative algorithms to refine the reconstruction. But even the atom probe detects only 60% of the atoms (see Section 7.4.1). Thus, the local chemistry is never known completely. The effect of disordered alloys becomes particularly clear if, in simulated measurements, the atom positions of a disordered alloy are compared to those of a pure tip of identical geometry. An example is shown in Figure 7.10. Owing to the different evaporation probabilities of the two species, the evaporation sequence is disturbed and the local fields are distorted in the case of an alloy in a different manner as compared to a pure specimen. Clearly, the reconstructed positions of the atoms do not agree. However, as can be seen from Figure 7.10, most positions agree within about one lattice constant. Only a minor fraction of atoms originating from zone line positions are shifted by much larger amounts, up to five lattice constants. In this way, the simulation indicates the lateral

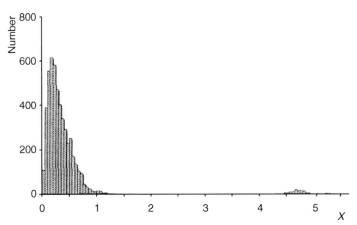

Figure 7.10 Statistics of reconstructed atom positions of an AB model alloy with respect to their ideal position in an homogeneous specimen (relative shift x in units of the lattice constant). (Reproduced with permission from Ref. [22].)

accuracy of the tomography to be slightly better than a lattice constant, as long as poles and zone lines are avoided for analysis.

7.4
Experimental Realization of Measurements

7.4.1
Position-Sensitive Ion Detector Systems

The rapid progress of APT in recent years has been made possible only by the remarkable evolution of spatially resolving detector systems with single-ion sensitivity. During the past two decades, several detector concepts have been proposed and put into operation. Those systems currently in use will be discussed at this point.

All available detector concepts are based on a stack of two to three multichannel plates (MCP). An MCP represents a secondary electron multiplier with many independent channels working in parallel. The device is composed of thousands of small glass tubes, each approximately 25 µm in diameter, and packed in parallel alignment to form a plate of about 1 mm thickness (see Figure 7.11). The front and reverse sides of the plate are coated with thin metallic films which serve as electrodes to supply a voltage in the 1 kV range. An ion which hits the inner wall of such a glass tube will produce a few secondary electrons that are accelerated by the supplied field. On their way towards the reverse side, they impact the glass wall several times and produce further secondary electrons. This cascade process finally produces a cloud of about 10^4 electrons per ion. If a consecutive stack of two or three MCPs is used instead of a single unit, the individual amplification factors will be multiplied so that a single ion hitting the front side with sufficient energy will produce finally a cloud of about

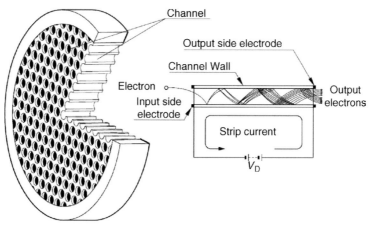

Figure 7.11 Functional principle of a multi-channel plate (MCP). The MCP is an arrangement of thousands of secondary electron multipliers working in parallel. Each of the electron multipliers is made from a tiny glass tube, 25 µm in diameter.

10^8 elementary charges, which is sufficient for further electronic evaluation. An MCP is a rather rapidly operating device; indeed, with optimized electronic circuitry the raise time of a single ion pulse is in the range of 100 ps. The spatial resolution of the MCP is determined by the dimension of the glass tubes. One important drawback here is that the detection efficiency of an MCP is way below 100%. Hence, only those charged particles which penetrate into one of the tiny tubes will induce the described avalanche process, while those hitting the massive front side are simply reflected. Due to mechanical requirements, channel plates cannot be produced with an open area fraction significantly larger than 60%. Attempts to improve the detection probability by placing additional electron mirrors in front of the channel plate have not been sufficiently successful so as to be used in modern-day instruments. Thus, based on principle, the presently available atom probes do count only half of the atoms of the analyzed volume.

For imaging purposes in FIM, it is sufficient to place a phosphorus screen behind the exit face of the MCP, so that each electron cloud produces a short light flash. Several attempts have been made to record these light flashes by means of a gated CCD camera in order to determine also quantitative positional information; examples include 'Optical PoSAP' [24] and 'Optical TAP' [25]. However, both systems suffered from a rather slow read-out of the camera, which severely limited the practical pulse frequency. Therefore, state-of-the-art instruments usually evaluate the charge clouds using methods mostly developed by nuclear physicists. For this, a multiple anode array is placed behind the MCP and connected to sensitive preamplifiers and fast converters in order to transform analogous charge information into digital data. Various concepts can be distinguished by their different layouts of the anode array and the complexity of the electronics. Historically, the first functioning system was the PoSAP, which was built around a 'wedge and strip anode' [26]. The name of this anode is self-explanatory, based on the sketch in Figure 7.12a. The geometry is designed in such a way that the relative fractions of the total charge measured on the three electrodes Q_X, Q_Y and Q_Z, vary with the position of the electron cloud. For the layout shown, the position may be calculated in a straightforward manner by

$$X \propto \frac{Q_X}{Q_X + Q_Y + Q_Z}; \quad Y \propto \frac{Q_Y}{Q_Y + Q_Z}. \tag{7.17}$$

Since only three independent anodes are used, the required electronics is reasonably simple. However, the layout has the important drawback that the total area of each electrode – and thus the respective capacities – are quite large and the drain of charges after the impact takes a considerable time. If a second ions hits the detector within this time gap, then both events cannot be separated; consequently, the operator is forced to use very low data rates.

Following a significant effort to improve the electronic instrumentation, a square array layout of many smaller electrodes was realized shortly afterwards, with the original TAP detector (see Figure 7.12b) [10]. By choosing the correct distance between the MCP and anode array, the electron cloud will always spread over at least three or four electrodes, so that its central position can be determined by charge

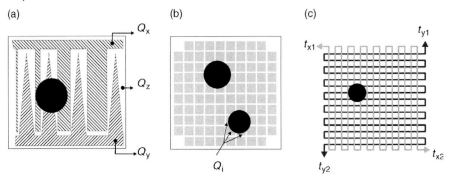

Figure 7.12 Anode layouts for the read-out of the positional information of a channel plate. (a) Wedge and strip detector; (b) TAP detector; (c) Delay line detector. Here, instead of the double-wire Lecher lines, only a single wire is shown to illustrate the principle. The extension of the electron cloud is symbolized by black circles.

weighting. Due to its parallel design the detector has some capability to separate multiple events, which allows much higher data rates than with the PoSAP detector.

The latest instruments (those constructed after 2003) mainly apply the delay line principle, which was first proposed in 1987 [27]. Here, instead of flat electrode areas, two independent double wire spirals are used that are wound along the X and Y axes of the detector, as shown schematically in Figure 7.12c. Each double wire represents a Lecher line, on which the pulse signal propagates with the velocity of light. As opposed to previous concepts, the spatial information is not determined from charge weighting but rather from time measurements, so that no expensive measurement of analogue signals is required. Each ending of the Lecher line is connected to a separate channel of a fast time-to-digital converter (TDC) with sub-nanosecond resolution. If the ion impacts at the center of the detector, the pulse signals will propagate symmetrically to both ends of the Lecher line, and will therefore reach the TDC at exactly the same time; in contrast, for an asymmetric impact position the two time signals will differ considerably. The sum and difference of the two time signals of such a Lecher line are presented in Figure 7.13; these were collected for many independent impacts on a circular detector of 120 mm diameter and a spire spacing of the anode of 1 mm. The sum of both time signals represents an instrumental constant, and corresponds approximately to the propagating time from one end of the line to the other. With 150 ns, this time interval is easily measurable. The time difference between both signals is proportional to the position according to

$$X = v_p \left(t_x^{(l)} - t_x^{(r)} \right), \qquad (7.18)$$

where the calibration parameter v_p denotes the propagation velocity along the spiral axis, which amounts to about 0.4 mm ns^{-1}. An analogous equation holds for the Y direction. With modern computer electronics, a time resolution below 100 ps – and

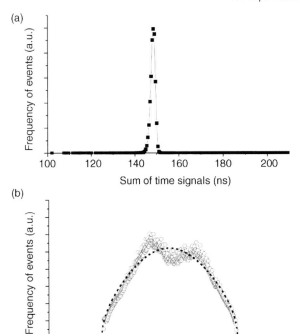

Figure 7.13 Evaluation of time signals of the delay line detector. (a) The sum of both signals is a constant which may be used to correlate time signals to individual events; (b) The difference is a direct measure of the position. The dashed line represents the spatial distribution expected for a circular detector in wide-angle configuration. Experimental data deviate due to anisotropic evaporation of the crystalline sample. (Illustration courtesy of P. Stender, University of Münster.)

thus a positional accuracy of better than 0.1 mm – is achievable. As a fast TDC is already required for ToF mass spectrometry, delay line detectors represent a very economic solution. Furthermore, the delay line principle allows very high data rates, so that evaporation pulses may be applied with frequencies of up to 300 000 Hz. In order to improve the multi-hit capability, a hexagon anode design [28] and quantitative evaluation of the pulse shape by use of fast oscilloscope electronics has been realized [29]. However, it is not yet clear whether this additional effort in instrumentation will provide an advantage in practical terms.

7.4.2
Instrumental Design of 3-D Atom Probes

In principle, a 3-D or tomographic atom probe consists of the same components as the FIM shown in Figure 7.2. Only the viewing screen must be replaced with one of the above-discussed position-sensitive detector systems, and the high-voltage supply

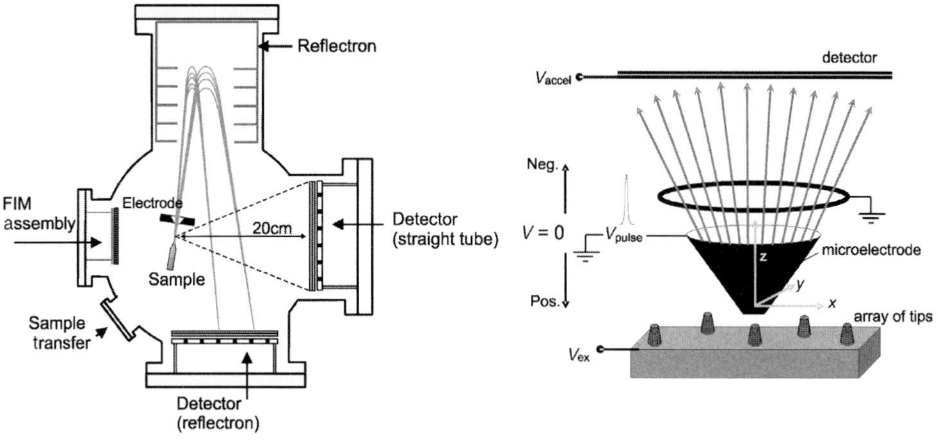

Figure 7.14 The design of the tomographic atom probe operated at the University of Münster [31]. The dedicated chamber layout allows switching between different geometries: (i) the straight flight tube yields a short flight path to optimize the open angle of the instrument; (ii) the reflectron arrangement yields an improved mass resolution; (b) The principle of a microelectrode atom probe. An extraction electrode with an open diameter of approximately 10 μm is placed close to a microtip to concentrate the electric field. (For an example, see Ref. [13].)

must be extended by the required voltage pulsing. By using the voltage pulse as start signal for an accurate time measurement, and the detector signal for stopping, the ToF of the ions can be measured in straightforward manner. However, the flight distance between the tip and detector must be adapted to obtain the desired mass resolution. During the past two decades, this flight distance has been steadily decreased in response to the improving time resolution of available measurement electronics. Recently, the flight path was reduced to 10–20 cm in so-called 'wide-angle instruments' [30, 31]. With detector diameters of up to 120 mm, this yields a geometrical aperture of $2\vartheta' \approx 44°$, which means an even larger effective opening in the range of $2\vartheta \approx 60°$ due to the curvature of the ion trajectories (as explained in Section 7.3.2). A schematic representation of a modern conventional 3-D atom probe is shown in Figure 7.14a. In contrast to the concepts discussed above, two modifications should be noted:

- Instead of a straight flight tube, a reflectron geometry is used with an ion mirror that leads to parabola-shaped ion trajectories. This geometry compensates for fluctuations in the kinetic energy of individual ions. A faster ion will penetrate deeper into the mirror field, and so will have a longer flight path. If the length and voltage of the reflectron are adjusted correctly, then ions of identical mass but with slightly varying initial velocity will hit the detector after identical flight times. In this way, the mass resolution is significantly improved. Originally, reflectrons were designed with a homogeneous mirror field. With such a device, a perfect time focusing is paid by trajectory deviations which can only be tolerated for small aperture angles. Therefore, a conventional reflectron is not compatible with the wide-angle concept. Very recently dedicated reflectrons have been designed with curved electrode shapes, which eliminate this problem [12]. When using a reflectron, the typical mass resolution of a 3-D atom probe can be expected to be $\Delta m/m = 1/500...1/2000$, whereas without such a device the resolution is limited to $\Delta m/m \approx 1/100$ [all data full width at half maximum (FWHM), determined at an effective mass of $m = 30$].

- The evaporation trigger is supplied as a negative pulse to an extraction electrode in front of the tip, which allows shorter and better-defined pulse shapes. Recently, the use of an extraction electrode has created important progress towards miniaturization. If a large number of atoms were to be evaporated and measured in a reasonable duration of the experiment, the pulse frequencies would need to be as high as possible. However, this strategy finds a natural limit, as no practicable means are available to produce a 5 kV nanosecond pulse with frequencies exceeding 20 kHz. A very intelligent method to circumvent this technical problem is to use a micrometer-sized electrode, as suggested in 1994 by Nishikawa and Kimoto [32]. By placing a tiny extraction electrode of some 10 µm bore size close to the tip, the β-parameter of Equation 7.1 is reduced by a factor of 2 to 3, as exemplified by the experimental data in Figure 7.15. In consequence, much lower voltage pulses are required, which today can be produced with repetition rates far in excess of 100 kHz. After solving any related technical problems, the concept has been put into operation during the past few years. Meanwhile, these instruments are functioning well and available commercially [33]. Their efficiency of analysis is impressive [34]; a data rate higher than 10 000 atoms per second is obtainable and data sets with more than 10^8 atoms have been routinely achieved. Beside the high

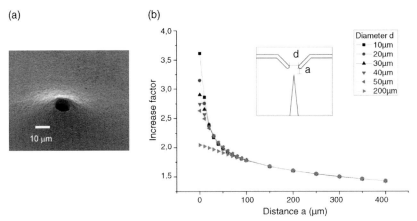

Figure 7.15 The increase of field by a microelectrode in front of the tip allows the voltage to be reduced by a factor of two to three at a constant tip radius: (a) SEM micrograph of an electro-plated Ni microelectrode (b) The relative increase of field at constant voltage for microelectrodes of 20 to 200 µm is plotted against the spacing between tip apex and electrode. (Illustration courtesy of Ralf Schlesiger, University of Münster [88].)

data rate, these instruments have an important advantage in specimen preparation. As indicated in Figure 7.14b, an array of microtips may be used rather than a single tip made from a supporting wire. As the requirement for a large aspect ratio is considerably relaxed by the microelectrode geometry, the tip array may be conveniently produced by sputtering through a suitable mask. Very recently, In As nanowires, grown naturally by using nanopatterned Au catalysts, were directly analyzed using a tomographic atom probe equipped with such a microelectrode [35].

7.4.3
Specimen Preparation

As APT requires a dedicated needle-shaped sample geometry, obtaining suitable specimens is a delicate matter. Although the art of preparation has undergone considerable progress, a large proportion of desired measurements fail due to this issue. Traditionally, the required needles were produced by the electropolishing of thin metallic wires under *in situ* observation by means of optical microscopy. At present, thin films and other complex nanostructure are the focus of interest, of which usually no conventional wire is available.

For studies on reactions in thin-film materials, layer systems have been deposited onto tungsten tips, which serve as a substrate. In order to achieve an optimum shape, the freshly prepared tungsten substrates are field-developed prior to deposition. As considerable stress is induced by the electrical field during the measurement, many investigations are prevented by insufficient mechanical stability of the interface between substrate and coating. Thus, this interface requires special care. Occasionally, an interlayer (often chromium) is used as an adhesion aid, while ion-beam cleaning of the substrates immediately before deposition has also been shown as advantageous [36]. In Figure 7.16a, a thin-film specimen produced in this way,

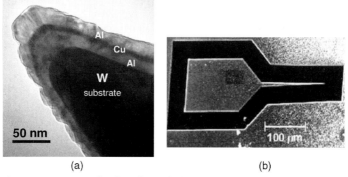

Figure 7.16 (a) Example of an Al/Cu/Al trilayer on a tungsten substrate tip, deposited using an ion beam sputtering technique. (Illustration courtesy of C. Ene, Göttingen.); (b) Scanning electron micrograph of a layer specimen prepared by electron-beam lithography. (Illustration courtesy of J. Schleiwies, Göttingen.)

namely an Al/Cu/Al trilayer deposited on tungsten, is presented. In this geometry, the main analysis direction is aligned normal to the interfaces, so that these specimens are especially suited for the investigation of reactive diffusion by local depth profiles of composition. However, when interpreting the analysis results it must be borne in mind that these films are deposited onto a curved surface. Usually, the curvature induces a rather small grain size, so that the microstructure is not directly comparable to that of thin films deposited onto conventional planar wafer substrates [37].

If this variation in microstructure cannot be tolerated – for example, because the properties of technical devices should be characterized – the tips may be cut by either lithography [38] or focused ion beam techniques. For the former a planar layer system is first coated with a suitable photo resist, and exposed to electron-beam lithography and developed chemically to obtain a suitable etching mask. A typical sample is shown in Figure 7.16b. The tip is attached to a 'handle' which is about 100 μm in length, with the wedge-shaped needle pointing to the right-side taper to a width below 100 nm at the apex. After etching by sputtering, the tips are removed from the substrate and glued to a supporting wire. In this geometry, the interesting interfaces are aligned parallel to the main analysis direction; thus, the method is well suited to investigations of interfacial roughness or pin holes in multilayers of small periodicity.

With the emergence of focused ion beam (FIB) facilities, equipped with a Ga beam of 5 to 30 keV, the preparation of difficult nanometric geometries has been revolutionized. Thus, it is no wonder that this technique is being used increasingly to prepare the required needles. Following an original proposition by Larson [39], thin films are deposited on top of flat-ended Si posts for that purpose. The width of the rectangular prism-shaped posts is chosen to be about $10 \times 10\,\mu m^2$ in cross-section – sufficiently large to mimic realistically the deposition conditions of a larger planar substrate, yet at the same time thin enough to limit the required beam time of the FIB. As an alternative, the cutting of a thin bar directly from a massive volume has also become usual practice as the intensity of available Ga beams has been further improved. Originating from transmission electron microscopy (TEM) studies [40], this 'lift-out' technique has been recently adapted to the needs of APT [34]. The essential preparation steps are shown in Figure 7.17. With any FIB method, the final step to produce a sharp tip is always an annular milling from the front face with a continuously decreasing size of the circular mask, as indicated in Figure 7.17c. It goes without saying that great care must be taken to avoid irradiation damage of the part which is to be analyzed. At 30 keV, the Ga ions of the cutting beam can be expected to penetrate at least 20 nm into the sample, and therefore suitable metallic coatings must be used to cover sensitive areas. Then, on completion of the procedure the energy of the beam should be reduced as much as is practicable.

The FIB has certainly revolutionized the preparation task, particularly if the tips are to be cut from chemically or mechanically difficult materials such as multilayers and semiconductors or ceramics. On the other hand, it cannot be overlooked that 'beam time' on these machines is still a rather limited resource. In this context, it should be noted that although many insulated measurements of FIB-prepared samples using

238 | 7 Nanoanalysis by Atom Probe Tomography

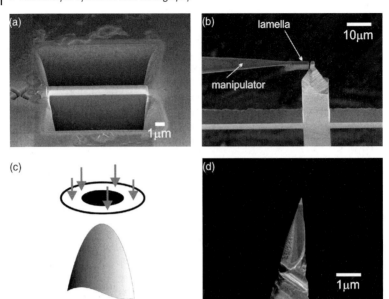

Figure 7.17 Stages of focused ion beam preparation using the lift-out procedure. (a) Cutting of lamella perpendicular to the substrate. The lamella edges are covered with deposited Pt; (b) Lamella moved to a supporting post (e.g. Cu). A part of the lamella had been fixed to the post and the remaining cut by the ion beam; (c) Annular milling from the front with continuously decreasing aperture size; (d) The final tip sample. (Illustration courtesy Ralf Schlesiger, University of Münster [86].)

APT are reported today, very few experimental series have been reported that characterize the different stages of a physical process. This may be seen as an indicator that the reliability and efficiency of the FIB procedures still require significant improvement.

7.5
Exemplary Studies Using Atom Probe Tomography

The application of APT to the physics of reactions is demonstrated here with some studies with nanosized, man-made geometries. As an analysis with an atom probe requires appreciable effort, the method should be used especially in those cases where its particular advantages can be best utilized. To summarize, the outstanding features of APT include:

- A spatial resolution of chemical analysis of a few Ångstroms. While theoretical performance data expect a similar resolution to be achieved with analytical TEM [electron energy loss spectroscopy (EELS) or energy-dispersive X-ray spectroscopy (EDS)], a comparison of composition profiles at interfaces reveals that APT has a significantly higher discriminating power.

- A standard-free chemical analysis with identical sensitivity of all elements across the periodic table. Whilst in TEM studies a variety of species cannot be measured due to physical limitation [e.g. low-mass elements in energy-dispersive X-ray (EDX) or peak overlap and low intensity in EELS], APT will always reveal a chemical contrast that even allows different isotopes to be separated.
- A real chemical mapping in three dimensions. While the 2-D image projection required in high-resolution TEM studies hinders the clear characterization of complex morphologies, APT is especially suited to investigate curved and buried interfaces in nanocrystalline matter.

The following examples have been selected to illustrate, in which way these advantages may be decisive in the success of experimental studies.

7.5.1
Nucleation of the First Product Phase

Owing to the technological trend towards miniaturization, the very early stages of reactive diffusion at thin-film interfaces have shifted into the focus of materials research. Frequently, it is argued that the thermodynamic driving force of forming a first reaction product is usually so high that the critical thickness of nucleation ranges down below the size of a lattice constant. Thus, nucleation should not be a rate-controlling step at all. However, initial evidence that this may not be true came from calorimetric studies of reactive diffusion in metallic thin films. Although only a single product forms, double-peaked heat releases were observed [41–43]. This experimental finding was interpreted as a two-stage mechanism [44]. In the first stage, nuclei form at the initial interface and grow quickly in lateral directions, while in the second stage heat release is attributed to parabolic thickness growth by volume diffusion. However, the process of nucleation remained quite unclear. Several mechanisms have been proposed to explain the apparent reduction of driving force in the presence of a sharp composition gradient [45–47]. In common, they predict that the composition gradient at the interface must first decrease by interdiffusion to a critical level before nucleation of the first intermetallic compound becomes possible. However, no clear experimental verification of this interpretation was provided.

A recent nanoanalytical study [48] was aimed at enlightening details of the early nucleation stages. For these experiments, bilayers of Co and Al, each 20–30 nm thick, were deposited on tungsten substrate tips, as described in Section 7.4.3. Two examples of 3-D reconstructions of the Co/Al interface are shown in Figure 7.18. Although the layers were deposited on curved surfaces, the initial interface appeared practically flat, as the radius of curvature was still significantly larger than the width of the analyzed volumes. This flat interface is preserved for short annealing so that the earliest reaction stages may be characterized by 1-D composition profiles determined normal to the interface, as shown in Figure 7.19. Due to the outstanding resolution, minor chemical modifications at the interface become noticeable. After a 5 min period of heat treatment at 300 °C, the zone of chemical transition at the interface

240 | 7 Nanoanalysis by Atom Probe Tomography

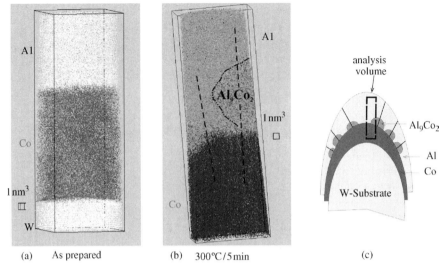

Figure 7.18 (a, b) Atomic reconstructions of the Al/Co interface in the as-prepared state (a) and after annealing at 300 °C for 5 min; (b) Positions of individual atoms are marked by gray coded dots; (c) Sketch of specimen geometry. (After [48]).

broadened to 3.5 nm, indicating a significant mixing of the components. However, the composition profile in this annealing state was well fitted by an error function. Thus, it follows that, up to this stage, only interdiffusion rather than formation of a new intermetallic product has taken place.

The first nucleation of a new phase is observed only in a small fraction of measurements at this annealing stage. In these cases, globular particles are detected at the interface towards the Al side (see Figure 7.18b). The fact that these particles

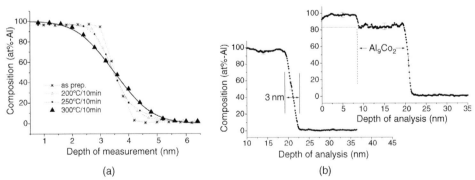

Figure 7.19 (a) Composition profiles determined perpendicular to the initial Al/Co interface after different annealing treatments; (b) Composition profiles determined along the left and right dashed lines in Figure 7.18b, demonstrating interdiffusion and phase formation, respectively. (After [48]).

appear only in part of the measurements after identical annealing conditions emphasizes the statistical nature of a nucleation process. Furthermore, it becomes clear from the volume reconstructions that nucleation takes place at heterogeneous sites at the interface, as sketched in Figure 7.18c. A composition profile across the newly formed phase identifies the product as Al_9Co_2 (see Figure 7.19b), while a profile determined across the remaining unreacted interface confirms again the interdiffusion on a depth of 3 to 4 nm as described previously.

The set of composition profiles in Figure 7.19, which could be obtained in this way only by APT, provide a clear demonstration that significant interdiffusion takes place before nucleation of the product. Furthermore, the experimental data quantify the critical diffusion depth before onset of nucleation to 3.5 nm (Al/Co at 300 °C).

If the theoretical nucleation thickness of the intermetallic product is estimated by the balance between the volume driving force and interfacial energy, a value of $d = 2\sigma/g_v = 0.2$ nm is expected. In view of this small value, it is very surprising, that the intermetallic Al_9Co_2 is only formed after the intermixed zone has already reached a thickness of 3.5 nm. However, based on the presented measurements, a theoretical study [49] was able to demonstrate that this behavior is clearly consistent with a polymorphic nucleation mechanism. Such a mechanism assumes that the nucleus of the new phase is produced by transforming the lattice structure, without modifying the local composition [45]. As any nucleus must have a minimum size in order to overcome the nucleation barrier, the ideal stoichiometric composition is only established in the center of the nucleus, whereas towards its boundaries the composition must deviate due to the existing concentration gradient. In other words, nucleation is only probable within a thin-layer fraction of the total diffusion zone, and the thickness of this layer shrinks with the chemical sharpness of the interface. In consequence, high concentration gradients will prevent nucleation.

In the cited theoretical study [49], the free energy ΔG required to form a nucleus under the constraint of a sharp concentration gradient was calculated. Numerical results are presented in Figure 7.20 for three different widths of the interdiffusion

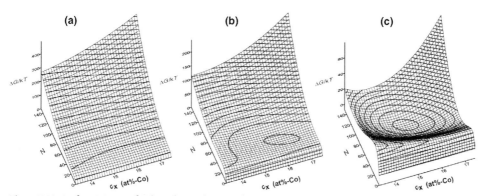

Figure 7.20 Surface $\Delta G(N, c_x)$ for the polymorphic transformation of a cubic volume into the Al_9Co_2 phase inside diffusion fields of width: (a) 3.0 nm; (b) 3.5 nm; and (c) 4.0 nm. Thermodynamic functions evaluated at a temperature of 573 K. (After [49]).

zone, namely 3.0, 3.5 and 4.0 nm. (In these plots, the size of the nucleus is expressed by the number N of atoms within the nucleus. The concentration c_x represents the composition at the center of nucleus.) For a diffusion width of 3.0 nm, the Gibbs free energy still increases monotonously with its size; thus, nucleation is forbidden. At 3.5 nm width, the situation has already slightly changed, with a weak local minimum appearing in the energy landscape. For 4.0 nm width, this minimum has become pronounced and its magnitude negative, which means that a product particle may now be formed under gain of energy. A nucleation barrier of $25\,kT$ is determined from the energy landscape, which is a quite realistic value to obtain reasonable nucleation rates. Thus, the critical interdiffusion width is predicted to be slightly larger than 3.5 nm, in remarkably good agreement with the experimental observation by APT.

The same considerations were also performed for other suggested nucleation mechanisms. For example, calculation of the so-called 'transversal mode' [46] yielded a critical diffusion width smaller than 1.0 nm, in obvious contrast to the measurements. In conclusion, the atom probe analysis yielded convincing evidence for the interdiffusion process taking place before nucleation of the first product, although the depth of mixing ranged down to only a few nanometers. By using APT it has been possible to determine the critical diffusion depth with better than 1 nm accuracy, which allows a distinction to be made between different suggested nucleation modes.

Other studies on reactive metallic model systems made particular use of the calibration-free analysis of the atom probe. This chemical accuracy becomes especially important if metastable phases that spread over only two to three lattice constants are to be identified. Under these circumstances, a structural image is rarely achieved by high-resolution (HR) TEM, although the phases may be already clearly characterized by the local level of composition determined with APT, as for example demonstrated in studies on Ni/Al [50] and Ag/Al [51].

7.5.2
Thermal Stability of Giant Magnetoresistance Sensor Layers

In recent years, reading heads based on the giant magnetoresistance (GMR) effect have made possible a dramatic increase in magnetic recording density. As the periodicity of the required multilayers ranges down to a few nanometers, their spatially resolved analysis is a challenge, even with APT. Hence, several groups have used APT to characterize the as-produced state of GMR sensor layers [52–55]. Co/Cu and Cu/Ni$_{79}$Fe$_{21}$ are two of the most often-used metallic systems. The soft magnetic alloy Ni$_{79}$Fe$_{21}$ (Permalloy, Py) is especially suited for position sensors, as the effect of hysteresis is very low, while Cu/Co is mostly used for reading heads in data storage. For many potential applications, including for example angular sensors in motor vehicles, the thermal stability of the device is an important issue. It is well known that the Cu/Py system is much more sensitive to a thermal load than is Cu/Co. Whereas for the latter the GMR amplitude remains stable up to 400 °C, in the former case the GMR effect begins to degenerate at temperatures of only 150 °C [56]. As both metallic systems are immiscible in a thermodynamic sense, and furthermore all diffusivities

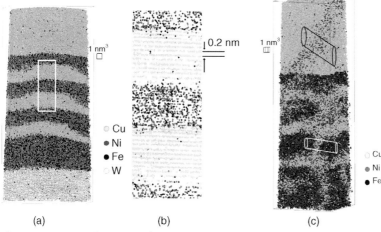

Figure 7.21 Atom probe tomography at a Cu/NiFe giant magnetoresistance structure. (a,b) In the as-prepared state; (c) After 30 min annealing at 400 °C. The detail (b) demonstrates lattice plane resolution and the outstanding chemical accuracy that allows the reproduction of sharp transitions in concentration from plane to plane. (After [58]).

are quite comparable, the low-temperature degradation of Cu/Py is surprising. In general, different mechanisms have been proposed as being responsible for GMR degradation: Van Loyen et al. [57] argued that at least two effects should contribute, namely grain boundary diffusion and inter- or demixing at the interface. In contrast, Hecker et al. [56] concluded that the alloying tendency of Ni and Cu above 250 °C controlled the decay of GMR in the Py systems.

In an extended APT study [58], $Cu_{2.5nm}/Py_{2.5nm}$ multilayers were deposited onto substrate tips and annealed in an UHV furnace. A typical volume reconstruction of an as-prepared state is shown in Figure 7.21a. The resolution is sufficient to distinguish individual (111) planes of Cu (Figure 7.21b), so that the dimensions of the reconstructed volume could be calibrated exactly. If the microstructures of the as-prepared state were compared to those after annealing up to 350 °C, then no remarkable difference was seen at first sight. Notably, the integrity of the multilayer was preserved. This was all the more striking as the magnetoresistivity vanished completely at the lower temperature of 150 °C. Only after annealing at even higher temperatures – at which the GMR effect also vanished in the more stable Cu/Co structure (400 °C) – was any clear indication of grain boundary transport observed (as shown in Figure 7.20c).

As an advantage of the 3-D experimental data, concentration profiles can be evaluated by defining analysis cylinders of smaller diameter and exactly aligning them perpendicular to the interfaces. In this way, artifacts due to local curvature or roughness of interfaces can be mostly excluded. An exemplary profile obtained after annealing at 350 °C is presented in Figure 7.22a, where the sharpness of the chemical

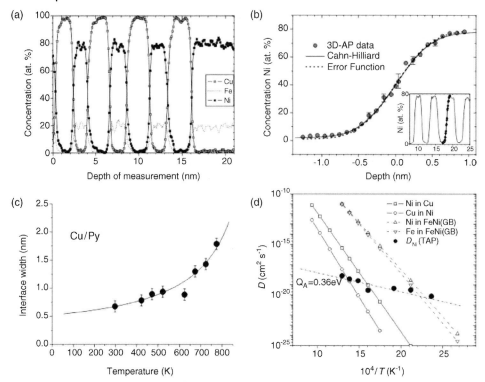

Figure 7.22 Nanoanalysis of Cu/Py multilayers. (a) Composition profile after 20 min annealing at 350 °C; (b) Error-function shape of chemical transition at interface; (c) Development of interfacial width with annealing temperature solid line represents prediction by Cahn–Hilliard theory. (d) Interdiffusion coefficient determined from this width in comparison to published data. (Reproduced with permission from Ref. [59]; © Carl Hanser Verlag 176.)

transition from almost 0% to 100 at% Cu on the length of two lattice plane distances, should be noted. Also of note, it is not possible to achieve such selectivity with analytical TEM.

Furthermore, the shape of the chemical transition can be characterized in detail and compared with expected model curves, as for example for interdiffusion (error-function) and interfacial thermodynamics (Cahn–Hilliard) shown in Figure 7.22b. Due to the outstanding sensitivity of APT, it is possible to note the smallest modifications of interfacial chemistry with temperature. The width of the chemical transition, defined exactly between 10% and 90% of the concentration amplitude, was used as a characteristic parameter. Clearly, although only on the order of 1 nm, the width is proven to increase significantly during low-temperature annealing (see Figure 7.22c). The formal description by an error function leads to the assumption that the width of the transition is controlled by kinetics of interdiffusion, as in the previously discussed study on Al/Co (see Section 7.5.1). However, the diffusion coefficients derived in accordance with this interpretation reveal a way too-low activation energy in comparison to reported data (see Figure 7.22d). Furthermore, interdiffusion would be difficult to understand from the viewpoint of thermodynamics, as the layer system is expected to be stable up to a temperature of about 1070 K. In an analysis of these data [59], it could be shown that the temperature dependence of the interface width is rather based on interfacial thermodynamics within the frame of Cahn–Hilliard theory. The experimentally observed temperature dependence of the width of the interface agrees nicely with the related model curve, as presented in Figure 7.22c. In this way, it is possible to determine, via direct measurement with APT, the required – but experimentally almost unknown – gradient coefficients that are important in the Cahn–Hilliard theory.

By combining wide field-of-view FIM and atom probe analysis, it was also possible [58] to exclude definitely grain boundary transport as a significant reason for GMR degradation in Cu/Py multilayers. As shown in Figure 7.21c, segregation – as an indicator for grain boundary diffusion – can be demonstrated in the volume reconstructions, but clearly not at temperatures relevant to GMR breakdown. Many individual grain boundaries were investigated after annealing at up to 250 °C, but no segregation was found in any case.

For the design of GMR devices, the APT study has important consequences. Clearly, a short-ranged broadening of interfaces on the depth of 1–2 nm due to interfacial thermodynamics (which has been neglected in previous studies) has an important influence on magnetoresistivity. In view of the small multilayer periodicity of only 4 nm, this does not come as a surprise. It is, furthermore, in complete agreement with measurements of total electrical resistivity [56], which had shown that the base resistance of the multilayer, indicating complete alloying of the layers, increased significantly only at temperatures higher than 250 °C. Thus, only a weak interfacial alloying could take place at the relevant conditions below that temperature. In order to develop temperature-stable devices, it is insufficient to select only a thermodynamic stable system. In addition, the critical temperature of the respective miscibility gap should be at least threefold higher than the application temperature to suppress the described interfacial broadening. As confirmation, a similar study with APT on Cu/Co (a system distinguished by a much higher critical temperature of about 1600 K) revealed no significant broadening of the interfaces up to a temperature of 450 °C. As a consequence, the degradation of the GMR effect was considered due to grain boundary diffusion [60].

7.5.3
Influence of Grain Boundaries and Curved Interfaces

In nanocrystalline matter with grain sizes down to about 10 nm, the volume fraction attributed to the grain boundary (GB) can easily exceed 50%. With further decreasing grain size an additional necessary topological feature of the boundary arrangement – the so-called 'triple line' – may affect atomic transport. Along a triple line, three GBs merge to form a line-shaped junction defect, the structure of which is expected to differ considerably from that of ordinary GBs. Most likely, it is more disordered, so that a rather high atomic mobility can be assumed. However, until now the measurement of triple line diffusion using analytical electron microscopy has been rare [61]. The difficulty of such measurement stems from the small effective cross-section of the defect, which requires chemical analysis of the highest possible resolution.

In atom probe experiments [62], both Au (15 nm thickness) and Cu layers (25 nm thickness) were deposited on top of tungsten tips. In order to slow down the intermixing of the soluble metals Cu and Au, a thin Co barrier (6 nm thickness) was inserted in between. Due to the strong substrate curvature, thin films deposited on the substrate tips tend to be very fine-grained, with grain sizes down to 5 nm [63]. Thus, these specimens are ideal candidates to observe the transport along topological singularities of the GB arrangement. A typical field ion micrograph of the upper Cu layer is shown in Figure 7.23a. Discontinuities in the structure of concentric rings mark grain boundaries (some of which are emphasized by dashed lines in the figure). Clearly, a polycrystalline structure with a grain size of about 15 nm has formed. Triple junctions are also seen, aligned approximately perpendicular to the surface so that they can be analyzed along the tip axis; in the figure a particularly clear junction is marked by a "T".

By selecting suitable areas for analysis, the local concentration field around the junction could be measured. The geometry of three GBs and an exemplary 2-D composition map is shown in Figure 7.23b. As the atom probe delivers 3-D data, 2-D composition maps can be evaluated in any arbitrary direction subsequent to the measurement, so that the penetration of solute into the triple line and the merging grain boundaries can be determined in a unique manner. The example 2-D map in the figure is aligned perpendicular to a triple line. Clearly, the line is locally enriched

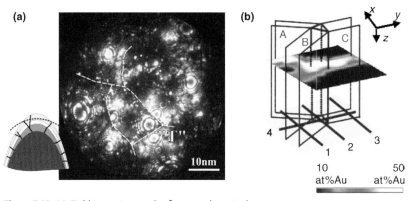

Figure 7.23 (a) Field ion micrograph of sputter-deposited Cu layer. The grain boundaries are marked by dashed lines, a triple junction by 'T'; (b) Two-dimensional composition map in gray-scale representation determined at a cross-section through a triple line and the related three boundaries (A, B, C). (After [62]).

in Au, and the three GBs are also distinguished by a measurable Au content, albeit of a somewhat lower level. The concentration fields around the triple line were evaluated by means of the approximate solution of transport equations proposed by Klinger et al. [64]. In the case of the triple line, the atomic transport may be understood by a three-level cascade process: (i) the material is transported along the triple line; (ii) leakage into the related three boundaries takes place; and (iii) atoms are drained from the grain boundaries into the bulk grain volume. Expressing this in a compact formula, the concentration field is described by:

$$c(x, y, z, t) = c_0 \cdot \exp\left(-\frac{\sqrt{3} \cdot \sqrt[4]{D_{GB} \cdot \delta \cdot s_{GB}} \cdot \sqrt[8]{4D_V/\pi t}}{\sqrt{D_{TL} \cdot q \cdot s_{TL}}} \cdot z\right)$$

$$\times \exp\left(-\frac{\sqrt[4]{4D_V/\pi t}}{\sqrt{D_{GB} \cdot \delta \cdot s_{GB}}} \cdot y\right) \quad (7.19)$$

$$\times \exp\left(1 - \mathrm{erf}\left[\frac{x}{2\sqrt{D_V t}}\right]\right)$$

(For a definition of the coordinates, see Figure 7.23b). If the volume diffusion coefficient D_V is known, the two other diffusion coefficients – or, more exactly, the respective transport products $p_{TL} := D_{TL} \cdot q \cdot s_{TL}$ and $p_{GB} := D_{GB} \cdot \delta \cdot s_{GB}$ – can be determined from logarithmic plots of composition versus penetration depth in the z- and y-directions. (δ, q, s_{TL} and s_{GB} define grain boundary width, effective cross-section of the triple line and the segregation factors of triple line and grain boundary, respectively.) Exemplary plots are presented in Figure 7.24, demonstrating the transport behavior in accordance with Equation 7.19. Indeed, by evaluating the slope of the straight lines in Figure 7.24, it was quantitatively found that D_{TL} is a factor 5600-fold larger than D_{GB} at a temperature of 295 °C, if potential segregation is neglected and $q = \delta^2$ is assumed.

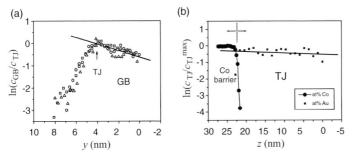

Figure 7.24 Au diffusion in Cu at 295 °C. Normalized concentration profiles along the (a) grain boundary (GB) and (b) triple junction (TJ), as determined with atom probe tomography. For details, see the text. (After [62]).

7.6
Approaching Nonconductive Materials: Pulsed Laser Atom Probe Tomography

On completing this survey of the atom probe method, an additional and quite important methodical aspect must be addressed. The following point could have been presented earlier, in Section 7.3, and frankly – from a logical standpoint – such an order would be more valid. However, by postponing it to the end of the chapter, our aim is to highlight the fact that pulsed laser evaporation represents the most active recent field in methodical research in FIM. Indeed, this technique promises not only to expose APT to a much broader range of materials, but also raises some interesting controversy on the fundamental mechanisms of field evaporation by using ultra-short light pulses of about 100 fs duration.

7.6.1
The Limitations of High-Voltage Pulsing

The conventional atom probe technique based on high-voltage pulsing has always been restricted to the investigation of sufficiently conductive materials, usually metals. Very few measurements of ceramic materials have been reported, and in all successful cases the nonconducting phases had the geometry of tiny particles or thin films embedded in a metallic matrix [65–67]. Besides the unfavorable mechanical properties of ceramics or semi-conductors, this situation is first and foremost due to the limited possibility of transferring a short pulse to the specimen surface. As illustrated in Figure 7.25, the arrangement of sample and detector represents nothing else but a resistive–capacitive (RC) oscillator which is naturally limited in its transfer frequency. A few lines of calculation are sufficient to estimate the required minimum conductivity of the tip material to achieve a sufficient band width of the transfer line. According to Poisson's law, the field in a vacuum is related to a density σ of charges at the surface of the metallic electrode:

$$\sigma = \varepsilon_0 \cdot F = \varepsilon_0 \frac{V}{\beta R}, \qquad (7.20)$$

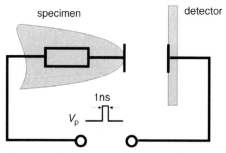

Figure 7.25 Electric equivalent circuit of the arrangement of sample and detector or vacuum chamber.

7.6 Approaching Nonconductive Materials: Pulsed Laser Atom Probe Tomography

where we have made use of Equation 7.1 to formulate the second equation. Approximating the effective electrode surface by a semi-sphere, the electrical capacity of the arrangement reads

$$C = \frac{2\pi R^2 \sigma}{V} = \frac{2\pi R \varepsilon_0}{\beta}. \tag{7.21}$$

With regards to the specimen geometry ($R \approx 50$ nm), this capacity amounts to about 10^{-18} F. Likewise, the resistivity $R_{el.}$ may be estimated by approximating the specimen shape with a truncated cone of length L and semi-shaft angle γ. So, one can formulate

$$R_{el.} = \frac{\rho}{\pi} \int_0^L \frac{dl}{(\gamma \cdot l + R)^2} \approx \frac{\rho}{\pi} \cdot \frac{1}{\gamma \cdot R} \tag{7.22}$$

in which ρ denotes the specific resistivity of the material. In order to transfer pulses properly, the limiting frequency $\omega = 1/(R_{el.} C)$ must reach 10^9 Hz, which requires the specific resistivity to be smaller than

$$\rho_c < \frac{\pi \gamma R}{\omega C} \approx 10^3 \, \Omega \, \text{cm}. \tag{7.23}$$

Clearly, the resistivity of metals is way below this limit, particularly at the low temperatures used for evaporation. However, already in the case of semi-conductors, sufficient conductivity is only achieved with very high doping. In the case of real insulators (such as oxides or nitrides) the condition of Equation 7.23 is failed by orders of magnitude in any case, leaving very little hope to perform successful measurements by electric pulsing.

Additional complications arise from the brittleness of the materials. If we multiply (for an estimate to order of magnitude) the charge density (Equation 7.20) by the field (Equation 7.1), then tensile stresses of up to GPa magnitude are predicted to be induced by the field. In the case of high-voltage pulsing, a significant fraction of this stress is supplied to the tip as a continuous sequence of mechanical shocks resembling a classical fatigue test. Considering that a stress of 1 GPa exceeds the fracture strength of many bulk materials, it is clear that specimens will fail frequently by fracture during the 'fatigue test' of the measurement, and that the risk of this failure increases with the brittleness of the tip material.

7.6.2
The Mechanism of Pulsed Laser Evaporation

With regards to Equation 7.6 (see Section 7.3), it is possible to imagine at least two alternative routes to trigger field evaporation. Either the numerator in the argument of the exponential could be reduced by a short increase of the electric field (as is done with conventional high-voltage pulsing), or the denominator may be increased by short rises in temperature. The latter can be achieved with the light flashes delivered by pulsed laser sources, so it is not surprising that attempts to utilize this effect date back to the 1980s. The early studies with pulsed laser sources were motivated by studies of the photoionization of adsorbed species [68, 69], but shortly afterwards the potential to control field evaporation was also acknowledged and the pulsed laser atom probe (PLAP) introduced by Kellog and Tsong [70]. In these early experiments, nitrogen lasers of 1 ns pulse width (comparable to the length of high-voltage pulses) were normally used. The authors stated the following as the main advantages of the laser evaporation mode:

- As no fast charge transport is required, a low conductivity of the sample no longer represents a bottleneck to the atom probe analysis. The controlled field evaporation of intrinsic Si could be demonstrated.

- As the remaining field is supplied in dc mode, the PLAP does not suffer from the energy spread which is unavoidable under high-voltage pulsing. Thus, a much better mass resolution could be achieved, even with straight flight tubes, without using any time-focusing devices.

- The level of mechanical stress is reduced by the fraction that has been formerly induced by the high-voltage pulses. The remaining stress is produced by the constant base voltage, and thus loaded to the tip permanently. This is a much more favorable situation for the specimen life time than the above-mentioned 'fatigue test' condition.

The time resolution – and thus the accuracy of the energy measurement – was improved by Tsong and coworkers to the order of 10^{-5}. So, by careful measurement of the energy deficits of evaporating species, the mechanism of desorption could be identified [71]. For laser pulses of nanosecond duration, it could be shown that with metals the laser effect was merely thermal in nature, whereas in the case of semiconductors (which were distinguished by a significant band gap) the additional influence of direct photoionization was indicated. As the laser acted predominantly as a pulsed heat source, the question immediately arose as to whether heating of the specimen surface might diminish the spatial resolution of the atom probe analysis. However, already Tsong and coworkers had already shown that it was possible to trigger the evaporation reliably by temperature pulses below 200 K, at which surface diffusion remained frozen, at least for the refractory metals being studied.

Until 1990, the advantages of the pulsed laser mode were exploited not only in investigations of Si, SiC and SiO but also of Group III–V semiconductor materials. However, as these studies were mostly restricted to verify mass spectra and to prove the average composition of the materials, the total number of evaluated atoms was consequently very low. Except for some interesting studies on the oxidation of Si [72], very few attempts were made at a spatially resolved analysis. In this context, it is quite remarkable that reports on the different versions of 3-D atom probes which emerged at that time made no mention of any attachment of a laser beam line. Apparently, only the group at Oxford University attempted to use the PoSAP instrument with laser assistance to perform for example, a study on GaInAs quantum well structures [73]. However also in this case the fact that no volume reconstructions were presented may be seen as an indication of the significant experimental problems that the group encountered. Compared to the vast amount of successful reports on conductive materials with high-voltage pulsing, the pulsed laser technique virtually disappeared during the 1990s. It might be speculated that this situation was essentially caused by the limited reliability of available laser systems and the involved alignment procedure.

Fortunately, this situation has changed remarkably since 2002, and many university research groups and commercial companies are now engaged in the development of atom probe instrumentation, both in terms of testing and offering instruments with the facility of laser-assisted evaporation. Today, commercial laser systems are much more stable, provide pulse frequencies, and have found vast improvements in the ease of operation. Moreover, continuous miniaturization in microelectronics has led to an enormous driving force to overcome any remaining barriers when developing new tools for high-resolution semiconductor analysis.

The first experiments with modern sub-picosecond laser pulses were reported by Gault et al. in 2005 [74]. These authors showed that the efficiency of laser pulsing depended on the orientation of light polarization with respect to the tip axis, and claimed this to be evidence of a direct, athermal influence of the optic wave. In other words, similar to high-voltage pulsing, the activation barrier in the numerator of Equation 7.6 should be temporarily decreased by the electrical field of the wave. This idea was very attractive, as in this way laser-triggered evaporation would become possible without heating the specimen. Moreover, since with a decreasing pulse width at constant pulse energy the electric field amplitude of the optical wave increased, this might indeed be the case. However, the average field strength of the laser beam in the experiments performed by Gault and coworkers was still in the range of only 0.1 V nm^{-1}, which was much less than the effect of typical high-voltage pulses. Nonetheless, one might expect a field enhancement due to polarization of the metallic tip. In theoretical studies, enhancement factors of up to three orders of magnitude have been predicted [75–77], with the fields becoming quite capable of having a significant effect on the evaporation barrier in Equation 7.6. A further conceptual difficulty arises from the fact that optical fields oscillate with a frequency of 10^{15} Hz, while the spectrum of atomic vibration is limited at about 10^{13} Hz. So, how might an atom be excited way off resonance? The group of Gault argued that the rapidly oscillating field could be converted into a directed field for the duration of the pulse by a nonlinear response, in the same way that a diode might be used to demodulate high-frequency radio signals. Indeed, in recent theoretical calculations proof was furthered that this 'optical rectification' would indeed produce field pulses of an appropriate order of magnitude and polarity [78]. In addition, optical pump probe experiments demonstrated a rapid sub-picosecond response, which has been interpreted in the same direction [79].

On the other hand, meanwhile, it became clear that the interpretation of measurements in terms of a clear athermal triggering by Gault et al. [74] was too euphoric, and their evidence less clear [80]. The dependence of evaporation rate on the polarization of the wave could be naturally explained by scattering at the nanometer-sized, cylindrically shaped tip, as had been shown earlier [81]. The measurement of the temperature rise produced by sub-picosecond pulses, although performed with a remarkable pump probe experiment that combined laser and voltage pulses [82], delivered an ambivalent result. Yet, the idea of achieving direct-field pulsing by using ultra-short laser pulses proved so attractive that considerable effort is currently being undertaken to clarify this aspect, notably by the group at the University of Rouen.

Today, several other teams are focusing on the practical consequences of laser pulsing to concrete analytical studies. With laser pulsing, several new parameters must be explored to define the optimum measurement conditions. An extended study [83] has investigated the influence of wavelength and energy density of the pulses, laser spot size, specimen geometry, and heat conductivity on the quality of mass spectra. Yet, interestingly, no significant influence of pulse width and wavelength on the quality of mass spectra was identified. The mass resolution achieved with laser pulsing is indeed significantly improved in comparison to high-voltage pulsing (up to $\Delta m/m = 1:1000$, without using a reflectron). However, significant thermal peak tails are observed in the mass spectra, with rather long laser pulses exceeding the important picosecond benchmark of electron phonon coupling, as well as with short pulses way below this threshold. Instead of the pulse width, the duration at which the ions are evaporated is essentially defined by the cooling period of the specimen after the pulse. Thus, samples of low heat conductance or of particularly long and thin geometry lead to poor mass resolution. With materials of low heat conductivity, such as stainless steels, a small heat spot achieved with an optimally focused laser beam has proved to be advantageous when achieving rapid cooling rates, and thereby good mass resolution. Remarkably, also in the case of a femtosecond laser, the tails of the mass peaks are correctly described by the principles of heat conduction, which proves that also with this type of laser a considerable temperature rise appears during the pulse.

In summary, it can be stated that today, a final conclusion of whether short pulses in the range of 100 fs provide any significant advantage in the analysis cannot be drawn. Although, in various reports, the impression is sometimes raised that femtosecond pulses are required for the reliable analysis of difficult materials (such as the measurement of oxides [84]), the opposite can be clearly demonstrated [85].

7.6.3
Application to Microelectronic Devices

Unaffected by the scientific controversy surrounding the mechanism of laser-assisted evaporation, the 3-D atom probe has advanced in recent years to a modern measurement tool, and is set to become an established component of those industrial laboratories involved with microelectronics. While the structural width of a transistor now ranges down to 35 nm, the measuring field of the atom probe has reached approximately 100 nm. This almost perfect matching of length scales has motivated the industrial application of the method. With the number of evaluated atoms exceeding 100 million, APT now permits the mapping of dopant distributions, in outstanding resolution, where the chemical sensitivity of analytical TEM is usually way too low (see e.g. [86]).

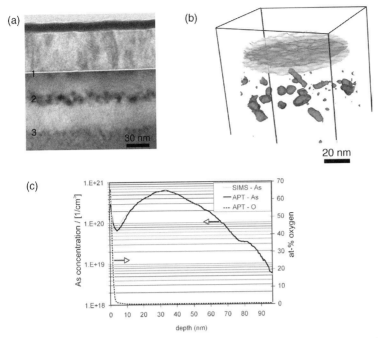

Figure 7.26 Analysis of As doping of Si by ion implantation. (a) Cross-section TEM image of the test structure showing an oxide layer (1), As-rich particles (2), and the end of damage range (3). The ion irradiation was apparent from the top; (b) Volume analysis by APT showing iso-concentration surfaces at 10 at.% oxygen (light gray) and 2 at.% As (dark gray); (c) Comparison of atom probe data with SIMS measurement. (Reproduced with permission from [87]).

These findings are illustrated by the results of the study shown in Figure 7.26. Here, a {001} Si wafer was implanted with 2×10^{15} As atoms per cm^2 with 30 keV kinetic energy. An oxide film of 2 nm thickness and 50 nm of undoped, polycrystalline Si were then deposited, to mimic the basic steps of microelectronics production. While forming the polycrystalline Si, the specimen had to be heated to 600 °C for about 30 min, so that the implanted As atoms could cluster. As shown in the TEM image (Figure 7.26a), a defect band was formed at approximately 30 nm beneath the substrate surface (this is localized in the figure by the thin light oxide layer). For atom probe analysis, tip-shaped samples were prepared by FIB milling. With pulsed laser assistance, the semiconductor sample could be properly evaporated, and from the volume reconstruction local composition maps calculated which allowed the compositional heterogeneities to be located by means of iso-concentration surfaces (Figure 7.26b). At a preset concentration level of 2 at.% As, these iso-surfaces mark As-rich clusters, which have formed by nucleation. The iso-surfaces at a level of 10 at.% oxygen were used to localize the interfaces to the oxygen layer. Quantitative composition profiles aligned perpendicular to the substrate surface could also be determined (as shown in Figure 7.26c) in

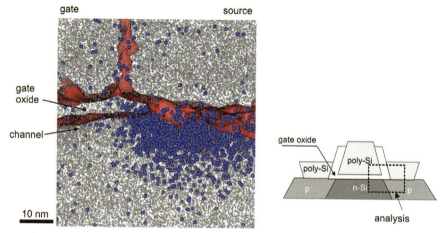

Figure 7.27 Three-dimensional reconstruction of a field effect transistor structure showing Si atoms (light gray dots) and boron dopant (blue). Iso-concentration surfaces at 3 at% oxygen localize the gate oxide film (red). (Illustration courtesy of D.J. Larson, Imago Scientific Instruments.) The schematic at the right clarifies the location of the analyzed region.

comparison to depth profiling by secondary ion mass spectrometry (SIMS), which represents the industrial standard. Clearly, the atom probe delivers quantitative data in agreement with SIMS, but more importantly the APT data proved to be more accurate. While the segregation of As to the substrate surface is indicated only faintly by SIMS, the corresponding composition peak appears very pronounced in the APT measurement, due to a significantly better spatial resolution of the latter technique. As the atom probe is equally sensitive to all chemical species, the oxygen content of the oxide is determined within the same measurement.

The final example refers to the preparation of tip samples of technical devices taken from industrial production. By using modern, dual-beam FIB, areas of interest can be selected for TAP measurement. In this way, the corner region of a field effect transistor – between the source contact and the channel beneath the gate contact – has been analyzed with regards to local dopant and oxygen distribution (see Figure 7.27). A few years ago, the achievement of such an analysis from a microchip was beyond thinking. Having mapped the spatial arrangement of the atoms, not only the depth profile beneath the source or drain contacts of the field effect transistor could be determined, but also the lateral distribution of the dopant. In this way, essential information can be obtained concerning dopant diffusion from the contact region into the channel under the gate oxide. As the structural width of transistors continues to decrease, the control of this sideward transport may in particular become a critical issue. It is likely that, in future, APT will become the most important tool for controlling this undesired process.

References

1 Müller, E.W. (1951) *Zeitschrift für Physik*, **131**, 136.
2 Kelly, T.F. and Miller, M.K. (2007) *Review of Scientific Instruments*, **78**, 031101.
3 Cerezo, A., Clifton, P.H., Galtrey, M.J., Humphreys, C.J., Kelly, T.F., Larson, D.J., Lozano-Perez, S., Marquis, E.A., Oliver, R.A., Sha, G., Thompson, K. and Zandbergen, M. (2007) *Materials Today*, **10**, 1.
4 Miller, M.K., Cerezo, A., Hetherington, M.G. and Smith, G.D.W. (1996) *Atom Probe Field Ion Microscopy*, Oxford Science Publications, Oxford.
5 Miller, M.K. (2000) *Atom Probe Tomography*, Kluwer Academic, New York.
6 Tsong, T.T. (1990) *Atom-Probe Field Ion Microscopy*, Cambridge University Press, Cambridge.
7 Müller, E.W. (1957) *Journal of Applied Physics*, **28**, 1.
8 Müller, E.W., Panitz, J.A. and McLane, S.B. (1968) *Review of Scientific Instruments*, **39**, 83.
9 Cerezo, A., Godfrey, T.J. and Smith, G.D.W. (1988) *Review of Scientific Instruments*, **59**, 862.
10 Deconihout, B., Bostel, A., Menand, A., Sarrau, J.M., Bouet, M., Chambreland, S. and Blavette, D. (1993) *Applied Surface Science*, **67**, 444.
11 Sijbrandij, S.J., Cerezo, A., Godfrey, T.J. and Smith, G.D.W. (1996) *Applied Surface Science*, **94–95**, 428.
12 Panayi, P. (2006) Great Britain Patent Application No. GB2426120A, November 15.
13 Kelly, T.F., Gribb, T.T., Olson, J.D., Martens, R.L. et al. (2004) *Microscopy and Microanalysis*, **10**, 373.
14 Tsong, J. (1978) *Journal of Physics F: Metal Physics*, **8**, 1349.
15 Bas, P., Bostel, A., Deconihout, B. and Blavette, D. (1995) *Applied Surface Science*, **87/88**, 298.
16 Al-Kassab, T., Wollenberger, H., Schmitz, G. and Kirchheim, R. (2003) High Resolution Imaging and Spectrometry of Materials (eds T. Ernst and M. Rühle), Springer, Berlin, p. 290.
17 Schmitz, G. and Howe, J.M. (2007) High Resolution Microscopy, in *Alloy Physics* (ed. W. Pfeiler), Wiley, pp. 774–860.
18 Hellman, O. and Seidman, D. (2000) *Microscopy and Microanalysis*, **6**, 437.
19 Warren, P.J., Cerezo, A. and Smith, G.D.W. (1998) *Ultramicroscopy*, **73**, 261–266.
20 Vurpillot, F., da Costa, G., Menand, A. and Blavette, D. (2001) *Journal of Microscopy*, **203**, 295.
21 Geiser, B.P., Kelley, T.F., Larson, D.J., Schneir, J. and Roberts, J.P. (2007) *Microscopy and Microanalysis*, **13**, 437–447.
22 Vurpillot, F., Bostel, A., Cadel, E. and Blavette, D. (2000) *Ultramicroscopy*, **84**, 213.
23 Vurpillot, F., Bostel, A. and Blavette, D. (2001) *Ultramicroscopy*, **89**, 137.
24 Cerezo, A., Godfrey, T.J., Hyde, J.M., Sijbrandij, S.J. and Smith, G.D.W. (1994) *Applied Surface Science*, **76/77**, 374.
25 Deconihout, B., Renaud, L., Da Costa, G., Bouet, M., Bostel, A. and Bavette, D. (1998) *Ultramicroscopy*, **73**, 253.
26 Cerezo, A., Godfrey, T.J. and Smith, G.D.W. (1988) *Review of Scientific Instruments*, **59**, 862.
27 Keller, H., Klingelhöfer, G. and Kankeleit, E. (1987) *Nuclear Instruments & Methods*, **A258**, 221.
28 Jagutzki, O., Cerezo, A., Czasch, A. et al. (2002) *IEEE Transactions on Nuclear Science*, **49**, 2477.
29 da Costa, G., Vurpillot, F., Bostel, A., Bouet, M. and Deconihout, B. (2005) *Review of Scientific Instruments*, **76**, 013304.
30 Deonihout, B., Vurpillot, F. and Gault, B. (2007) *Surface and Interface Analysis*, **39**, 278.
31 Stender, P., Oberdorfer, C., Artmeier, M. and Pelka, P. (2007) *Ultramicroscopy*, **107**, 726–733.
32 Nishikawa, O. and Kimoto, M. (1994) *Applied Surface Science*, **76/77**, 424.

33 Kelly, T.F., Gribb, T.T., Olson, J.D., Martens, R.L. et al. (2004) *Microscopy and Microanalysis*, **10**, 373.

34 Miller, M.K. and Russel, K.F. (2007) *Ultramicroscopy*, **107**, 761–766.

35 Perea, D.E., Allen, J.E., May, S.J., Wessels, B.W., Seidman, D.N. and Lauhon, L.J. (2006) *Nano Letters*, **6**, 181.

36 Schleiwies, J. and Schmitz, G. (2002) *Materials Science and Engineering A*, **327**, 94.

37 Lang, C. and Schmitz, G. (2003) *Materials Science and Engineering A*, **353**, 119.

38 Hono, K., Hasegawa, N., Okano, R., Fujimori, H. and Sakurai, T. (1993) *Applied Surface Science*, **67**, 407.

39 Larson, D. (2001) *Microscopy and Microanalysis*, **7**, 24.

40 Giannuzi, L.A. and Stevie, F.A. (1999) *Micron*, **30**, 197.

41 Michaelsen, C., Barmak, K. and Weihs, T.P. (1997) *Journal of Physics D - Applied Physics*, **30**, 3167.

42 Roy, R. and Sen, S.K. (1992) *Journal of Materials Science*, **27**, 6098.

43 Bergmann, C., Emeric, E., Clugnet, G. and Gas, P. (2001) *Defect and Diffusion Forum*, **194–199**, 1533.

44 Coffey, K.R., Clevenger, L.A., Barmak, K., Rudman, D.A. and Thompson, C.V. (1989) *Applied Physics Letters*, **55**, 852.

45 Gusak, A.M. (1990) *Ukrainian Journal of Physics*, **35**, 725.

46 Desré, P.J. and Yavari, R. (1990) *Physical Review Letters*, **64**, 1533.

47 Hodaj, F. and Desré, P.J. (1996) *Acta Materialia*, **44**, 4485.

48 Vovk, V., Schmitz, G. and Kirchheim, R. (2004) *Physical Review B - Condensed Matter*, **69**, 104102.

49 Pasichnyy, M.O., Schmitz, G., Gusak, A.M. and Vovk, V. (2005) *Physical Review B - Condensed Matter*, **72**, 014118.

50 Jeske, T. and Schmitz, G. (2001) *Scripta Materialia*, **45**, 555.

51 Schleiwies, J. and Schmitz, G. (2002) *Materials Science and Engineering A*, **327**, 94.

52 Schleiwies, J., Schmitz, G., Heitmann, S. and Hütten, A. (2001) *Applied Physics Letters*, **78**, 3439.

53 Zhou, X.W. et al. (2001) *Acta Materialia*, **49**, 4005.

54 Larson, D.J., Cerezo, A., Clifton, P.H., Petford-Long, A.K., Martens, R.L., Kelly, T.F. and Tabat, N. (2001) *Journal of Applied Physics*, **89**, 7517.

55 Larson, D.J. et al. (2003) *Physical Review B - Condensed Matter*, **67**, 144420.

56 Hecker, M., Tietjen, D., Wendrock, J., Schneider, C.M., Cramer, N. and Malinski, L. (2002) *Journal of Magnetism and Magnetic Materials*, **247**, 62.

57 van Loyen, L., Elefant, D., Tietjen, D., Schneider, C.M., Hecker, M. and Thomas, J. (2000) *Journal of Applied Physics*, **87**, 4852.

58 Ene, C.B., Schmitz, G., Kirchheim, R. and Hütten, A. (2005) *Acta Materialia*, **53**, 3383.

59 Stender, P., Ene, C.B., Galinski, H. and Schmitz, G. (2008) *International Journal of Materials Research*, **99**, 480.

60 Vovk, V. and Schmitz, G., *Ultramicroscopy* (inpress).

61 Bokstein, B., Ivanov, V., Oreshina, O., Pteline, A. and Peteline, S. (2001) *Materials Science and Engineering A*, **302**, 151.

62 Schmitz, G., Ene, C., Lang, C. and Vovk, V. (2006) *Advanced Science and Technology*, **46**, 126.

63 Lang, C. and Schmitz, G. (2003) *Materials Science and Engineering A*, **353**, 119.

64 Klinger, L.M., Levin, L.A. and Petelin, A.L. (1997) *Defect and Diffusion Forum*, **143–147**, 1523.

65 Shashkov, D.A. and Seidman, D.N. (1995) *Physical Review Letters*, **75**, 268.

66 Kluthe, C., Al-Kassab, T. and Kirchheim, R. (2002) *Materials Science and Engineering A*, **327**, 70.

67 Kuduz, M., Schmitz, G. and Kirchheim, R. (2004) *Ultramicroscopy*, **101**, 197.

68 Tsong, T.T., Block, J.H., Nagasaka, M. and Viswanathan, B. (1976) *Journal of Chemical Physics*, **65**, 2469.

69 Nishigaki, S., Drachsel, W. and Block, J.H. (1979) *Surface Science*, **87**, 389.

70 Kellog, G.L. and Tsong, T.T. (1980) *Journal of Applied Physics*, **51**, 1184.

71 Tsong, T.T. (1984) *Physical Review B - Condensed Matter*, **30**, 4946.

72 Grovenor, C.R.M. and Cerezo, A. (1989) *Journal of Applied Physics*, **65**, 5089.

73 Liddle, J.A., Norman, A., Cerezo, A. and Grovenor, C.R.M. (1989) *Applied Physics Letters*, **54**, 1555.

74 Gault, B., Vurpillot, F., Bostel, A., Menand, A. and Deconihout, B. (2005) *Applied Physics Letters*, **86**, 094101.

75 Novotny, L., Bian, R.X. and Xie, X.S. (1997) *Physical Review Letters*, **79**, 645.

76 Martin, O. and Girard, C. (1997) *Applied Physics Letters*, **70**, 705.

77 Martin, Y., Haffmann, H.F. and Wickramasinghe, H.K. (2001) *Journal of Applied Physics*, **89**, 5774.

78 Vella, A., Deconihout, B., Marrucci, L. and Santamato, E. (2007) *Physical Review Letters*, **99**, 046103.

79 Vella, A., Gilbert, M., Hideur, A., Vurpillot, F. and Deconihout, B. (2006) *Applied Physics Letters*, **89**, 251903.

80 Cerezo, A., Smith, G.D.W. and Clifton, P.H. (2006) *Applied Physics Letters*, **88**, 154103.

81 Robins, E.S., Lee, M.J.G. and Langlois, P. (1986) *Canadian Journal of Physics*, **64**, 111.

82 Vurpillot, F., Gault, B., Vella, A., Bouet, M. and Deconihout, B. (2006) *Applied Physics Letters*, **88**, 094105.

83 Bunton, J.H., Olson, J.D., Lenz, D.R. and Kelly, T.F. (2007) *Microscopy and Microanalysis*, **13**, 418.

84 Gault, B., Menand, A., de Geuser, F. and Deconihout, B. (2006) *Applied Physics Letters*, **88**, 114101.

85 Oberdorfer, C., Stender, P., Reinke, C. and Schmitz, G. (2007) *Microscopy and Microanalysis*, **13**, 342.

86 Kelley, T.F., Larson, D.J., Thompson, K., Alvis, R.L., Bunton, J.H., Olson, J.D. and Gorman, B.W. (2007) *Annu. Rev. Mater. Res*, **37**, 681–727.

87 Thompson, K., Flaitz, P.L., Ronsheim, P., Larson, D. and Kelley, T. F. (2007) *Science*, **317** 1370.

88 Schlesiger, R. (2008) Design und Aufbau einer Mikroelektroden Atomsonde, Dipl. Thesis, University of Münster.

8
Cryoelectron Tomography: Visualizing the Molecular Architecture of Cells*

Dennis R. Thomas and Wolfgang Baumeister

8.1
Introduction

In order to reveal the networks of macromolecular interactions which underlie higher cellular functions on a systems level, new techniques are needed. As a starting point, it is crucial to have a validated and quantified list of all components, which can be provided by using mass spectrometry-based proteomics techniques [1]. The next challenge is to analyze the interaction patterns *in situ* with increasing degrees of complexity, ranging from that of supramolecular modules, to organelles or even whole cells. Some of these systems are tightly integrated complexes, robust enough to withstand the isolation and purification procedures that are used traditionally in biochemistry. These functional modules are amenable to detailed studies with the established tools of structural biology [2]. Other complexes interact transiently and weakly to form supramolecular networks that are designed to associate or dissociate in response to specific signals. Many such putative cellular complexes have been identified through affinity-based isolation methods. The components of these complexes can be identified using mass spectrometry to provide valuable information regarding the composition of such interacting complexes. In this manner interactions can be detected, whether they are direct or indirect [3]. However, this type of approach is prone to error and does not provide information about the order or molecular details of how the components interact. As the complexity of interacting systems increases, the component lists and affinity-based interaction data no longer suffice to describe the architecture of networks [4]. Ideally, the target would be a 'snapshot' of the system in action, acquired in a nondisruptive, noninvasive manner, avoiding perturbations to the system. Thus, we can hope to obtain a detailed three-dimensional (3-D) image of the system in action at molecular resolution.

*This chapter is a modified and updated version of a previously published article: Baumeister, W. (2005) From proteomic inventory to architecture. *FEBS Letters* **579**(4), 933–7.

Electron tomography is uniquely suited to studying large pleomorphic structures and visualizing macromolecules in their functional cellular context. The development and implementation of automated low-dose data-acquisition procedures, has made it possible to study biological samples embedded in vitreous ice. Electron cryotomography provides a 3-D picture of complex systems preserved in as near to a native state as can be achieved at molecular resolution [5]. Today, we have the tools to bridge the (resolution) gap that currently exists between cellular and high-resolution molecular structural techniques. The reconstructed tomograms of organelles or cells at molecular resolution represent snapshots of their proteomes. These 3-D snapshots can be interpreted using advanced pattern recognition techniques, revealing the molecular architecture present at the instant the sample was frozen. The fitting of tomographic maps of cells with high-resolution structures of their components should ultimately enable us to generate pseudo-atomic models of large – and otherwise elusive – assemblies and networks in the act of performing their cellular role [6].

8.2
Basic Principles and Challenges of Electron Tomography

An electron micrograph is a two-dimensional (2-D) projection of the sample present on the specimen support. Much like a medical X-ray, all structural features present in three dimensions are imaged, but are seen superimposed in the 2-D image. As a consequence, the images are difficult to interpret directly. In electron tomography, the specimen holder is tilted incrementally around an axis perpendicular to the electron beam, and images are taken at each position. These images are projections of the same object as viewed from different angles. Based on principles first described by Radon [7], these projections can be combined to produce a 3-D density map (Figure 8.1). Before such a density map is calculated – most commonly by using a 'weighted back-projection' algorithm – the projection images must be all be aligned to a common origin.

In order to successfully apply electron tomography to ice-embedded specimens, it is necessary to overcome two conflicting sets of limitations which have, for more than two decades, stood in the way of its widespread application to radiation-sensitive biological samples:

- In order to obtain the most detail with the least distortion in the reconstruction, it is necessary to sample as large an angular range as possible while tilting with increments as small as possible. This strategy calls for maximizing the number of projection images used for the reconstruction.

- Biological specimens embedded in vitreous ice, are extremely sensitive to radiation damage; therefore, it is very important to minimize exposure to the beam. To acquire a tilt series of perhaps more than a hundred images, the dose per image must be limited in order to prevent radiation damage from destroying the finer details of the structure or, worse yet, rendering reconstructions useless. The problem is, as one lowers the dose per image, the signal-to-noise ratio (SNR)

Figure 8.1 (a) Single axis tilt tomographic data acquisition. The unknown object is represented by a 'flexible knot' to emphasize the fact that electron tomography can reconstruct structures with unique topologies. A set of projection images is recorded as the object is tilted incrementally; (b) Following alignment of the projection images, the object is reconstructed generally by weighted back-projection. The 2-D projections are recombined to generate the 3-D density distribution of the object – the tomogram. The implementation of algorithms such as algebraic reconstruction techniques (ARTs) and simultaneous iterative reconstruction technique (SIRT) provide means for refining reconstructions [8, 9].

decreases. Thus, it is necessary to choose between having a better SNR but fewer projections, or more projections with a lower SNR.

Given the above considerations, we must consider how to 'spend' the allowable electron dose. Early theoretical considerations [10], corroborated by computer simulations [11], have suggested that, in principle, the electron dose needed to visualize structural features at a particular resolution limit is the same for 2-D and 3-D images containing equivalent information. In principle, if one were to average tomograms (3-D images), the dose could be distributed over as many projections as required to achieve the desired resolution, at the expense of lowering the SNR of the individual 2-D images. A consequence of combining the information from the projection images is an improvement in the SNR, similar to the improvement obtained by averaging statistically noisy images of repetitive structures. There is, nevertheless, a practical limitation, namely that the SNR of the 2-D images must be sufficient so as to permit the accurate alignment needed to establish a common framework of coordinates for all projection images.

If a constant exposure or dose per image were to be used throughout the tilt series then, as the tilt angles increased, images would be obtained with increasingly worse SNRs. This is because, as the tilt angle increases, the thickness of the specimen through which the electron beam passes also increases. One way to solve this problem would be to distribute the dose such that the dose or exposure time increased with the increasing tilt. The intended effect would be to keep the signal content of the images more or less constant. There are variations which can be used with this approach. One is to collect increasingly finer tilt increments at higher tilt angles without changing dose or, alternatively, to combine the previous two ideas and both increase the dose at higher tilt and collect finer increments. This latter approach has the drawback of spending a great deal of the total allowed dose at high tilt angles which, due to increased specimen thickness and potential beam damage, may or may not contribute much to the information in the resulting tomogram.

Projection image alignment can be achieved using cross-correlation methods that may include area matching or feature tracking, or by the alignment of high-contrast fiducial markers, such as gold nanoparticles. The electron-dense gold particles have a relatively high SNR, even in very low-dose projections. However, gold – while not subject to radiation damage itself – is affected by the beam because the ice matrix in which it is embedded is subject to radiation-induced changes. If one compares the image of the untilted specimen at acquired at the start of the tilts series with an untilted image of the same area recorded after the tilt series has been completed, this can be observed as a movement of the gold relative to the specimen over the course of collecting a tilt series. Like radiation damage, this effect is dependent on the total dose and other factors such as specimen thickness. Thus, whether using cross-correlation only methods, or using fiducial marker-based alignment, the total electron dose which can be used is limited.

8.3
Automated Cryoelectron Tomography

Computer-controlled transmission electron microscopes first became commercially available during the late 1980s. This development, combined with the improving quality and availability of large-area charge-coupled device (CCD) cameras, made possible the development of sophisticated software that could be used to control the microscope and image-acquisition in a fully automated manner [12–15]. This software maintains the specimen centered in the field of view, by controlling the stage movement or beam shift, and determines image defocus automatically (Figure 8.2). An important development is the ability to track (center) and focus on areas some distance along the tilt axis away from the sample being imaged, thus minimizing the exposure of the area of interest. The fraction of the dose that is spent on overhead [search, centering, (auto) focusing] with automation has thus been reduced to as little as 3% of the total dose – something which is utterly impossible to achieve with manual operation [16].

These developments have greatly improved the status of electron tomography of cryopreserved specimens from that of a technique with potential to that of one beginning to produce exciting results. Starting with 'phantom cell' experiments – that is, liposomes encapsulating macromolecules [17, 18], and more recently with viruses, prokaryotic [19, 20] and eukaryotic cells [21] – we can now apply the potential of 3-D imaging to samples preserved in a close-to-life state. Vitrification by rapid freezing not only preserves the native molecular and cellular structures, but also allows 'snapshots' to be taken of dynamic events, thus freezing moments in time; an example would be trapping the short-lived open state of the acetylcholine receptor [22–24]. Vitrified samples do not suffer from the artifacts traditionally associated with chemical fixation and staining, nor with the dehydration of cellular structures. Tomograms of frozen-hydrated structures have a natural density distribution (albeit noisy), whereas staining and preservation reactions

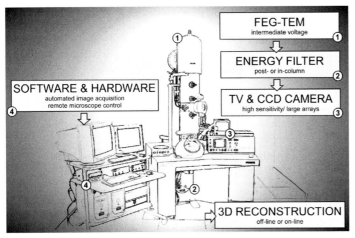

Figure 8.2 Cartoon showing where the advances in electron microscopy come into play. (1) The development of highly coherent intermediate-voltage electron sources [field emission guns (FEGs)] results in images with good signal-to-noise ratios extending to higher resolutions; (2) The application of energy filters to exclude inelastically scattered electrons from the image reduces noise in the image; (3) The acquisition of digital images on CCD cameras allows data acquisition to be automated; (4) The ability to control specimen movement, focusing, image tracking and image acquisition with computer software increases the efficiency of data collection and minimizes the electron dose spent on non-data-acquisition steps.

tend to produce artificial contrast from intricate mixtures of positive and negative staining. Unfortunately, the artifacts of staining and preservatives make the molecular interpretation of tomograms from such techniques very problematic [25].

8.4
Resolution, Signal-to-Noise Ratio and Visualization of Tomograms

Today, the development of automated procedures has made the recording of low-dose tilt series a routine procedure, with user-friendly software available for downstream processing [26, 27]. It is in fact now significantly less cumbersome and time-consuming to obtain a cryotomogram than to go through the conventional procedures of plastic embedding and sectioning biological samples. There is, however, one caveat – that sample thickness is a limitation. The thickness of whole prokaryotic cells, isolated organelles or thin (<1 μm) eukaryotic cells limits the resolution of cryotomograms to at best a range of 4–5 nm, although the prospects for improvement are good (see Refs. [26, 28]). The obstacle presented by sample thickness makes the development of reliable protocols for the sectioning of frozen-hydrated material a priority. While progress is clearly being made towards obtaining thin, artifact-free sections from samples preserved in vitreous ice, tomographic studies of thin cryosections are still not routine [29]. However, on-going improvements in instrumentation can be expected to make significant improvements in data quality. Better still, more efficient CCD detectors, in particular, would improve the resolution by retaining the higher resolution signal that currently is lost to the modulation transfer function (MTF) of present-day CCDs. These improvements would allow tomography to enter the realm of molecular resolution (2–3 nm). In addition, with dual-axis tilting schemes, the effects of missing data (due to the restricted tilt range; usually $\pm 70°$) could be reduced and resolution become more isotropic.

In electron tomography, as in the processing of single particles or 2-D crystals, an attainable resolution and the SNR are intricately linked. A signal may be present in a tomogram at high resolution, but detection of the information is limited by the limited degree of averaging which is inherent in the tomographic reconstruction process. As a consequence of averaging, the noise cancels (and thus is reduced in the averaging process) while the signal is summed. However, the signal can only be detected at frequencies where an acceptable SNR is achieved [30]. Electron microscopic single-particle analysis benefits greatly from the stratagem of combining a very large number of images (i.e. >10 000) of a structure viewed in random orientations, into a 3-D reconstruction, where averaging significantly improves the SNR [31]. The noise reduction resulting from averaging during reconstruction is quite limited in electron tomography (<150 images), given the uniqueness of cellular tomograms. Therefore, other means of noise reduction must be applied to the analysis of tomograms. For example, if the tilt series has been acquired in such a way that information does not extend past a certain resolution limit (say 5 nm), the tomogram may be Fourier-filtered with a cut-off at that resolution, thereby eliminating what can only be noise at frequencies higher than the cut-off frequency. Sophisticated 'de-noising' algorithms are also available; these are based on the same basic principle but employ 'diffusion' criteria based on local continuity of density [32]. Although the gain in SNR obtained from such methods is not very large, this does help in visualizing the underlying structure present in the tomograms. Nonetheless, new algorithms are required to achieve a greater noise reduction.

The interpretation of a tomogram at an ultrastructural level requires the identification of structural components – that is, membrane-bound organelles, cytoskeletal filaments or large macromolecular complexes. In the past, manual assignment has been commonly used as human pattern recognition is often superior to available segmentation algorithms; on the other hand, machine-based segmentation should, in principle, be more objective. Continuous structures are relatively easy to recognize and delineate, in spite of the low SNR present in cryotomograms. For example, visualizing the organization of the cytoskeleton in both *Spiroplasma melliferum* and *Dictyostelium discoideum*, was possible at the level of individual filaments, without the need for extensive post-processing (Figure 8.3) [21, 33].

Although averaging can obviously not be applied to tomograms of unique structures such as individual cells or organelles, such tomograms may nevertheless contain multiple copies of components such as ribosomes, chaperones or proteases. Small regions of the tomogram containing isolated complexes can be extracted from the tomogram *in silico*, and these so-called 'subtomograms' can be subjected to classification against a library of known structures (see Figure 8.4). The subtomograms can then be aligned to a common orientation and averaged within the appropriate class. The result should be an average with a better SNR and a higher

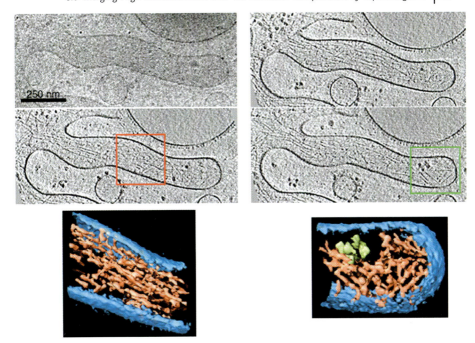

Figure 8.3 Actin filament organization in filopodia [34]. The upper left panel is a projection image from the tilt series. The next three images are sequential sections taken from the tomogram. Two segments of the filopod have been surface- rendered to reveal the organization of the actin filaments and interactions with the membrane. The red box indicates the region shown in the rendered image (lower left); the green box indicates the region rendered (lower right).

resolution. The original low-resolution tomographic image can be replaced by the average or by the higher-resolution template itself, if available. The result is a 'synthetic' tomogram with a much improved, local SNR. Such a procedure has been used in a tomographic study of enveloped *herpes simplex* virions [19] (Figure 8.5), and to also visualize nuclear pore complexes in intact nuclei [35, 36] (Figure 8.6).

8.5
Merging High Resolution with Low: The Molecular Interpretation of Cryotomograms

In tomographic reconstructions of vitrified samples, the macromolecular content of an organelle or cell is present in its native state, thereby making interpretation at the molecular level less problematic. Although the resolution of an individual tomogram may be limited, the advantage is that everything can be seen in its native context. A vast amount of information is available, as tomograms are 3-D images of the entire proteome, and should ultimately enable the spatial relationships of macromolecules to be mapped in an unperturbed cellular environment. However, the retrieval of this information is faced with major problems. First, although everything can be seen, the identification of what is seen may be difficult in the crowded macromolecular

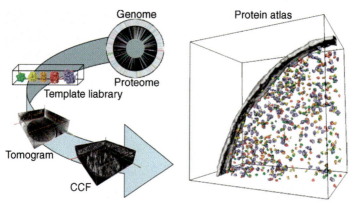

Figure 8.4 Using template matching to generate a protein atlas of a cell. Proteomics, nuclear magnetic resonance, X-ray crystallography and electron microscopy have produced a wealth of structural templates which can be used as a library for analyzing tomograms. In this example, templates from the library have been chosen as probes for the tomogram. Each template, in a number of different orientations, is cross-correlated against the tomogram. The positions of complexes appear as peaks in the 3-D cross correlation function (CCF). A box or subvolume is extracted centered on each of these peaks, resulting in a dataset of 3-D images or subvolumes. These subvolumes are then aligned in three dimensions against the library of complexes. Each volume is assigned to the class to which it correlates best, and each of the classes is averaged, increasing the signal-to-noise ratio and improving resolution. Finally, the class average or original template for the class can be substituted for the original complex in the tomogram, in the determined orientation. This produces a 3-D map of the cell with the contents identified. It is now possible to consider how the various complexes might be related in the cellular context.

environment where complexes literally touch each other [36]. Therefore, the interpretation of low-SNR cryo-tomograms is difficult and can be tedious. Furthermore, because it is not possible to tilt a full 180°, the tomographic reconstructions are distorted by missing data, and this results in a nonisotropic resolution. There are essentially only two options for identifying macromolecules in tomograms, namely *specific labeling* or *pattern recognition methods*, where complexes are matched against a library of known structures. Of course, the two approaches are not mutually exclusive.

8.5.1
Specific Labeling

Proteins exposed on the surface of cells or organelles can be labeled with antibodies or, specific ligands bound to gold nanoparticles. These labels provide indicators for the presence of a specific molecule within a broad molecular landscape. Intracellular labeling is more problematic and requires innovative approaches. These approaches could be based on noninvasive genetic manipulations generating covalent fusions with a protein such as metallothionein, which has the potential to bind heavy metals such as gold [38, 39]. Ideally, the aim would be to introduce a label in a time-resolved experiment (thereby identifying a particular event), and subsequently to remove the background, as is achieved with the ReAsH compound in fluorescence microscopy [40]. However, the achievement of labeling that can be statistically quantified is a

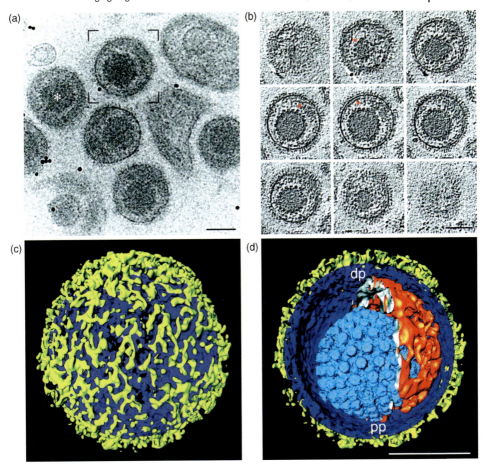

Figure 8.5 (a, b) Tomographic reconstruction of HSV-1 virions in vitreous ice (based on Ref. [19]). (a) The untilted projection from a tilt series. The black dots are 10 nm gold particles used as fiducial markers; (b) Gallery of sections from the tomogram taken from the virion framed in panel (a). Each section is an average of seven planes from the tomogram, and represents a slab which is 5.2 nm thick. The sections are separated by 15.5 nm. Red arrowheads mark filaments in the tegument. All scale bars are 100 nm. (c, d) The surface-rendered tomogram from the same virion after denoising; (c) Outer surface showing the distribution of glycoprotein spikes (yellow) protruding from the membrane (blue); (d) Cutaway view of the virion interior, showing the capsid (light blue) and the tegument 'cap' (orange) inside the envelope (blue and yellow). pp = proximal pole; dp = distal pole.

daunting task. It is also difficult to imagine that labeling can be developed such that it becomes a high-throughput technology capable of mapping entire proteomes. In order to identify every molecule of interest, the entire procedure of labeling, as well as data acquisition and reconstruction of the tilt series, must be repeated. Moreover, the unique nature of cellular tomograms makes the direct correlation of the different maps impossible, which in turn poses a major problem when deriving such maps from the molecular interaction patterns.

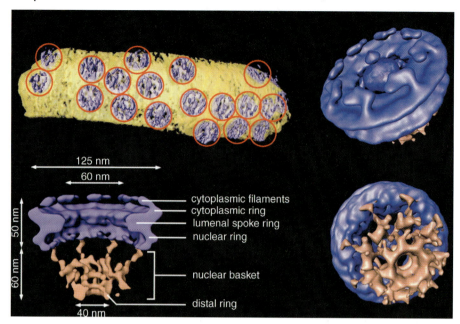

Figure 8.6 The structure of the nuclear pore complex (NPC) obtained from averaging subvolumes [35]. A surface-rendered representation of a segment of nuclear envelope (NPCs in blue, membranes in yellow). The dimensions of the rendered nuclear volume are 1680 nm × 984 nm × 558 nm. The upper right diagram shows the cytoplasmic face of the NPC, and the lower right the inward (nuclear) face. The distal ring of the basket is connected to the nuclear ring by the nuclear filaments. The lower left diagram shows a cutaway view of the NPC. The dimensions of the main features are indicated. The nuclear basket is shown in brown.

An alternative strategy is to combine fluorescent light microscopy and cryo-tomography. Specific fluorescent labels can be engineered into proteins of interest, or alternatively fluorescent labels can be applied and taken up by the cells. Progress has been made in the development of the cryogenic light microscope [41, 42], which can visualize fluorescence in vitrified specimens. This allows a fluorescence image of the frozen hydrated specimen to be recorded on the grid, and for any targets of interest for tomography to be identified. The specimen is then transferred directly to the electron microscope, where tomographic tilt series can be recorded of the target areas. Software-based methods can be used to align the fluorescence image with a low magnification image from the electron microscope, thus locating the desired areas in the electron microscope overview image.

8.5.2
Pattern Recognition

The identification of a single 26S proteasome in the cytoplasm of a *Dictyostelium* cell suggested that a template-matching approach could be used for mapping cellular proteomes [21]. Given that one can see 'everything' in a tomogram, it makes sense to probe the tomogram with a library of templates derived from known structures, using intelligent pattern recognition algorithms. The intent is to locate known structures and determine the molecular context in which complexes are organized in organelles or cells, with less emphasis on the discovery of novel molecular features and more on determining whether there is some correlation in the spatial arrangement of complexes, relative to one another. To achieve this, a 'template-matching' strategy is being pursued [43, 44]. Given the array of high- to medium-resolution relevant structures that are available to be included in a template library, the systematic analysis of tomograms by scanning for the presence of these known templates is feasible. The procedure is computationally intensive, as not only must the positions matching a given template be determined, but also their spatial orientations. The tomogram must be probed with every template, and the best match determined for each complex in the tomogram. Ideally, such a multi-template search would result in a 3-D map with each low-resolution complex that was identified replaced by the higher-resolution template that it matched, in the orientation determined (Figure 8.4). Simulations and experiments with 'phantom cells' have shown that such an approach is feasible [44], and that the search results can be validated. At present, only large complexes such as the ribosome have been identified with an acceptable accuracy (>95%) at the routinely attained resolution of 4–5 nm. Yet, an improvement in resolution to 2–3 nm would allow the accurate identification of smaller complexes [45].

8.6
Creating Template Libraries

Once the challenges of obtaining a sufficiently good resolution have been met, the next target will be to expand the libraries of available templates. Many approaches can contribute to achieve this goal. Structural genomics efforts will increase the pace at which high-resolution structures of domains, subunits or larger entities become available, and eventually will provide a comprehensive structural dictionary. Hybrid methods, combining information from different sources and of variable quality, will also play an important role [2, 4]. Electron microscopic single-particle analysis will undoubtedly continue to provide many medium-resolution structures of complexes. Clearly, the prospects for accelerating throughput by means of automated data collection and analysis show great promise [15].

Nonetheless, sample preparation continues to be the major bottleneck that slows progress in the field. Today, improvements are required in the methods used to isolate and purify proteins. This is especially true for labile macromolecular assemblies (which are probably more abundant in cells than stable complexes), with novel strategies being required. Whilst traditional biochemical methods tend to optimize for yield and/or purity, they are frequently time-consuming, and labile or transient complexes are generally lost during a traditional course of isolation. Given the modest requirements of single-particle analysis – where only very small quantities of material are needed and impurities can be removed computationally – there is no reason to purify labile complexes to exhaustion. It has been shown recently, by using lipid monolayer techniques and taking advantage of His tags, that it is possible specifically to pick up only the desired His-tagged protein from a crude extract and directly freeze and image the specimen in vitreous ice [46]. There is, therefore, plenty of scope for innovative approaches aimed either at minimizing the purification steps, or even avoiding purification altogether by taking advantage of optimized *in vitro* translation systems.

8.7
Outlook

With cryoelectron tomography providing 3-D images at molecular resolution, and with image analysis tools at our disposal for interpreting the tomograms, we are now poised to integrate structural data into pseudo-atomic maps of organelles or cells. Whilst these maps will provide unprecedented insights into the molecular architecture that underlies cellular behavior, they will also pose new challenges. It will not be a trivial task to extract generic features from the maps, nor to derive general rules regarding the principles that govern supramolecular organization, given the stochastic nature of cellular systems as well as their dynamics. Most importantly, systems analysis will need to start taking this into account, and there will also be a need to develop statistical methods similar to those used when analyzing macroscopic social systems. Alternatively, sophisticated molecular dynamics software could be expanded to model large-scale systems.

References

1 Aebersold, R. and Mann, M. (2003) Mass spectrometry-based proteomics. *Nature*, 422 (6928), 198–207.
2 Sali, A., Glaeser, R., Earnest, T. and Baumeister, W. (2003) From words to literature in structural proteomics. *Nature*, 422 (6928), 216–225.
3 Aloy, P., Boettcher, B., Ceulemans, H. *et al.* (2004) Structure-based assembly of protein complexes in yeast. *Science*, 303 (5666), 2026–2029.
4 Robinson, C.V., Sali, A. and Baumeister, W. (2007) The molecular sociology of the cell. *Nature*, 450 (7172), 973–982.
5 Baumeister, W., Grimm, R. and Walz, J. (1999) Electron tomography of molecules and cells. *Trends in Cell Biology*, 9 (2), 81–85.
6 Baumeister, W. (2004) Mapping molecular landscapes inside cells. *Biological Chemistry*, 385 (10), 865–872.
7 Radon, J. (1917) Über die Bestimmung von Funktionen durch ihre Integralwerte längs gewisser Mannigfaltigkeiten. Berichte über die Verhandlungen der Königlich Sächsischen Gesellschaft der Wissenschaften zu Leipzig. *Mathematisch Physikalishe Klasse*, 69, 262–277.
8 Gordon, R., Bender, R. and Herman, G.T. (1970) Algebraic reconstruction techniques (ART) for three-dimensional electron microscopy and X-ray photography. *Journal of Theoretical Biology*, 29 (3), 471–481.
9 Gilbert, P. (1972) Iterative methods for the three-dimensional reconstruction of an

object from projections. *Journal of Theoretical Biology*, **36** (1), 105–117.
10 Hegerl, R. and Hoppe, W. (1976) Influence of electron noise on three-dimensional image reconstruction. *Zeitschrift für Naturforschung*, **A314**, 1717–1721.
11 McEwen, B.F., Downing, K.H. and Glaeser, R.M. (1995) The relevance of dose-fractionation in tomography of radiation-sensitive specimens. *Ultramicroscopy*, **60** (3), 357–373.
12 Typke, D., Dierksen, K. and Baumeister, W. (1991) Automatic electron tomography (ed. W. Bailey) *Proceedings, 49th Annual Meeting of the Electron Microscopy Society of America*, San Francisco Press, San Francisco, CA, pp. 544–545.
13 Dierksen, K., Typke, D., Hegerl, R., Koster, A.J. and Baumeister, W. (1992) Towards automatic electron tomography. *Ultramicroscopy*, **40**, 71–87.
14 Dierksen, K., Typke, D., Hegerl, R. and Baumeister, W. (1993) Towards automatic electron tomography. II. Implementation of autofocus and low-dose procedures. *Ultramicrosopy*, **49**, 109–120.
15 Carragher, B., Fellmann, D., Geurra, F. *et al.* (2004) Rapid routine structure determination of macromolecular assemblies using electron microscopy: current progress and further challenges. *Journal of Synchrotron Radiation*, **11** (Pt 1) 83–85.
16 Koster, A.J., Grimm, R., Typke, D. *et al.* (1997) Perspectives of molecular and cellular electron tomography. *Journal of Structural Biology*, **120** (3), 276–308.
17 Dierksen, K., Typke, D., Hegerl, R., Walz, J., Sackmann, E. and Baumeister, W. (1995) Three-dimensional structure of lipid vesicles embedded in vitreous ice and investigated by automated electron tomography. *Biophysical Journal*, **68** (4), 1416–1422.
18 Grimm, R., Barmann, M., Hackl, W., Typke, D., Sackmann, E. and Baumeister, W. (1997) Energy filtered electron tomography of ice-embedded actin and vesicles. *Biophysical Journal*, **72** (1), 482–489.
19 Grunewald, K., Desai, P., Winkler, D.C. *et al.* (2003) Three-dimensional structure of herpes simplex virus from cryo-electron tomography. *Science*, **302** (5649), 1396–1398.
20 Grimm, R., Singh, H., Rachel, R., Typke, D., Zillig, W. and Baumeister, W. (1998) Electron tomography of ice-embedded prokaryotic cells. *Biophysical Journal*, **74** (2 Pt 1), 1031–1042.
21 Medalia, O., Weber, I., Frangakis, A.S., Nicastro, D., Gerisch, G. and Baumeister, W. (2002) Macromolecular architecture in eukaryotic cells visualized by cryoelectron tomography. *Science*, **298** (5596), 1209–1213.
22 Dubochet, J., Adrian, M., Chang, J.J. *et al.* (1988) Cryo-electron microscopy of vitrified specimens. *Quarterly Reviews of Biophysics*, **21** (2), 129–228.
23 Berriman, J. and Unwin, N. (1994) Analysis of transient structures by cryo-microscopy combined with rapid mixing of spray droplets. *Ultramicroscopy*, **56** (4), 241–252.
24 Unwin, N. (1995) Acetylcholine receptor channel imaged in the open state. *Nature*, **373** (6509), 37–43.
25 Baumeister, W. (2002) Electron tomography: towards visualizing the molecular organization of the cytoplasm. *Current Opinion in Structural Biology*, **12** (5), 679–684.
26 Plitzko, J., Frangakis, A.S., Nickell, S., Förster, F., Gross, A. and Baumeister, W. (2002) *In vivo* veritas: electron cryotomography of cells. *Trends in Biotechnology*, **20**, s40–s44.
27 Nickell, S., Forster, F., Linaroudis, A. *et al.* (2005) TOM software toolbox: acquisition and analysis for electron tomography. *Journal of Structural Biology*, **149** (3), 227–234.
28 Gruska, M., Medalia, O., Baumeister, W. and Leis, A. (2008) Electron tomography of vitreous sections from cultured mammalian cells. *Journal of Structural Biology*, **161** (3), 384–392.
29 Al-Amoudi, A., Norlen, L.P. and Dubochet, J. (2004) Cryo-electron microscopy of

vitreous sections of native biological cells and tissues. *Journal of Structural Biology*, **148** (1), 131–135.

30 Unser, M., Trus, B.L., Frank, J. and Steven, A.C. (1989) The spectral signal-to-noise ratio resolution criterion: computational efficiency and statistical precision. *Ultramicroscopy*, **30** (3), 429–433.

31 Frank, J. (2002) Single-particle imaging of macromolecules by cryo-electron microscopy. *Annual Review of Biophysics and Biomolecular Structure*, **31**, 303–319.

32 Frangakis, A.S. and Hegerl, R. (2001) Noise reduction in electron tomographic reconstructions using nonlinear anisotropic diffusion. *Journal of Structural Biology*, **135** (3), 239–250.

33 Kurner, J., Medalia, O., Linaroudis, A.A. and Baumeister, W. (2004) New insights into the structural organization of eukaryotic and prokaryotic cytoskeletons using cryo-electron tomography. *Experimental Cell Research*, **301** (1), 38–42.

34 Medalia, O., Beck, M., Ecke, M. et al. (2007) Organization of actin networks in intact filopodia. *Current Biology*, **17** (1), 79–84.

35 Beck, M., Forster, F., Ecke, M. et al. (2004) Nuclear pore complex structure and dynamics revealed by cryoelectron tomography. *Science*, **306** (5700), 1387–1390.

36 Beck, M., Lucic, V., Forster, F., Baumeister, W. and Medalia, O. (2007) Snapshots of nuclear pore complexes in action captured by cryo-electron tomography. *Nature*, **449** (7162), 611–615.

37 Grunewald, K., Medalia, O., Gross, A., Steven, A.C. and Baumeister, W. (2003) Prospects of electron cryotomography to visualize macromolecular complexes inside cellular compartments: implications of crowding. *Biophysical Chemistry*, **100** (1–3), 577–591.

38 Mercogliano, C.P. and DeRosier, D.J. (2006) Gold nanocluster formation using metallothionein: mass spectrometry and electron microscopy. *Journal of Molecular Biology*, **355** (2), 211–223.

39 Mercogliano, C.P. and DeRosier, D.J. (2007) Concatenated metallothionein as a clonable gold label for electron microscopy. *Journal of Structural Biology*, **160** (1), 70–82.

40 Gaietta, G., Deerinck, T.J., Adams, S.R. et al. (2002) Multicolor and electron microscopic imaging of connexin trafficking. *Science*, **296** (5567), 503–507.

41 Sartori, A., Gatz, R., Beck, F., Rigort, A., Baumeister, W. and Plitzko, J.M. (2007) Correlative microscopy: bridging the gap between fluorescence light microscopy and cryo-electron tomography. *Journal of Structural Biology*, **160** (2), 135–145.

42 Schwartz, C.L., Sarbash, V.I., Ataullakhanov, F.I., McIntosh, J.R. and Nicastro, D. (2007) Cryo-fluorescence microscopy facilitates correlations between light and cryo-electron microscopy and reduces the rate of photobleaching. *Journal of Microscopy*, **227** (Pt 2), 98–109.

43 Bohm, J., Frangakis, A.S., Hegerl, R., Nickell, S., Typke, D. and Baumeister, W. (2000) Toward detecting and identifying macromolecules in a cellular context: template matching applied to electron tomograms. *Proceedings of the National Academy of Sciences of the United States of America*, **97** (26), 14245–14250.

44 Frangakis, A.S., Bohm, J., Forster, F. et al. (2002) Identification of macromolecular complexes in cryoelectron tomograms of phantom cells. *Proceedings of the National Academy of Sciences of the United States of America*, **99** (22), 14153–14158.

45 Ortiz, J.O., Forster, F., Kurner, J., Linaroudis, A.A. and Baumeister, W. (2006) Mapping 70S ribosomes in intact cells by cryoelectron tomography and pattern recognition. *Journal of Structural Biology*, **156** (2), 334–341.

46 Kelly, D.F., Dukovski, D. and Walz, T. (2008) Monolayer purification: a rapid method for isolating protein complexes for single-particle electron microscopy. *Proceedings of the National Academy of Sciences of the United States of America*, **105** (12), 4703–4708.

9
Time-Resolved Two-Photon Photoemission on Surfaces and Nanoparticles

Martin Aeschlimann and Helmut Zacharias

9.1
Introduction

Occupied and unoccupied states of solid-state materials are traditionally investigated using photoemission and inverse photoemission spectroscopies. Due to the limited and short path length of electrons in the 20 to 100 eV range, an excellent surface specificity is achieved with these methods, enabling differentiation to be made between signals from the bulk of a material and from layers in the vicinity of the surface. The advent of ultra-short laser pulses among those laboratories investigating surface sciences has extended the field of research, as this technique provides the possibilities both to investigate occupied states with photon energies sufficiently high to surpass the work function of the material, and to study unoccupied states by using resonantly enhanced two-photon photoemission (2PPE). In this process, electrons are promoted by a first laser photon to an unoccupied state; this may be a volume, surface, an adsorbate state or an image potential state (IS) [1] of the material. Before the electron can re-equilibrate, a second photon from the same (or a different) laser beam is absorbed by the excited state. This promotes the electron of the excited state above the vacuum level, leading to an emission of photoelectrons which thus have been interacting with two laser fields. In this way, the spectral resolution for the intermediate unoccupied states is greatly improved compared to inverse photoemission [2].

The two-pulse scheme of photoemission bears another extremely important advantage, in that it allows the delay of one pulse against the other. This so-called time-resolved two-photon photoemission (TR-2PPE) allows investigation of the dynamics of the intermediate electronically excited state on a femtosecond time scale. Moreover, energy and momentum transfer processes in the excited state can also be studied, while scattering phenomena can also be addressed. These time-resolved photoemission experiments have been conducted for about 25 years. Initially, the thermalization dynamics of band edge carriers in semiconductors and electron–phonon (e–ph) dynamics in metals were studied using lasers of typically a few tens of picoseconds pulse duration [3–6]. More recently, however, by using lasers with femtosecond time resolution, the lifetimes of image potential states on silver surfaces [7] and questions regarding electron–electron (e–e) scattering dynamics in bulk metals have been addressed [8–10].

During the past two decades, photoemission techniques have been coupled with electron imaging by means of photoemission electron microscopy (PEEM), and this has allowed material-specific microscopic images of nanostructures on surfaces to be obtained with sub-100 nm resolution [11]. Important information concerning various new and unexpected phenomena, including the nonlinear behavior of surface reaction dynamics [12] and of layered magnetic materials, has been obtained using this method. These imaging methods, when combined with time resolution in the femtosecond regime [11, 13], will become increasingly important in emerging areas such as molecular electronics, self-assembled and self-organized functional layers, organic solar energy converters and photocatalytic reaction centers.

In this chapter we will provide examples of the imaging methods, together with their recent application in the preparation of metallic nanostructures. Such nanostructures will become increasingly important as fields of molecular switching and electronics, plasmonics and functional molecular assemblies evolve. Moreover, as the dimensions of structures shrink to the submicrometer level, novel properties and functions will undoubtedly emerge. The spectral tuning of the emission of entities and controlled charge transport in organized nanostructured environments [14–16] represent just two aspects of functional modification in molecular electronics. It is further envisioned that the optical phase control of charge transport [17] may become very important in such systems. It follows that a precise knowledge of the basic electronic properties of these nanostructured systems is, therefore, of major importance.

9.2
Theoretical Background

The electron dynamics in metallic nanostructures are governed by a few elementary processes which will now briefly be described. Besides the fundamental charge screening, which takes place on an attosecond time scale, the decoherence of excited electrons and electron–electron scattering processes are primary processes leading to a rapid redistribution of electron energy and momentum. Thereby, a hot electron gas is created. On nanostructures a coherent excitation of the whole electronic system – the creation of particle plasmons – is of fundamental nature and a specific element to be considered in these systems. The decay processes of particle plasmons are presently under intense investigation. The hot electrons created by various processes then couple to the phonon system, eventually dissipating the energy to heat. A concise theoretical description of the basic processes involved, which is beyond the scope of this chapter, can be found in recent reports [18–20].

9.2.1
Electron–Electron Interaction

The simplest description of electronic interaction in metals can be derived from the Drude theory of metals [21]. Although based on currently outdated assumptions, this theory provides a good estimate of the magnitude of electron scattering for various classes of metals. The electron mean free path, divided by their velocity at the Fermi energy, yields the collision free time which denotes the lifetime of an electron at a certain energy. At room temperature, this lifetime ranges from a few femtoseconds in transition metals to a few tens of femtoseconds in noble metals.

A more quantitative description of the electron–electron scattering rate is obtained from the Fermi liquid theory [22, 23] for a free electron gas (FEG FLT). In brief, phase space arguments are invoked to describe the interaction of an electron with kinetic energy above the Fermi level with unexcited electrons. Due to the same mass of the interacting electrons, and the fact that all states below the Fermi level are occupied, the inelastic scattering of a hot electron with the cold Fermi gas yields two electrons, both with kinetic energies now above the Fermi level. For a single excited electron this inelastic process yields a new kinetic energy and momentum, thus limiting the mean free path of an electron at a given energy. It is clear that electrons far above the Fermi level have a larger phase space available for scattering than those close to the Fermi level. Therefore, the scattering rate strongly decreases as the energy of the hot electron relaxes towards the Fermi level.

These inelastic processes continue, and finally a hot electron gas is rapidly created. When an intense laser pulse creates the primary excitation, the excitation density is high and therefore also the density of the hot electron gas. Scattering events with adsorbed molecules then become probable, and processes such as desorption induced by electronic transitions (DIET) and desorption induced by multiple electronic transitions (DIMET) or electronic friction induced adsorbate excitation my become observable.

Within the relaxation time approximation, and invoking Boltzmann transport theory with the Fermi–Dirac distribution for the occupation, the Fermi liquid theory yields in three dimensions for the scattering rate [24]

$$\tau^{-1} = \tau_0^{-1}(E-E_F)^2 + b(k_B T)^2 \qquad (9.1)$$

At excitation energies above 200 meV, the thermal contribution can usually be neglected at all practical temperatures. The prefactor τ_0 depends only on the electron density in the metal under consideration:

$$\tau_0 = \frac{64}{\sqrt{3}\pi^{5/2}} \sqrt{\frac{m_e}{ne^2}} \; E_F^2 \qquad (9.2)$$

The inelastic lifetime of an electron above the Fermi level, given by the inverse scattering rate, is therefore inversely proportional to the square of the energy difference to the Fermi level and, taking the dependence of E_F on the electron density into account, proportional to $n^{5/6}$. At excitation energies of about 3 eV this amounts to a few femtoseconds, whereas close to the Fermi level, $E<0.2$ eV, it may approach the picosecond range, depending on the metal under consideration [25, 26].

Usually, the FEG FLT serves as a benchmark for comparison of the observed hot-electron inelastic lifetimes τ with the theory, and therefore electron relaxation dynamics has first been investigated for noble metals, where a reasonable agreement with the FEG FLT was expected [10, 27–30]. For all other metals, the key role in the low-energy electron relaxation dynamics is played by the electronic structure of the system close to the Fermi level. For example, in transition metals the high density of d bands in the proximity of the Fermi level leads to a very fast electron relaxation. In this context, the exact energy position and shape of the d bands must be considered in detail in order to achieve a complete understanding of the relaxation processes. During the past decade, several TR-2PPE experiments have been performed for the transition metals Ta [31], Ru [32], Mo and Rh [33], for ferromagnetic $3d$ metals Fe, Co and Ni [34], and for high-T_C superconductors [35] and $4f$- rare earth metal Yb [36].

Theoretical calculations of excited electron lifetimes have been performed in the past within the GW approximation (GWA) for electron self-energy for bulk noble [37, 38] and $4d$ transition metals [39–42], as well as for the $5d$ transition metal Ta [31]. In contrast to noble metals, which show qualitatively similar band structure and density of states (DOS) [43], the electronic structure of $4d$ metals varies strongly on moving from the start of the $4d$ series to the end [43]. Calculations performed by Zhukov *et al.* [41] and Bacelar *et al.* [42] have shown that the evaluated lifetimes also vary widely along the $4d$ series, following trends in electronic structure. The extension, $GW + T$, of the GW approximation by inclusion of multiple electron-hole scattering within a T-matrix approximation [44–48] results in a decrease of the GW lifetime value that brings theory and experiment to better agreement [47, 48].

9.2.2
Plasmonic Processes

In noble metal nanoparticles collective electronic oscillations – so-called particle plasmons or localized surface plasmons (LSPs) – can be excited by electromagnetic waves. Therefore, they are detectable as pronounced resonances in the scattering and absorption cross-section, for the noble metals Ag and Au commonly located in the visible or ultraviolet (UV) region of the spectrum [49]. The resonance frequency of the plasma oscillation is determined by the dielectric properties of the metal and the surrounding medium, as well as by the particle size and shape [49–51]. The collective oscillation can be interpreted as a displacement of the electrons in the particle against the positively charged background of the atomic nuclei. Resonant excitation of this collective charge oscillation causes a large enhancement of the local field inside and near the particle [52] which dominates the linear and nonlinear responses of the particles to the light field. The field enhancement caused by the electron oscillation (see Figure 9.1) is thought to be responsible for the enhancement of nonlinear optical effects such as surface-enhanced Raman scattering (SERS) [53], surface second harmonic generation [54, 55] and multiphoton photoemission [56]. In recent years, the promising research field of plasmonics and ultrafast nano-optics has emerged, exploiting the high potential of plasmons to concentrate and channel light into subwavelength structures of nanoscopic circuits [57].

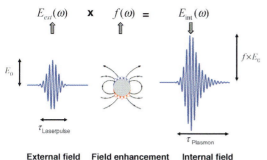

Figure 9.1 In metallic nanostructures collective electronic oscillations – local surface plasmons – can be excited by light; this results in a strong enhancement of the local field both inside and near the surface structure.

Although, the spectral positions of the resonances of particle plasmon excitations as a function of particle size, shape and dielectric properties are well understood [49–52], the ultrafast dynamics of these collective electronic excitations have remained a highly interesting topic to be studied in more detail. In order to understand the dynamics, it is essential to investigate those mechanisms relevant to the loss of phase coherence between the electrons contributing to the collective excitation – that is, the dephasing of the plasmonic state.

9.2.3
Two-Temperature Model

When a strong femtosecond laser pulse produces a large primary population of hot electrons, the electron–electron scattering leads to the formation of a hot electron gas. The formation and decay of this not-equilibrated hot electron distribution was first observed by Bokor and coworkers in thin gold films. The equilibration process of this electron gas can, in general, be described fairly well by the Fermi liquid theory, together with the Boltzmann transport equation in the relaxation time formulation [8, 9]. However, deviations from this description were noted at an early stage of these investigations. An important (and, at first sight, counterintuitive) finding was that the relaxation to a heated thermal distribution occurred faster if a more intense laser excitation was applied, and thus the density of hot electrons was higher.

The hot electron dynamics is usually modeled using a two-temperature model [58], where the electronic and phonon systems assume different temperatures; this is justified due to the wide differences in the heat capacities.

$$C_{el}\frac{\partial T_{el}}{\partial t} = \frac{\partial}{\partial z}\left(\kappa\frac{\partial T_{el}}{\partial z}\right) - g_{\infty}(T_{el} - T_{ph}) + S(z,t) \tag{9.3}$$

$$C_{ph}\frac{\partial T_{ph}}{\partial t} = g_{\infty}(T_{el} - T_{ph}) \tag{9.4}$$

$S(z,t)$ represents the exciting optical intensity as it penetrates into the bulk of the material. The lateral dimensions are usually large, and thus uniform compared to the large gradients in z direction. $C_{el} = \gamma\, T_{el}$ represents the electronic heat capacity, and C_{ph} that of the phonon system. The coupling between the electronic system and the phonons is described by g_∞. With this system of equations, a good estimate of the magnitude and time dependence of electronic and phonon temperatures is usually achieved. This can easily be extended by frictional coupling to an adsorbate [59, 60], which thereby acquires internal energy and also serves as a cooling heat bath for the excited surface layers. Similarly, diffusive and ballistic electron transport from the surface layers represents an important sink for the deposited laser energy (see for example, Ref. [32]).

9.2.4
Electron–Phonon Coupling

The hot electron gas created by electron–electron collisions then couples to the phonons of the substrate, which eventually leads to a dissipation of the energy into a second heat bath. A theoretical description of this process has only in recent years become possible. Usually, the structure of a solid and the motion of its atomic constituents is very successfully described by the Debye model using quasi-particles, the phonons. For the electronic system, on the other hand, Bloch waves describe the motion and energetics of the quasi-particles. In both descriptions they are considered independent of each other, and this constitutes the usual Born–Oppenheimer approximation, which is well known from molecular physics. In this picture an electron–phonon interaction cannot take place, and therefore one must go beyond the Born–Oppenheimer approximation and formulate a description of these nonadiabatic processes.

As a formal derivation of the approach is beyond the scope of this chapter, the reader is advised to consult the appropriate literature, which is based on the Fröhlich description [61] of the electron–phonon (e–ph) interaction [62]. Physical insight into the problem is gained with the introduction of the so-called Eliashberg function, $\alpha^2 F(\omega)$, the product of the phonon density of states $F(\omega)$ and the electron–phonon interaction strength α^2 [63]. For the emission of a phonon it is given by

$$\alpha^2 F_{i,k_i}(\omega) = \int d^2q \sum_{f,v} \left|g_{q,v}^{i,f}\right|^2 \delta(E_{i,k_i} - E_{f,k_f} \pm \omega_{q,v}) \delta(\omega - \omega_{q,v}) \quad (9.5)$$

where $g_{q,v}^{i,f}$ denotes the electronic part of the interaction [62]. For the evaluation of $g_{q,v}^{i,f}$ one has to enter the screening potential. In order to illustrate this Eliashberg function, its dependence on phonon frequency is shown in Figure 9.2 for Be [64]. It transpires that the simple Thomas–Fermi screening yields results which are well compatible with experimental data. The electron–phonon coupling parameter $\lambda(E_i, k_i)$ is then obtained from

$$\lambda(E_i, k_i) = 2 \int_0^{\omega_{max}} \frac{\alpha^2 F_{i,k_i}(\omega)}{\omega} d\omega \quad (9.6)$$

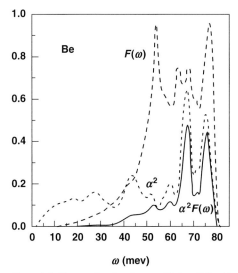

Figure 9.2 Phonon density of states $F(\omega)$ and Eliashberg function $\alpha^2 F(\omega)$ for Be(0001). (From Ref. [64].)

Any phonon-induced interaction should be proportional to the occupation number $n_B(\omega)$ of phonons. The spectral broadening due to e–ph coupling is then

$$\Gamma(E_i, k_i) = 2\pi \int \alpha^2 F_{i,k_i}(\omega)[(2n_B(\omega) + 1) + f(E_{k_i} + \omega) - f(E_{k_i} - \omega)]d\omega \quad (9.7)$$

For increasing temperature this occupation number follows

$$n_B(\omega) \to k_B T / \omega, \quad (9.8)$$

and thus shows at high temperatures a linear dependence on T [65]. The spectral broadening of a state due to e–ph coupling can then be written as

$$\Gamma(E_i, k_i) = 2\pi\lambda(E_i, k_i)k_B T. \quad (9.9)$$

In order to calculate this width, it is necessary to know the phonon density of states $F(\omega)$ and the e–ph coupling strength $|g_{q,n}^{i,f}|^2$, given by [18]

$$g_{q,\nu}^{i,f} = \frac{1}{\sqrt{2M\omega_{q\nu}}} \langle \psi_{k_i} | \hat{\varepsilon}_{q\nu} \cdot \nabla_R V_{sc} | \psi_{k_f} \rangle \quad (9.10)$$

where $\nabla_R V_{sc}$ is the gradient of the screened potential, $\hat{\varepsilon}_{q\nu}$ the polarization of the phonons, and M the atomic mass. ψ_{k_i,k_f} denotes the electronic wave functions of the initial and final states.

Experimentally, this e–ph coupling manifests itself via the spectral width Γ of states. An ideal situation to study this coupling arises therefore from surface states located in projected bulk bandgaps, as on the {111} surfaces of noble metals or of quantum-well states [66]. In this case, e–e scattering and electron–defect scattering may also contribute to the total linewidth Γ, and hence it is necessary to seek alternative ways to separate these contributions. The e–ph coupling is also important when describing the cooling of a hot, nonequilibrium electron gas created by an intense laser pulse, as described above. Here, a connection between the microscopic constant λ and the phenomenologically used value g_∞ is given by [67]

$$g_\infty = \frac{3\hbar\gamma}{\pi k_B}\lambda\langle\omega^2\rangle, \tag{9.11}$$

where $\langle\omega^2\rangle$ denotes the second moment of the phonon spectrum.

9.3
Experimental

The investigation of electronic dynamics on nanostructures requires first an *in situ* preparation of the nanostructured system. For this purpose, a surface science ultra-high-vacuum machine with standard techniques for cleaning and preparation, like low-energy electron diffraction (LEED) and Auger electron spectroscopy, must also be equipped with instruments for either surface structuring or controlled growth techniques. Further, an *in situ* control and manipulation of the sample is suggested, which can be achieved by using scanning electron microscopy (SEM), scanning tunneling microscopy (STM) [68] or atomic force microscopy (AFM) [69] or by photoelectron emission spectroscopy [70] and microscopy [11]. Each of these techniques has in the past been described extensively, and the reader is referred to respective review articles.

In order to study the ultrafast electron dynamics, an appropriate laser source must be added, as well as a means of detecting the laser-generated photoelectrons. This can be achieved by a conventional dispersive electron spectrometer, equipped with two-dimensional (2-D) signal detection, or by a time-of-flight (ToF) detector, which makes use of the multiplexing advantage, especially when a multi-anode assembly is added. Figure 9.3 shows, in schematic form, a typical experimental set-up. The standard laser system consists of a Ti : sapphire laser oscillator operating in the spectral vicinity of 800 nm and with pulse durations of typically 12–25 fs. As the work functions of typical metals range from about 4 to 6 eV, frequency doubling and tripling in optically nonlinear crystals, such as β-BaB_2O_4 (BBO) and $LiBO_3$ (LBO), is often employed. For an optimal time resolution the frequency doubling crystal should be thin, with about 100–200 µm thickness depending on the spectral width that is to be converted. Efficiencies to produce 400 nm radiation in the range of 15–20% can be achieved. The third harmonic of the 800 nm fundamental radiation is produced by sum frequency mixing the fundamental with its second harmonic in a second BBO crystal. The frequency-converted pulses should be recompressed to compensate for the material dispersion in both the doubling and mixing crystals as well as the chamber windows, or any other optics in the beam path from the laser to the sample (see Figure 9.3). Besides simply reducing the number of photons required to reach the vacuum level of a system, the use of two colors – one from the fundamental and one from a

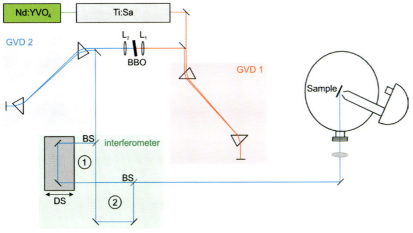

Figure 9.3 Schematic set-up for space and time-resolved two-photon photoemission experiments. The sketched dispersive hemispherical electron energy analyzer can be replaced by a time-of-flight detector or an imaging PEEM. Pulse compressors (GVD 1 and 2) (pre-)compensate a pulse stretching in optical elements in the path of both the fundamental and an harmonic beam.

frequency-doubled or -tripled beam – leads to a significant increase in the analytical power of the experiment.

Amplified Ti : sapphire radiation is required when an optical parametric oscillator is to be pumped in order to obtain radiation that is tunable across the visible spectrum. This allows one specifically to tune to and excite unoccupied resonance states of the sample. The dispersion of a state can also be followed. Amplified Ti : sapphire radiation with pulse energies in the 1 mJ regime and durations of 20–35 fs can also be used to produce extreme ultraviolet (XUV) radiation by high harmonic generation, with tunable photon energies up to about 100 eV and beyond [71–74]. Besides accessing more strongly bound states, or even shallow core levels, this radiation also allows the momentum space of a state to be addressed in the whole Brillouin zone. In addition, high-lying plasmonic states of all metals can, in principle, be reached directly. By shortening the pump pulse duration in the sub- 10 fs regime, using nonlinear optical techniques, it may be possible to produce not only high harmonics in the water window spectral range ($\lambda \sim 4.4$ to 2.3 nm) [71, 72]. By employing carrier envelope phase-stabilized pulses and correctly selecting the XUV spectral range, pulse trains of approximately 100 as [75–78] or even single pulses with durations as short as 80 as [79] can be produced. Such an approach will undoubtedly open a future window for investigating the electron dynamics of screening and dephasing processes.

For time-resolved experiments, care must be taken to properly delay and recombine both pulses on the sample. For this, either a Mach–Zehnder or a Michelson interferometer set-up may be employed. A temporal delay of both pulses is achieved with different path lengths for the two interferometer arms; this difference must be controlled at better than 150 nm when a time step resolution of 1 fs is to be achieved, as the pulse will travel twice through the delayed arm. When the two pulses are collinearly overlapped, interferometric correlation signals with a periodicity of the optical wavelength are obtained. Due to a cycle duration of about 1.3 fs for 400 nm light, a much smaller step size (20 nm) and higher stability and reproducibility is required. Then, instead of using a conventional motorized translation stage, a piezoelectrically driven stage and an interferometer that is actively controlled with a frequency-stabilized HeNe laser beam are employed [80]. It is also important to control the polarization (s, p or circular) of both beams applied to the sample. The performance of both phase-resolved and phase-averaged 2PPE has been tested on a polycrystalline tantalum film (see Figure 9.4; photon energy 3.1 eV) [81]. In this way, the oscillation fringes due to the interference between pump- and probe- pulse are clearly resolved; that is, the accurate reproduction and periodicity of these measurements over the entire temporal delay proves the position stability of the set-up employed.

Due to increasing interest in the specific hot-electron dynamics of spatially heterogeneous systems such as metallic nanostructures, an extension to a space- and time-resolved 2PPE set-up, using time-resolved photoemission electron microscopy (TR-PEEM) for 2-D-electron detection has been established [13]. This method is capable of high spatial resolution in the 20 nm regime, which enables the focus to be set on the details of an individual nanoparticle. A typical TR-PEEM experiment is shown schematically in Figure 9.3 which is, except for the detector, identical to that of a TR-2PPE set-up. The TR-PEEM method has been well reviewed [11]. For a full pump-probe scan, the delay between the two pulses is varied in small steps (typically $\Delta t = 1$ fs), and for each step a PEEM image is taken

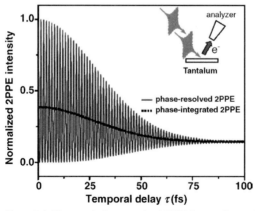

Figure 9.4 Time- and phase-resolved 2PPE from polycrystalline tantalum measured at an electron kinetic energy of 6 eV. The gray line is the 2PPE interferogram; the black dotted line shows the data for a conventional phase-averaged 2PPE measurement. (From Ref. [81].)

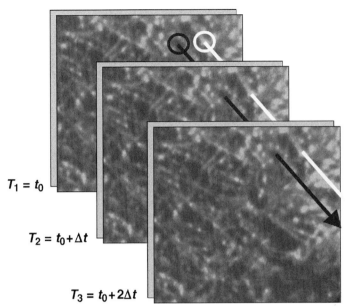

Figure 9.5 Schematics of measuring time-resolved 2PPE pump–probe images obtained with a PEEM.

(see Figure 9.5). This results in a series of images that contains a correlation trace for each pixel (Figure 9.6); these can then be plotted in a lifetime map deduced from a pixel-wise analysis of the cross-correlation traces of a TR-PEEM scan, as shown in Figure 9.7. The lifetime map contains information on the dynamic behavior of the electron system at the sample surface (decay time T_1 of the intermediate state) with the spatial resolution of the PEEM.

Nanostructured samples are, in general, prepared using either electron beam lithography (EBL) for larger structures (<40 nm), or cluster deposition for nanostructures as small as <1 nm. EBL, as a lift-off process, allows the controlled and

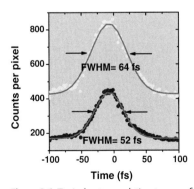

Figure 9.6 Typical autocorrelation traces of individual pixels of the PEEM images of Figure 9.5.

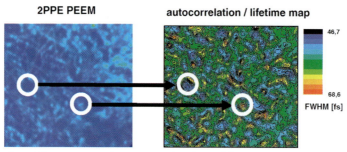

Figure 9.7 2-D autocorrelation map (right) obtained from time-resolved PEEM images taken according to Figure 9.5.

flexible design of metallic nanoparticles with regards to their shape and size. The optical properties – and especially the position and width of the LSP-resonance – depend critically on the shape and size of the particles, and this allows the characteristic LSP resonance frequencies to be tuned to the wavelength regime accessible by the available femtosecond laser system, e.g., the fundamental and second harmonic of a Ti:sapphire laser. Figure 9.8 shows SEM images of different silver nanostructures deposited on indium tin oxide (ITO) -covered glass substrates [82]. The dimensions of the elliptically shaped silver nanoparticles in Figure 9.8a are 140 nm (long axis), 60 nm (short axis) and 50 nm (height). These constitute versatile samples for investigating variations in the LSP decay in respect of resonant or off-resonant excitation. The silver nanodot array (Figure 9.8b; diameter 200 nm, height 50 nm) and the silver nanowire array (Figure 9.8c; length 1.6 µm, width 60 nm, height 50 nm) can be used to illustrate the potential of the time-resolved PEEM technique to map retardation effects associated with a plasmon excitation at nanometer resolution. Studies of the plasmon-induced coupling between neighboring nanoparticles are possible with nanodot pairs of varying center-to-center spacing. Figure 9.8d shows an example of 50 nm dimers (height 40 nm) at an interparticle spacing of 130 nm (grating constant 740 nm).

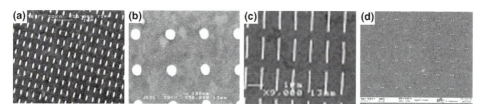

Figure 9.8 SEM images of different Ag nanostructures on an indium tin oxide (ITO) substrate. (a) 140 nm × 60 nm × 50 nm height; (b) Ø 200 nm × 50 nm height; (c) l = 1400 nm, w = 60 nm, h = 50 nm; (d) pairs with Ø 50 nm × 40 nm height and a separation of 130 nm.

For small cluster deposition onto surfaces, an UHV gas-aggregation cluster source is required including a quadrupole mass selector [83]. This allows the flexible *in situ* preparation of monodisperse cluster distributions over a broad size range. An alternative possibility would be to produce a defined density of atomic defects in a graphite surface, known as the Hövel-method [84]. This involves using a focused ion beam (FIB) technique that allows for the etching of well-defined nanopits into highly ordered pyrolytic graphite (HOPG) substrates and a subsequent oxidation procedure. The pits have a statistical variation in their depth between one and three monolayers (MLs). Further details on the FIB technique and oxidation procedure are available in Ref. [85]. The evaporation of about four MLs of silver at room temperature results in the condensation of near-monodisperse silver clusters in the native and artificially created defects in the HOPG surface.

9.4
Relaxation of Excited Carriers

During the past two decades, the study of image potential states has evolved as a paradigm for the investigation of electron dynamics at surfaces. This field has recently extensively been reviewed [86], and therefore only selected aspects will be discussed here. The main decay channel of these states isolated in a directional bandgap is the overlap of their wave function with bulk electronic wave functions. This overlap increases as the energy of the states approach the band edges. Then also the phonon contribution to the decay rate increases. For well-isolated image states, such as the $n = 1$ state on Cu(100), this contribution to the linewidth is expectedly very low, with a coupling parameter of $\lambda \sim 0.01$ and a contribution to the spectral width of $\Gamma < 1$ meV [87, 88], while for Cu(111) λ increases to 0.06 [89].

On the other hand, interband and intraband electron scattering also constitute important scattering mechanisms for image potential states. Steps and adatoms on the surface play an important role in these scattering phenomena. For example, on stepped surfaces scattering within the $n = 1$ image state results in a broadening of the level and thus in a shortening of the lifetime [90]. On vicinal Cu surfaces with (111) terraces, Roth *et al.* [91] found that for electrons with a downwards momentum the steps show a greater decay rate than in the opposite direction, and this results in lifetime differences within $n = 1$ of up to 4fs. Copper and cobalt adatoms induce intraband and interband scattering between the image states [92, 93]. An inelastic interband scattering of $n = 2$ electrons with bulk electrons populates the bottom of the $n = 1$ band. When probing levels within the $n = 1$ band above the bottom of the $n = 2$ level, a resonant (quasi-)elastic interband scattering occurs from $n = 2$ to $n = 1$, with strongly increasing probability as the energy increases further. On the other hand, the adatoms do not have any significant influence on the intraband scattering rate. It can be envisioned that a lateral confinement of nanostructures might also lead to an increased coupling between image states and bulk bands due to a shift of the band edges.

An especially illustrative example for a hole decay has been provided by Berndt and coworkers, who studied the occupied surface states of noble metals by using

Table 9.1 Experimental and theoretical hole lifetimes of occupied surface states on noble metals. (From Ref. [94].)

Metal	Δ (meV)	β	τ (fs)	Experiment		Theory	
				Γ_{STM} (meV)	Γ_{PES} (meV)	Γ_{old} (meV)	Γ_{new} (meV)
Ag	8	0.89	120	6	20[a]	5.3	7.2
Au	23	0.82	35	18	60[b]	8.6	18.9
Cu	30	0.80	27	24	21[c]	10.2	21.7

[a]Ref. [70].
[b]Extrapolated to $T = 0\,\text{K}$ from Ref. [95].
[c]Ref. [96].

STM [94]. The width Δ of the onset of dI/dV spectra can be transformed into lifetimes τ via

$$\tau = \beta h/(4\Delta), \tag{9.12}$$

where h denotes Planck's constant and β is a scale factor close to unity. Using this method, it can be ensured that the surface area studied is indeed clean, and therefore defect scattering can be excluded. The results obtained, together with a comparison of values derived from photoelectron spectroscopy [70, 95, 96], and a comparison with theoretical calculations, is shown in Table 9.1. The agreement between these measurements and the latest theoretical calculations is excellent. The results show that the holes created in the surface state are filled by electron–electron scattering from the 2-D electron gas of the still-occupied surface state band, rather than from the underlying 3-D bulk electron gas.

The global relaxation dynamics of a hot, nonequilibrium electron gas can be assessed by measuring the energy distribution function as a function of delay time after creating the hot distribution. This may be exemplified for hot electrons in Ru as studied by Wolf and coworkers [32]. The group investigated the relaxation dynamics at different pump pulse fluences. The relaxation dynamic is then modelled using a modified two-temperature model, where the electron distribution was parametrized by splitting it into a thermalized component at low temperature, while a second component was also thermalized, albeit at a higher temperature and with a lower population. This procedure describes the nonthermal electron energy distribution quite well. Both electron distributions couple to the phonon heat bath (for details, see Ref. [32]). Besides the expected localization of energy at the surface by electron–phonon coupling (as discussed above), it has also been found that ballistic transport out of the surface region has a significant influence on the temperatures and their time dependencies in the surface region. This leads in turn to notably lower temperatures (as shown impressively in Figure 9.9), with corresponding consequences for electronic coupling to adsorbates and laser-induced desorption dynamics. Such transport effects have also been observed for copper surfaces [97].

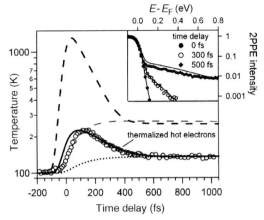

Figure 9.9 Time evolution of the electron and phonon temperatures according to the two-temperature model (thick and thin dashed lines). The extended heat bath model yields significantly lower values (shown by the thick and dotted lines) for the electrons and phonons, respectively. The open dots are experimental results. The inset shows measured electron kinetic energy distributions at different delay times. Data and calculations are for a Ru surface with a pump fluence of about 500 µJ cm^{-2}. (From Ref. [32].)

As mentioned above, on a microscopic level the electron–phonon coupling phenomena are best studied via isolated states in the band gap of materials. By using high-resolution photoelectron spectroscopy, the occupied surface states of noble metals have been studied [96, 98]. Figure 9.10 shows the experimental linewidth of the surface states, together with theoretical calculations [99], whereby an excellent agreement is obtained. This figure also shows the theoretically determined contribution of e–e scattering (Γ_{ee}) to the total width of the states. At low binding energies of the hole, the contribution of the Rayleigh phonon mode is dominant, while the overall spectral width becomes small. A similarly good agreement between theory and experiment was obtained previously for the very isolated surface state at Γ on the Be (0001) surface [100]. Here, the main contribution comes again from intraband scattering.

When defects are present on the surface, an additional broadening occurs. Such an effect has recently been reported for the Au (111) [101, 102] and Al(111) occupied surface states [102]. By monitoring the deviation of broadening from the expected linear temperature dependence, an activation energy for the creation of surface defects can be derived

$$\Gamma_{ed} = C \exp\{-[E_a/k_B T]\}. \tag{9.13}$$

On aluminum, the experimentally determined value for the activation energy of about $E_a = 170$ meV compares well with theoretical expectations for the creation of defects by kinks on a step edge, while the creation of adatoms requires about 300–600 meV. The corresponding value for Au(111) amounts to $E_a = 81$ meV, which suggests that kinks at step edges might well be responsible for the observed broadening.

In image states such a defect scattering can be isolated from the e–ph scattering, because these states are located in front of the surface and therefore are detached

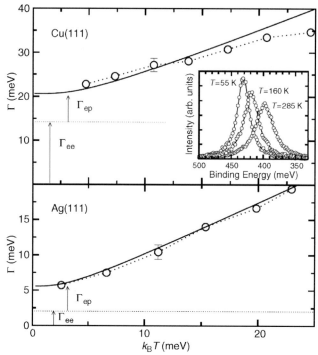

Figure 9.10 Temperature dependence of the experimental and theoretical line width of the occupied surface state on Cu(111) and Ag(111). Γ_{ee} denotes the e–e scattering contribution to the line width, independent of temperature. The inset shows experimental photoemission spectra at the Γ point at selected temperatures. (From Ref. [99].)

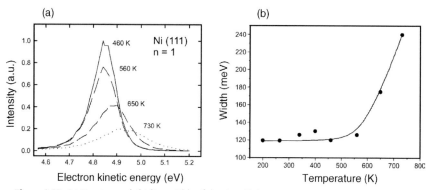

Figure 9.11 (a) Spectra and (b) line width of the ($n=1$) image state on Ni(111) as a function of temperature. The state is resonantly excited from the occupied surface state at $E_b = 0.2$ eV.

from the motion of the atomic surface constituents. A similar strong temperature dependence of the width of an isolated state has been observed for the ($n=1$) image potential state on Ni(111) (H. Zacharias and R. Paucksch, unpublished results). In this experiment, one photon at $\lambda = 263$ nm ($h\nu = 4.7$ eV) resonantly excites the image state from the occupied surface state at $E_b = 0.2$ eV [103], while a second photon of the same energy liberates the excited electron. Figure 9.11 shows the resonant spectra of $n=1$ and the observed widths as a function of temperature. Above 450 K, an exponential growth of the spectral width is observed. Based on this temperature dependence, an activation energy of about $E_a = 130$ meV was derived, this again being in agreement with the creation of kinks at step edges.

9.5
Volume Excitation in Metallic Nanostructures Investigated by TR-PEEM

The potential of TR-PEEM as a versatile tool for mapping the electron dynamics of metallic nanoparticles was introduced by Schmidt et al. [13]. Figure 9.12a shows the investigated lifetime map of a patterned silver film (hexagon-shaped patches) on a silicon substrate. The mapped color variations correspond to a change in the full-width half-maximum (FWHM) of the correlation trace for each pixel, which becomes more evident from a statistical analysis (histogram) of the investigated regions, as shown in Figure 9.12b.

Figure 9.13a shows a PEEM image of a Ag nanoparticle array (as shown in Figure 9.8b) at resonance excitations. The 400 nm laser light used for the TR-2PPE experiment couples almost resonantly to the in-plane mode of the particle. The 2PPE image shows distinct interparticle brightness variations which are dominated by defect-induced indirect transitions rather than by differences in the collective electron response [104]. Although the properties of the particles' plasmon resonances are not affected by this, they must of course be included in the analysis of the time-resolved data. The lifetime map, as shown in Figure 9.13b, visualizes the lateral

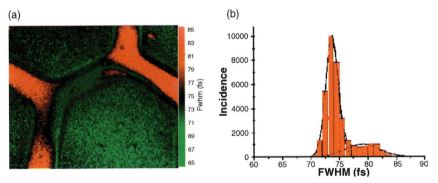

Figure 9.12 (a) Lifetime map of a hexagonal silver nanoparticle; (b) Lifetime map distribution showing a most probably value of 73.5 fs with a distribution width of ±1.5 fs.

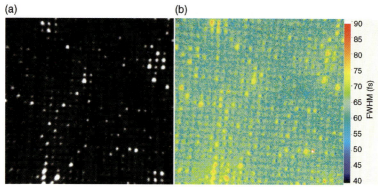

Figure 9.13 (a) PEEM image and (b) corresponding lifetime map of the nanostructure from Figure 9.8b. The FWHM of the autocorrelation curves ranges from 60 to 90 fs.

variations in the electron dynamics in color coding. Here, the red tones correspond to high FWHM values, indicating long decay times, while blue tones are associated with a faster decay. There is certain correlation between brightness in the 2PPE image and the FWHM values of the lifetime map: those particles which appear bright in the 2PPE image tend to exhibit longer average decay times in the lifetime image. This effect must obviously be related to the defect induced transitions. Whilst on the one hand the involved intermediate single-electron states give rise to a higher overall transition probability and a higher photoemission yield, on the other hand they show longer decay times.

The dynamic processes associated with a plasmon excitation in a single particle can be studied in further detail when the phase-resolved set-up is employed. The image

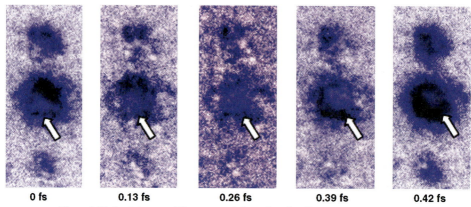

Figure 9.14 Spatiotemporal femtosecond dynamics of a silver nanoparticle with a diameter of 200 nm. A modulation of the lateral photoemission distribution as a function of phase delay between two identical exciting femtosecond laser pulses for 2PPE is observed. This is assigned to a phase propagation of a plasmon through the nanoparticle.

sequence in Figure 9.14 shows the plasmon dynamics in a single particle, where the time interval between the two images is 0.13 fs. A clear variation in the contrast within the area of the nanoparticle in the sub-femtosecond time scale is detectable. The result can be explained in the following way: The electric field amplitude is determined by the phase delay $\Delta\varphi(\tau)$ between the pump and the probe laser pulse, as adjusted by the Mach–Zehnder interferometer, as well as the polarization field of the particle plasmon which is oscillating at its resonance frequency. Due to oblique incidence from the right, the laser light would be expected to couple first to the LSP-mode at the right edge of the particle. Here, the external (laser) field and internal (plasmon) field attain a fixed phase relation. As the propagation velocity of the external and internal fields vary, a position-dependent phase lag between the two field components is acquired as the plasmon excitation travels through the particle. The particle internal structure visible in a single PEEM image of Figure 9.14 is a residual of the varying interference between the external light field and the particle internal LSP-field, directly connected to the plasmon phase. The parallel data acquisition by the PEEM allows any systematic errors to be excluded.

These results are in good agreement with the findings by Kubo *et al.* [105, 106], who also combined ultrafast laser spectroscopy and electron microscopy in order to image the quantum interference of localized surface plasmon polariton (SPP) waves with sub-wavelength spatial resolution and sub-femtosecond temporal precision. The sample used was based on a 400 nm-thick silver film perforated by an array of 100 nm-wide slits with a period of 780 nm. This approach resulted in a silver grating, which had the properties of an optical band-pass transmission filter. The polycrystalline grating creates nanoscale roughness, in which localized plasmon modes can be excited. So, by scanning the time delay between identical, phase-correlated pump and probe pulses in 174 optical cycle steps, and recording the resulting change in the polarization interference pattern, a movie of the SPP propagation wave packet at the Ag–vacuum interface can be created, as shown in Figure 9.15. As the driving pulse wanes, the coherent polarization excited at each dot shifts to its own resonant frequency. For instance, as shown in Figure 9.16, the phase of dots A, B and D (dot C) is retarded (advanced), causing the intensity maxima to rise later (sooner) with respect to the phase of the driving field. The circled hotspots in Figure 9.15 indicate the change in the intensity maxima (constructive interference) in five cycle intervals due to the phase slip of the surface plasmon modes with respect to the driving field.

The SPP wave packet propagation length – and hence coherent control studies – can be improved by using single-crystal nanostructures, as demonstrated by L.I. Chelaru *et al.* [107]. Figure 9.17 shows an example of the appearance of SPP waves in Ag single-crystal particles that were formed by self-assembly [108]. In Figure 9.17a, a SPP wave is started at the marked edge of the triangular island, and travels across the island during the time of observation. The striped pattern on the otherwise dark islands is a representation of the SPP wave by means of a beat pattern formed between the propagating plasmon wave and the laser pulse used to probe the structure [109]. Modulation of the local near-field in the surrounding of the island is caused by diffraction of the laser pulse. In Figure 9.17b, the two independent SPP waves are created at the edges of a hexagonal Ag island. The SPP waves superpose,

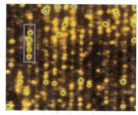

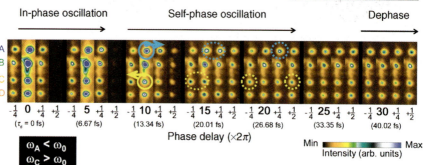

Figure 9.15 Interferometric TR-PEEM image of four localized surface plasmons on a Ag grating. The delay time between pump and probe pulses is advanced from −0.33 fs to +40.69 fs in π/2 steps (0.33 fs) of the carrier wavelength of 400 nm. During excitation (up to 5½ × 2π) all dots oscillate in phase; later the phase in dots A, B and D is retarded, but in dot C it is advanced compared to the exciting laser field. (Reproduced from Ref. [105].)

modulate the overall local electric field, and are reflected at the end of the structure. This is different in panel Figure 9.17c, where the angle between the two overlapping SPPs is smaller and the modulation of the electric field strength is much more pronounced. The field strength at the edge of the particle is strongly modulated and

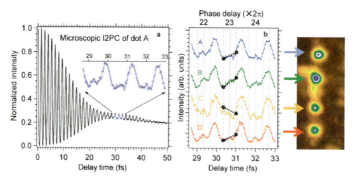

T_1: 43–65 fs (population decay time)
T_2: 4.8–5.9 fs (dephasing life time)

Figure 9.16 Phase slip between the LSP and the advancing light field of dots A to D in Figure 9.15. (From Ref. [105].)

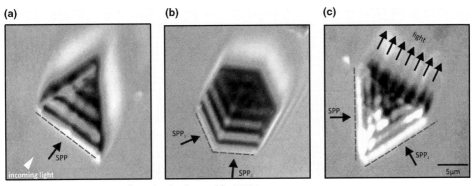

Figure 9.17 SPP waves on silber islands observed by PEEM illuminated under θ = 75°. (a) Triangular shape; (b) Hexagonal islands with SPP interferences; (c) Interference between two SPP waves where the island acts as beam splitter for the SSP wave. The scale bar represents 5 μm. (From Ref. [108].)

the SPP is converted back into light only at those positions where the field strength is particularly high. Ultimately, the island in panel Figure 9.17c acts as a beam-splitter. For particles such as those shown in Figure 9.17, propagation of the plasmon with ∼60% of the speed of light is observed as a systematic shift of the beat pattern as a function of the delay time between pump pulse and probe pulse [106].

9.6
Long-Lived Resonances in Adsorbate/Substrate Systems

Adsorbate atoms may not only serve as general scattering centers on well-prepared surfaces but also shorten the lifetime of well-defined surface states. When the electrons do scatter resonantly into unoccupied states of adsorbates, or when such states are directly optically excited, the lifetimes may be significantly prolonged compared to those expected from a simple Drude or Fermi liquid picture for electrons at the same excitation energy. A prominent example is the adsorption of alkali atoms on noble metal surfaces. In the low coverage limit an adsorbate-induced antibonding (A) state around 2.5–3 eV above Fermi is found on noble metals. The experimentally observed relatively long lifetimes of excited adsorbate states of up to 50 fs at low temperatures for Cu(111) [110, 111] and Ag(111) [112, 113] have initiated a number of theoretical investigations of this effect.

Although the decay of such states could, in principle, be viewed as a one-electron resonant charge transfer, which would yield very short lifetimes in the sub-femtosecond regime, the directional band gap at these surfaces hinders a fast decay, because the overlap of the state with bulk wave functions is small. With one-electron wave packet propagation calculations this reasoning could be supported [114–117]. Figure 9.18 shows the differences of the wave packet propagation for the Cs 6s state adsorbed on a jellium and a Cu(111) surface. Initially, the wave packet propagates in

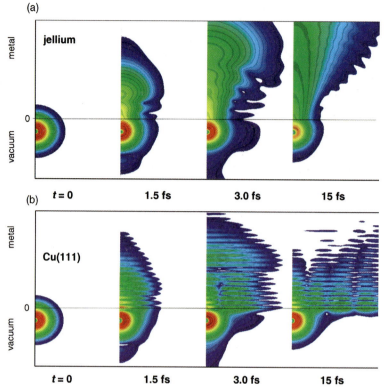

Figure 9.18 Dynamics of the Cs (6s) wave packet excited in front of (a) a jellium and (b) a Cu(111) surface. The Cs atom is placed at z = 10 a.u. from the image plane. A propagation along the surface normal into the bulk initially observed for both substrates continues only for the jellium surface. For Cu(111), propagation stops after only 3 fs, followed by a propagation in a high-$\vec{k}_{\parallel}$ direction along the surface. A high intensity of the 6s wave packet remains in the vicinity of the surface. (After Ref. [114].)

both cases into the bulk. On jellium (Figure 9.18a), this propagation continues with minimal lateral spreading of the wave packet, but on Cu(111) (Figure 9.18b) the movement of the wave packet into the bulk has come to a halt after only 3 fs, followed by a propagation in high-$\vec{k}_{\parallel}$ directions parallel to the surface, with the highest intensity remaining in the vicinity of the surface. Such behavior causes a relatively long resonance lifetime. Whilst these long-lived states were unexpected in the adsorbate/substrate systems, an understanding of the origin of long-lived states has opened some extremely interesting new possibilities towards the control of reactions at surfaces. Then, other processes may also come into play, including inelastic e–e scattering with bulk electrons, where the adsorbate atom begins to move on the new electronically excited potential energy surface. The excited electron then couples to the nuclear motion, and processes as in DIET or DIMET become important [118]. This type of motion has been observed directly and spectroscopically using TR-2PPE [119].

By taking into account the bandwidth of the ultrashort laser pulse and the nuclear motion of the adsorbate after excitation, an excellent agreement between experimental and theoretical lifetimes of the excited state for all alkalis studied has been achieved [114–117]. In differing from many other antibonding states of adsorbate/substrate systems, for the alkalis on noble metal surfaces this state can be populated by a direct dipole-allowed transition from the occupied surface state, also located in the same directional bandgap. This provides the opportunity to initiate the excitation with temporally and spectrally shaped pulses in order to control the desorption process [120]. For excitation laser pulses shorter than the lifetime of the excited A state, a complete population transfer to the A state can be achieved when dissipation is neglected. Including dissipation, the populations transfer falls to about 40% for a Gaussian pulse of 20 fs duration and a peak intensity of $0.5\,TW\,cm^{-2}$, while for an optimal pulse shape a transfer of 95% can still be achieved. In the case of longer and less intense pulses, the optimally controlled pulses are even more successful. Using again 20 fs-long Gaussian pulses, but now at only $10^{10}\,W\,cm^{-2}$ (which represents an experimentally feasible intensity at metal surfaces), only 3% population transfer is to be expected (in theory). For controlled pulses of 60 fs duration – that is, much longer than the lifetime of the A state – a transfer of 75% may still be achieved. In the control cases pulses are used which show the highest intensities at the extreme end of the pulse, although it will be difficult to produce such pulses experimentally. Nevertheless, even when taking the nuclear motion of the Cs atom on the excited potential into account, the use of shaped pulses might lead to an increased yield also for dissipative surface reactions.

With organic adsorbates, modified and new properties of nanostructures are obtained, the aim being to create functional materials in a controlled manner [121]. The organic constituents of such adsorbate layers provide a wide variability in the desired properties, which may range from sensor and molecular recognition to molecular electronics, photo-switches and molecular magnets. In most of these envisioned applications a nanostructured assembly of the functional materials is necessary. Many of the systems' properties rely on an heterogeneous electron transfer between the adsorbate and substrate and, in the case of sensor or recognition devices, between the active layer and the incoming molecules. The dynamic properties of these organic adsorbate films are therefore of fundamental interest for such functional materials.

Basic investigations of the action of alkane layers on the electronic properties of the underlying metals, as well as the electron localization in organic overlayers, have been carried out by Harris and coworkers [122, 123]. In the context of this chapter, the photo-induced electron transfer dynamics from (self-) organized and specifically bound organic molecules are of particular interest, because here the dynamics can be probed directly in the time domain. These processes also relate to light-harvesting applications, either for reactions or for the direct production of an electrical current.

Willig and coworkers investigated the influence of various anchor groups between a perylene-derived chromophore and the (110) surface of rutile TiO_2 on electron transfer [124]. These anchor groups were seen to serve two purposes: (i) to separate the electronic states of the chromophore and the solid surface; and (ii) to enable a control of the electronic coupling strength between both systems. The anchor groups also stabilize the complex and provide the option of definite binding angles and sites. The electron transfer is initiated by a direct femtosecond laser excitation of the singlet state of the chromophore. In the case of one conjugate anchor group – acrylic acid, which may be viewed as a molecular wire – the electron transfer time is found to be about fourfold faster ($\tau \sim 13.5$ fs) than for an insulating group of the same length, propionic acid, which acts as a tunneling barrier. For an even wider barrier which consists of a three-ring structure, where the central ring is aliphatic and rigid, the transfer times are prolonged to tenfold that of the first type [124].

When the electron transfer time is short – which can be achieved by directly binding the organic molecule to the substrate – the electron migration inside the substrate can be monitored using TR-2PPE. The experimentally observed times for the electron to escape from the surface layers (or, more specifically, from the detection depth of 2PPE) agrees well with rates obtained from density functional theory (DFT) calculations for alizarin adsorbates [125]. For the system investigated, it turned out that in rutile the energy does not relax on a 200 fs time scale, which is tentatively assigned to the population of new interface states created as the bonds with the organic adsorbates are formed. Such electron escape to the bulk states of a substrate is important in terms of the total energy transfer from an organic adsorbate to unoccupied states of a solid. When this process is rapid, charge conduction across the interface is favored; however, when it is slow an accumulation of charge in excited states of the organic adsorbate, with the possibility of a radiative deactivation, may occur.

A somewhat unexpected feature of organic adsorbates derives from the existence of very long-lived triplet states with lifetimes in the microsecond range. As yet, such states have not been identified for small inorganic adsorbates or for bare surfaces. Here, the dynamics in unoccupied states of ordered C_{60} on metal surfaces will be discussed as a specific example of an organic adsorbate. Due to the large work function of C_{60} of 6.8 eV above the highest occupied molecular orbital (HOMO), and depending on the work function of the metal, the lowest unoccupied molecular orbital (LUMO) state of the first C_{60} ML may be either below or above the metal Fermi level. Therefore, the character of the C_{60} film may be either metallic or semiconducting [126]. For thicker layers above about 5 ML the properties of C_{60} crystals are generally assumed, in which the interaction energies of the individual C_{60} constituents are comparatively weak. Hence, the properties of the film resemble those of the C_{60} clusters. Such layer dependency already provides the possibility of tuning the properties of a functional film, whilst in addition a substitution at a C–C bond by functional groups enables further variability [127, 128].

The symmetry properties of C_{60} allow optical transitions from the HOMO to the LUMO + 1, and from the HOMO-1 to the LUMO states – which are actually comparatively broad bands rather than sharply defined single states. With the third harmonic of a Nd-laser at photon energies of about $h\nu = 3.5$ eV ($\lambda \sim 355$ nm), a direct excitation via both transitions is possible. The population both in these states and in the energetically lower lying excitonic states, which are populated via internal energy transfer, can be probed with the fifth harmonic of a Nd-laser at 210 nm ($h\nu \sim 5.88$ eV).

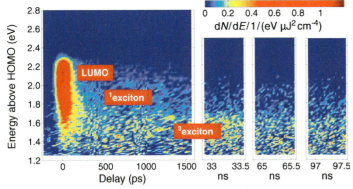

Figure 9.19 2-D lifetime plot for the LUMO, ^{1}exciton, and ^{3}exciton states on C_{60} up to a delay time of 100 ns. On this time scale the intensity of the ^{3}exciton decreases only marginally.

This second photon has insufficient energy to directly emit photoelectrons from the HOMO state. Figure 9.19 shows, in a 2-D plot, the kinetic energy distribution of emitted photoelectrons as a function of time delay between a pump and a probe laser pulse. In this case, the LUMO state has been directly populated in a dipole-forbidden transition, and it is evident that, besides the LUMO, lower-lying states are populated. Even for delay times exceeding 100 ns after pumping a signal intensity is observed for these energetically lower states. This long-lived state is assigned as a signature of the triplet exciton. In energetic terms, the singlet exciton is identified somewhat higher and below the directly excited LUMO.

An analysis of the intensity dependence at certain kinetic energies revealed for the LUMO a lifetime of about 84 ps – shorter than a previously reported value of 134 ps [129]. For the singlet exciton a lifetime of about 1050 ps was found, and was in good agreement with earlier reports [129]. In order to measure the lifetime of triplet excitons a second pump laser was used such that the probe laser could be electronically delayed with respect to the pumping laser. For these states, lifetimes between 22 μs for free and up to 200 μs for bound excitons were observed, depending also on the thickness of the C_{60} film (A. Rosenfeldt, B. Göhler and H. Zacharias, unpublished results). This was significantly longer than had been observed for a photo-polymerized C_{60} film [130], and in good agreement with lifetimes of these states in solution [131]. When the density of triplet excitons becomes high, the comparatively fast (spin-allowed) process of triplet–triplet annihilation can be observed (Figure 9.20).

It should be noted that these longlived states may be chemically active. Following the adsorption of NO molecules onto a thick, ordered C_{60} film, UV laser excitation of the system led to a desorption of the NO, with delay times of up to 200–400 μs after pumping. An analysis which yielded a chemically active state with a decay time of about 160 μs [132] was recently confirmed by directly measuring the velocity of the late-arriving molecules. Their velocity was found to be much greater than the corresponding arrival time at the detecting laser, which meant that these molecules

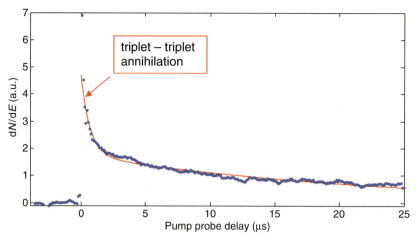

Figure 9.20 Lifetime of a C_{60} triplet excitonic state with $\tau = 22\,\mu s$ in a 14 ML film. Also indicated is an exciton–exciton annihilation occurring on a time scale of 720 ns.

were not desorbed promptly but rather at a later time (T. Hoger *et al.* unpublished results). Hence, these excitonic states in organic thin films may serve as efficient energy reservoirs for photochemical activity and reactions [133].

9.7
Outlook: Spatial and Temporal Control of Nano-Optical Fields

The optical response of nanostructures creates a variety of fascinating properties, including sub-wavelength variation of the field, local field enhancement and local fields with vector components perpendicular to those of the incident field. Moreover, the combination of ultrafast laser spectroscopy (i.e. illumination with broadband coherent light sources) and near-field optics continue to open new realms for nonlinear optics on the nanoscale. As an example, the spectral phase of the incident light will influence the peak intensity of the local field distribution [134], thus providing a means to manipulate the nonlinear response of a nanostructure. More recently, it has been shown theoretically that the interaction of polarization-shaped laser pulses with a nanostructure allows the simultaneous control of spatial and temporal evolution of the optical near-field distribution [135].

A first step towards the experimental demonstration of simultaneous spatial and temporal field control used adaptive control, combining multi-parameter pulse shaping with a learning algorithm, and demonstrated the generation of user-specified optical near-field distributions in an optimal and flexible fashion. Shaping of the polarization of the laser pulse provides a particularly efficient and versatile nano-optical manipulation method [136]. Additionally, pump-probe sequences can be generated, in which excitations occur not only at different times but also at different

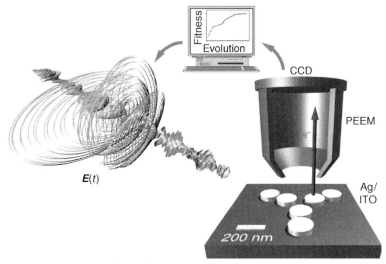

Figure 9.21 Experimental set-up for applying polarization-shaped femtosecond pulses to a nanostructure. A PEEM was used to measure the 2PPE signal.

positions that are separated by less than the diffraction limit. This in turn opens a route towards space–time-resolved spectroscopy with potential for the direct observation of nanoscopic energy or electron transport [135]. In a first experiment [137], femtosecond polarization shaping, adaptive optimization and two-photon PEEM were combined (see Figure 9.21). The pulse shaper contains a two-layer, 128-element liquid crystal display (LCD) spatial light modulator in the Fourier plane of a zero-dispersion compressor in folded 4f configuration [136]. Each LCD layer modulates the spectral phase of one of the two transverse polarization components, leading to a spectral variation of polarization state and phase. Hence, in the time domain the intensity, instantaneous oscillation frequency and polarization state (elliptical eccentricity and orientation) can be controlled to vary within a single laser pulse. Pulse-shape optimization occurs for both LCD layers independently. The data in Figure 9.22 show that adaptive polarization pulse shaping allows the optimization of a particular emission pattern. In Figure 9.22b and c, the ratio A/B between the emission yields integrated over the two rectangles A and B was optimized using an evolutionary algorithm. The experimental results showed that the local interference of the optical near-fields generated by the two orthogonal incident polarization components could be utilized to manipulate the local field distribution.

Currently, in these experiments two-photon PEEM is used to qualitatively monitor the local field distribution, with the interpretation of the acquired images being based on the assumption that the highest local field intensity will produce the highest photoemission yield in a 2PPE process. However, this yield is influenced by additional parameters such as the intermediate state lifetime in the metal, or the field component perpendicular to the surface. A detailed modeling of the emission process, together with a comparison with the experimentally observed emission

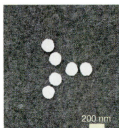
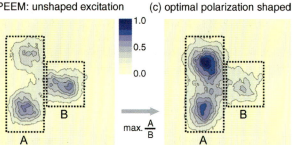

Figure 9.22 Control of nanoscopic photoelectron emission. (a) Scanning electron microscopy image of a single nanostructure; (b) An unshaped laser pulse is used as a reference; (c) Maximization of the integrated emission ratio A/B by optimal polarization shaping leads to emission patterns that exhibit a high contrast between region A versus region B. Hence, the photoelectron emission can be controlled experimentally on a nanoscopic length scale.

pattern, should provide an improved understanding of two-photon photoemission from nanostructured objects.

Acknowledgments

The authors enjoyed fruitful and stimulating discussions and support with and by many colleagues, notably E.V. Chulkov and P.M. Echenique (San Sebastian), J. P. Gauyacq (Paris), F.-J. Meyer zu Heringdorf (Duisburg), W. Pfeiffer (Bielefeld) and M. Bauer (Kiel). They also wish to thank their coworkers and graduate students. Financial support from the Deutsche Forschungsgemeinschaft via the priority programs SPP 1093 'Dynamics of electron transfer processes at interfaces', through projects Za 110/21 and AE 19/3, and SPP 1153 'Cluster in contact with surfaces' through project AE 19/12, is gratefully acknowledged.

References

1 Echenique, P.M. and Pendry, J.B. (1978) *The Journal of Physics C*, **11**, 2065.
2 Giesen, K., Hagen, F., Himpsel, F.J., Riess, H.J. and Steinmann, W. (1985) *Physical Review Letters*, **55**, 300.
3 Yen, R., Liu, J.M., Bloembergen, N., Yee, T.K., Fujimoto, J.G. and Salour, M.M. (1982) *Applied Physics Letters*, **40**, 185.
4 Williams, R.T., Boyt, R.R., Rife, J.C., Long, J.P. and Kabler, M.N. (1982) *Journal of Vacuum Science and Technology*, **21**, 509.
5 Bokor, J. (1989) *Science*, **246**, 1130.
6 Haight, R. (1995) *Surface Science Reports*, **21**, 275.
7 Schoenlein, R.W., Fujimoto, J.G., Eesley, G.L. and Capehart, T.W. (1988) *Physical Review Letters*, **61**, 2596.
8 Fann, W.S., Storz, R., Tom, H.W.K. and Bokor, J. (1992) *Physical Review Letters*, **68**, 2834.
9 Fann, W.S., Storz, R., Tom, H.W.K. and Bokor, J. (1992) *Physical Review B - Condensed Matter*, **46**, 13592.

10 Schmuttenmaer, C.A., Aeschlimann, M., Elsayed-Ali, H.E., Miller, R.J.D., Mantel, D.A., Cao, J. and Gao, Y. (1994) *Physical Review B - Condensed Matter*, **50**, 8957.

11 Schönhense, G., Elmers, H.J., Nepijko, S.A. and Schneider, C.M. (2006) *Advances in Imaging and Electron Physics*, **142**, 159.

12 Rotermund, H.H. (1997) *Surface Science Reports*, **29**, 267.

13 Schmidt, O., Bauer, M., Wiemann, C., Porath, R., Scharte, M., Andreyev, O., Schönhense, G. and Aeschlimann, M. (2002) *Applied Physics B: Lasers and Optics*, **74**, 223.

14 Clark, T.D., Tien, J., Duffy, D.C., Paul, K.E. and Whitesides, G.M. (2001) *Journal of the American Chemical Society*, **123**, 7677.

15 Hurst, S.J., Payne, E.K., Qin, L.D. and Mirkin, C.A. (2006) *Angewandte Chemie - International Edition*, **45**, 2672.

16 Popovic, Z., Otter, M., Calzaferri, G. and De Cola, L. (2007) *Angewandte Chemie - International Edition*, **46**, 6301.

17 Güdde, J., Rohleder, M., Meier, T., Koch, S.W. and Höfer, U. (2007) *Science*, **318**, 1287.

18 Chulkov, E.V., Borisov, A.G., Gauyacq, J.P., Sanchez-Portal, D., Silkin, V.M., Zhukov, V.P. and Echenique, P.M. (2006) *Chemical Reviews*, **106**, 4160.

19 Schöne, W.-D. (2007) *Progress in Surface Science*, **82**, 161.

20 Ueba, H. and Gumhalter, B. (2007) *Progress in Surface Science*, **82**, 193.

21 Ashcroft, N.W. and Mermin, N.D. (1976) *Solid State Physics*, Holt, Rinehart & Winston, New York.

22 (a) Landau, L.D. (1956) *Journal of Experimental and Theoretical Physics (USSR)*, **30**, 1058; (b) Landau, L.D. (1957) *Journal of Experimental and Theoretical Physics (USSR)*, **32**, 59; (c) Landau, L.D. (1958) *Journal of Experimental and Theoretical Physics (USSR)*, **35**, 97.

23 Pines, D. and Nozieres, P. (1989) *The Theory of Quantum Liquids*, Addison-Wesley, Reading.

24 Quinn, J.J. and Ferrell, R.A. (1958) *Physical Review*, **112**, 812.

25 Aeschlimann, M., Bauer, M. and Pawlik, S. (1996) *Chemical Physics*, **205**, 127.

26 Hertel, T., Knoesel, E., Wolf, M. and Ertl, G. (1996) *Physical Review Letters*, **76**, 535.

27 Ogawa, S. and Petek, H. (1996) *Surface Science*, **357**, 585.

28 Knoesel, E., Hotzel, A. and Wolf, M. (1998) *Physical Review B - Condensed Matter*, **57**, 12812.

29 (a) Petek, H., Nagano, H. and Ogawa, S. (1999) *Physical Review Letters*, **83**, 832; (b) Petek, H., Nagano, H. and Ogawa, S. (1999) *Applied Physics B: Lasers and Optics*, **68**, 369.

30 Aeschlimann, M., Bauer, M., Pawlik, S., Knorren, R., Bouzerar, G. and Bennemann, K.H. (2000) *Applied Physics A: Materials Science and Processing*, **71**, 485.

31 Zhukov, V.P., Andreyev, O., Hoffmann, D., Bauer, M., Aeschlimann, M., Chulkov, E.V. and Echenique, P.M. (2004) *Physical Review B*, **70**, 233106.

32 Lisowski, M., Loukakos, P.A., Bovensiepen, U., Stähler, J., Gahl, C. and Wolf, M. (2004) *Applied Physics A: Materials Science and Processing*, **78**, 165.

33 Mönnich, A., Lange, J., Bauer, M., Aeschlimann, M., Nechaev, I.A., Zhukov, V.P., Echenique, P.M. and Chulkov, E.V. (2006) *Physical Review B - Condensed Matter*, **74**, 035102.

34 Knorren, R., Bennemann, K.H., Burgermeister, R. and Aeschlimann, M. (2000) *Physical Review B - Condensed Matter*, **61**, 9427.

35 Nessler, W., Ogawa, S., Nagano, H., Petek, H., Shimoyama, J., Nakayama, Y. and Kishio, K. (1998) *Physical Review Letters*, **81**, 4480.

36 Marienfeld, A., Cinchetti, M., Bauer, M., Aeschlimann, M., Zhukov, V.P., Chulkov, E.V. and Echenique, P.M. (2007) *Journal of Physics - Condensed Matter*, **19**, 496213.

37 Echenique, P.M., Pitarke, J.M., Chulkov, E.V. and Rubio, A. (2000) *Chemical Physics*, **251**, 1.

38 Zhukov, V.P., Aryasetiawan, F., Chulkov, E.V., de Gurtubay, I.G. and Echenique,

P.M. (2001) *Physical Review B - Condensed Matter*, **64**, 195122.
39 Ladstädter, F., de Pablos, P.F., Hohenester, U., Puschnig, P., Ambrosch-Draxl, C., de Andres, P.L., García-Vidal, F.J. and Flores, F. (2003) *Physical Review B - Condensed Matter*, **68**, 085107.
40 Ladstädter, F., Hohenester, U., Puschnig, P. and Ambrosch-Draxl, C. (2004) *Physical Review B - Condensed Matter*, **70**, 235125.
41 Zhukov, V.P., Aryasetiawan, F., Chulkov, E.V. and Echenique, P.M. (2002) *Physical Review B - Condensed Matter*, **65**, 115116.
42 Bacelar, M.R., Schöne, W.-D., Keyling, R. and Ekardt, W. (2002) *Physical Review B - Condensed Matter*, **66**, 153101.
43 Papaconstantopoulos, D.A. (1986) *Handbook of the Band Structure of Elemental Solids*, Plenum, New York.
44 Springer, M., Aryasetiawan, F. and Karlsson, K. (1998) *Physical Review Letters*, **80**, 2389.
45 Karlsson, K. and Aryasetiawan, F. (2000) *Physical Review B - Condensed Matter*, **62**, 3006.
46 Nechaev, I.A. and Chulkov, E.V. (2005) *Physical Review B - Condensed Matter*, **71**, 115104.
47 Nechaev, I.A. and Chulkov, E.V. (2006) *Physical Review B - Condensed Matter*, **73**, 165112.
48 Zhukov, V.P., Chulkov, E.V. and Echenique, P.M. (2005) *Physical Review B - Condensed Matter*, **72**, 155109.
49 Mie, G. (1908) *Annalen der Physik*, **25**, 377.
50 Kreibig, U. and Vollmer, M. (1995) *Optical Properties of Metal Clusters, Springer Series in Materials Science, Vol. 25*, Springer, Berlin.
51 Bohren, C.F. and Huffmann, D.R. (1983) *Absorption and Scattering of Light by Small Particles*, Wiley, New York.
52 Kottmann, J.P. and Martin, O.J.F. (2001) *Optics Letters*, **26**, 1096.
53 Moskovits, M. (1985) *Reviews of Modern Physics*, **57**, 783.
54 Simon, H.J. and Chen, Z. (1989) *Physical Review B - Condensed Matter*, **39**, 3077.
55 Bouhelier, A., Beversluis, M., Hartschuh, A. and Novotny, L. (2003) *Physical Review Letters*, **90**, 013903.
56 Scharte, M., Porath, R., Ohms, T., Aeschlimann, M., Krenn, J.R., Dittelbacher, H., Aussenegg, F.R. and Liebsch, A. (2001) *Applied Physics B: Lasers and Optics*, **73**, 305.
57 Salerno, M., Krenn, J.R., Lamprecht, B., Schider, G., Ditlbacher, H., Félidj, N., Leitner, A. and Aussenegg, F.R. (2002) *Opto-Electronics Review*, **10**, 217.
58 Anisimov, S.I., Kapeliovich, B.L. and Perel'man, T.L. (1974) *Soviet Physics – JETP*, **39**, 375.
59 Budde, F., Heinz, T.F., Kalamarides, A., Loy, M.M.T. and Misewich, J.A. (1993) *Surface Science*, **283**, 143.
60 Brandbyge, M., Hedegard, P., Heinz, T.F., Misewich, J.A. and Newns, D.M. (1995) *Physical Review B - Condensed Matter*, **52**, 6042.
61 Fröhlich, H. (1950) *Physical Review*, **79**, 845.
62 (a) Eiguren, A., Hellsing, B., Chulkov, E.V. and Echenique, P.M. (2003) *Physical Review B - Condensed Matter*, **67**, 235432; (b) Eiguren, A., de Gironcoli, S., Chulkov, E.V., Echenique, P.M. and Tosatti, E. (2003) *Physical Review Letters*, **91**, 166803.
63 Grimvall, G. (1981) *The Electron–Phonon Interaction in Metals*, North-Holland, Amsterdam.
64 Sklyadneva, I.Yu., Chulkov, E.V., Schöne, W.-D., Silkin, V.M., Keyling, R. and Echenique, P.M. (2005) *Physical Review B - Condensed Matter*, **71**, 174302.
65 McDougall, B.A., Balasubramanian, T. and Jensen, E. (1995) *Physical Review B - Condensed Matter*, **51**, 13891.
66 Mathias, S., Wiesenmayer, M., Aeschlimann, M. and Bauer, M. (2006) *Physical Review Letters*, **97**, 236809.
67 Brorson, S.D., Kareroonian, A., Moodera, J.S., Face, D.W., Cheng, T.K., Ippen, E.P., Dresselhaus, M.S. and Dresselhaus, G. (1990) *Physical Review Letters*, **64**, 2172.

68 Fuchs, H., Hölscher, H. and Schirmeisen, A. (2005) *Encyclopedia of Materials: Science and Technology*, Elsevier, pp. 1–12.

69 Schirmeisen, A., Anczykowski, B. and Fuchs, H. (2007) *Handbook of Nanotechnology II* (ed. B. Bushan), Wiley, pp. 737–765.

70 Matzdorf, R. (1998) *Surface Science Reports*, **30**, 153.

71 Chang, Z., Rundquist, A., Wang, H., Murnane, M.M. and Kapteyn, H.C. (1997) *Physical Review Letters*, **79**, 2967.

72 Spielmann, Ch., Burnett, N.H., Sartania, S., Koppitsch, R., Schnürer, M., Kan, C., Lenzner, M., Wobrauschek, P. and Krausz, F. (1997) *Science*, **278**, 661.

73 Siffalovic, P., Drescher, M., Spieweck, M., Wiesenthal, T., Lim, Y.C., Weidner, R., Elizarov, A., Heinzmann, U. (2001) *Review of Scientific Instruments*, **72**, 30.

74 Tsilimis, G., Benesch, C., Kutzner, J. and Zacharias, H. (2003) *Journal of the Optical Society of America B - Optical Physics*, **20**, 246.

75 Goulielmakis, E., Uiberacker, M., Kienberger, R. et al. (2004) *Science*, **305**, 1267.

76 Sansone, G., Benedetti, E., Calegari, F. et al. (2006) *Science*, **314**, 443.

77 Paul, P.M., Toma, E.S., Breger, P., Mullot, G., Augé, F., Balcou, Ph., Muller, H.G. and Agostini, P. (2001) *Science*, **292**, 1689.

78 Mauritsson, J., Johnsson, P., Gustafsson, E., L'Huillier, A., Schafer, K.J. and Gaarde, M.B. (2006) *Physical Review Letters*, **97**, 013001.

79 Goulielmakis, E., Schultze, M., Hofstetter, M. et al. (2008) *Science*, **320**, 1640.

80 Ogawa, S., Nagano, H., Petek, H. and Heberly, A. (1997) *Physical Review Letters*, **78**, 1339.

81 Lange, J., Bayer, D., Rohmer, M., Wiemann, C., Gaier, O., Aeschlimann, M. and Bauer, M. (2006) *Proceedings of SPIE*, **6195**, 61950.

82 Bayer, D., Wiemann, C., Gaier, O., Bauer, M. and Aeschlimann, M. (2008) *Journal of Nanomaterials*, **00**, 249514.

83 Schlipper, R., Kusche, R., von Issendorff, B. and Haberland, H. (1998) *Physical Review Letters*, **80**, 1194.

84 Becker, Th., Hövel, H., Bettac, A., Reihl, B., Tschudy, M. and Williams, E.J. (1997) *Journal of Applied Physics*, **81**, 154.

85 Rohmer, M., Galeh, F., Aeschlimann, M., Bauer, M. and Hövel, H. (2007) *European Journal of Physics D*, **45**, 491.

86 Echenique, P.M., Berndt, R., Chulkov, E.V., Fauster, Th., Goldmann, A. and Höfer, U. (2004) *Surface Science Reports*, **52**, 219.

87 Weinelt, M. (2002) *Journal of Physics - Condensed Matter*, **14**, R1099.

88 Eiguren, A., Hellsing, B., Chulkov, E.V. and Echenique, P.M. (2003) *Journal of Electron Spectroscopy and Related Phenomena*, **129**, 111.

89 Knoesel, E., Hotzel, A. and Wolf, M. (1998) *Journal of Electron Spectroscopy and Related Phenomena*, **88–91**, 577.

90 Berthold, W., Höfer, U., Feulner, P., Chulkov, E.V., Silkin, V.M. and Echenique, P.M. (2002) *Physical Review Letters*, **88**, 056805.

91 Roth, M., Weinelt, M., Fauster, T., Wahl, P., Schneider, M.A., Diekhöner, L. and Kern, K. (2004) *Applied Physics A: Materials Science and Processing*, **78**, 155.

92 Boger, K., Weinelt, M., Wang, J. and Fauster, T. (2004) *Applied Physics A: Materials Science and Processing*, **78**, 161.

93 Hirschmann, M. and Fauster, T. (2007) *Applied Physics A: Materials Science and Processing*, **88**, 547.

94 Kliewer, J., Berndt, R., Chulkov, E.V., Silkin, V.M., Echenique, P.M. and Crampin, S. (2000) *Science*, **288**, 1399.

95 LaShell, S., McDougall, B.A. and Jensen, E. (1996) *Physical Review Letters*, **77**, 3419.

96 Theilmann, F., Matzdorf, R., Meister, G. and Goldmann, A. (1997) *Physical Review B - Condensed Matter*, **56**, 3632.

97 Lisowski, M., Loukakos, P.A., Bovensiepen, U. and Wolf, M. (2004) *Applied Physics A: Materials Science and Processing*, **79**, 739.

98 Reinert, F., Nicolay, G., Schmidt, S., Ehm, D. and Hüfner, S. (2001) *Physical Review B - Condensed Matter*, **63**, 115415.

99 Eiguren, A., Hellsing, B., Reinert, F., Nicolay, G., Chulkov, E.V., Silkin, V.M., Hüfner, S. and Echenique, P.M. (2002) *Physical Review Letters*, **88**, 066805.

100 Silkin, V.M., Balasubramanian, T., Chulkov, E.V., Rubio, A. and Echenique, P.M. (2001) *Physical Review B - Condensed Matter*, **64**, 085334.

101 LaShell, S., McDougall, B.A. and Jensen, E. (2006) *Physical Review B - Condensed Matter*, **74**, 033410.

102 Fuglsang Jensen, M., Kim, T.K., Bengio, S., Sklyadneva, I.Yu., Leonardo, A., Eremeev, S.V., Chulkov, E.V. and Hofmann, Ph. (2007) *Physical Review B - Condensed Matter*, **75**, 153404.

103 Kutzner, J., Paucksch, R., Jabs, C., Zacharias, H. and Braun, J. (1997) *Physical Review B - Condensed Matter*, **56**, 16003.

104 Wiemann, C., Bayer, D., Rohmer, M., Aeschlimann, M. and Bauer, M. (2007) *Surface Science*, **601**, 4714.

105 Kubo, A., Onda, K., Petek, H., Sun, Z., Jung, Y.S. and Kim, H.K. (2005) *Nano Letters*, **5**, 1123.

106 Kubo, A., Pontius, N. and Petek, H. (2007) *Nano Letters*, **7**, 470.

107 Chelaru, L.I., Horn von Hoegen, M., Thien, D. and Meyer zu Heringdorf, F.-J. (2006) *Physical Review B - Condensed Matter*, **73**, 115416.

108 Meyer zu Heringdorf, F.-J., Buckanie, N.M., Chelaru, L.I. and Raß, N. (2008) in *EMC 2008*, vol. 1 (eds. M. Luysberg, K. Tillmann and T. Weirich), Springer, Berlin, pp. 737.

109 Chelaru, L.I. and Meyer zu Heringdorf, F.-J. (2007) *Surface Science*, **601**, 4541.

110 Bauer, M., Pawlik, S. and Aeschlimann, M. (1997) *Physical Review B - Condensed Matter*, **55**, 10040.

111 Ogawa, S., Nagano, H. and Petek, H. (1999) *Physical Review Letters*, **82**, 1931.

112 Bauer, M., Pawlik, S. and Aeschlimann, M. (1999) *Physical Review B - Condensed Matter*, **60**, 5016.

113 Petek, H., Nagano, H., Weida, M.J. and Ogawa, S. (2001) *Journal of Physical Chemistry B*, **105**, 6767.

114 Borisov, A.G., Kazansky, A.K. and Gauyacq, J.P. (2001) *Physical Review B - Condensed Matter*, **64**, 201105.

115 Borisov, A.G., Gauyacq, J.P., Kazansky, A.K., Chulkov, E.V., Silkin, V.M. and Echenique, P.M. (2001) *Physical Review Letters*, **86**, 488.

116 Gauyacq, J.P. and Kazansky, A. (2005) *Physical Review B - Condensed Matter*, **72**, 045418.

117 Gauyacq, J.P., Borisov, A.G. and Bauer, M. (2007) *Progress in Surface Science*, **82**, 244.

118 Frischkorn, C. and Wolf, M. (2006) *Chemical Reviews*, **106**, 4207.

119 Petek, H., Weida, M.J., Nagano, H. and Ogawa, S. (2000) *Science*, **288**, 1402.

120 Kröner, D., Klamroth, T., Nest, M. and Saalfrank, P. (2007) *Applied Physics A: Materials Science and Processing*, **88**, 535.

121 Tans, S.J., Verschueren, A.R.M. and Dekker, C. (1998) *Nature*, **393**, 49.

122 Harris, C.B., Ge, N.-H., Lingle, G.R., Jr McNeill, J.D. and Wong, C.M. (1997) *Annual Review of Physical Chemistry*, **48**, 711.

123 Szymanski, P., Garrett-Roe, S. and Harris, C.B. (2005) *Progress in Surface Science*, **78**, 1.

124 (a) Gundlach, L., Ernstorfer, R. and Willig, F. (2007) *Applied Physics A: Materials Science and Processing*, **88**, 481; (b) Gundlach, L., Ernstorfer, R. and Willig, F. (2007) *Progress in Surface Science*, **82**, 355.

125 Duncan, W.R., Stier, W.M. and Prezhdo, O.V. (2005) *Journal of the American Chemical Society*, **127**, 7941.

126 see e.g. Tzeng, C.-T., Lo, W.-S., Yuh, J.-Y., Chu, R.-Y. and Tsuei, K.-D. (2000) *Physical Review B - Condensed Matter*, **61**, 2263, and references therein.

127 Maggini, M., Scorrano, G. and Prato, M. (1993) *Journal of the American Chemical Society*, **115**, 9798.

128 Maggini, M. and Prato, M. (1998) *Accounts of Chemical Research*, **31**, 519.

129 Jacquemin, R., Kraus, S. and Eberhardt, W. (1995) *Solid State Communications*, **105**, 449.

130 Long, J.P., Chase, S.J. and Kabler, M.N. (2001) *Physical Review B - Condensed Matter*, **64**, 205415.

131 See e.g. Wasielewski, M.R., O'Neil, M.P., Lykke, K.R., Pellin, M.J. and Gruen, D.M. (1991) *Journal of the American Chemical Society*, **113**, 2774.

132 Hoger, T., Grimmer, D. and Zacharias, H. (2007) *Applied Physics A: Materials Science and Processing*, **88**, 449.

133 O'Regan, B. and Grätzel, M. (1991) *Nature*, **353**, 737.

134 Stockman, M.I., Faleev, S.V. and Bergman, D.J. (2002) *Physical Review Letters*, **88**, 067402.

135 Brixner, T., García de Abajo, F.J., Schneider, J. and Pfeiffer, W. (2005) *Physical Review Letters*, **95**, 093901.

136 Brixner, T., Garcia Abajo, S.J., Schneider, J., Spindler, C. and Pfeiffer, W. (2006) *Physical Review B - Condensed Matter*, **73**, 125437.

137 Aeschlimann, M., Bauer, M., Bayer, D., Brixner,T., Garcia de Abajo, F.J., Pfeiffer, W., Rohmer, M., Spindler, C. and Steeb, F. (2007) *Nature*, **446**, 301.

10
Nanoplasmonics
Gerald Steiner

10.1
Introduction

The basic foundation of nanoplasmonic involves interactions between light and metal particles. Such particles, which are often referred to as *clusters*, have been the subject of a large number of investigations. Gustav Mie's theory about the study of optical properties of small gold particles, published in 1908, was the starting point for the new field of plasmonics. During the past half-century the nature of interactions between photons and electrons in metal clusters has been the focus of many research fields, including not only physics but also chemistry, materials science, medicine, biology and the environmental sciences. Clearly, metal cluster photonics is important in many fields, even in nanotechnology. Although the term 'cluster' is often used for a number of unspecified particles, there is no precise definition of a cluster. Hence, in this chapter clusters are defined as particles composed of a certain number of atoms that form a spherical or elliptical particle with the dimension in the range between 5 and 100 nm, corresponding to 500...10^7 atoms. The range from single atoms to bulk material is illustrated in Figure 10.1. It should be noted, that some publications use the term cluster just for very small clusters consisting of less than 100 atoms. The defined cluster region covers a rather wide range in relation to their optical properties. Metal clusters are subjected to variations not only in size but also in shape (some frequently identified forms are shown in Figure 10.2), but they are rarely spherical. In particular, clusters adsorbed onto a surface are ellipsoid in shape. Nanorods are tiny rod-shaped particles, less than 100 nm in diameter, but often with a length in excess of a few micrometers. The advantage of nanorods lies in their high aspect ratio (the ratio of length to diameter), as this permits them to show certain properties not seen with spherical or elliptical clusters. The optical properties of metal clusters depend heavily on the shape, the environment of the cluster, and the type of metal. Nonetheless, the two general features of all clusters are their ability to interact with light and to produce a localized electromagnetic field, the details of which are discussed in the following sections [1–3].

Nanotechnology. Volume 6: Nanoprobes. Edited by Harald Fuchs
Copyright © 2009 WILEY-VCH Verlag GmbH & Co. KGaA, Weinheim
ISBN: 978-3-527-31733-2

10 Nanoplasmonics

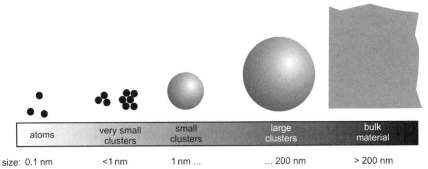

Figure 10.1 Dimensions of metal clusters on the scale from single atoms to bulk material.

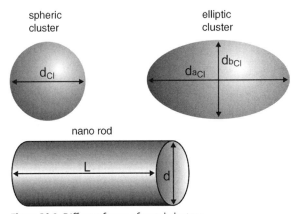

Figure 10.2 Different forms of metal clusters.

10.2
Single Clusters

In general, the description of the interaction between light and metal clusters is obtained by the solution of Maxwell's equation and imposing the boundary conditions. A simple approach to understand the optical response of metal clusters is to consider the free electrons. The positive charges are assumed to be immobile; then, if the cluster is illuminated by light the electric vector of the light wave will displace the free electrons. Under consideration of the boundary condition the cluster exhibits polarization; this effect is illustrated in Figure 10.3.

The internal field E_i of a spherical cluster is given by

$$E_i = E_0 \frac{3\varepsilon_1}{(\varepsilon r_{Cl} + \varepsilon i_{Cl}) + 2\varepsilon_1} \tag{10.1}$$

where E_0 is the electric field of the incident light. The dielectric function of the cluster material is given by the real part εr_{Cl} and the imaginary part εi_{Cl}. The surrounding

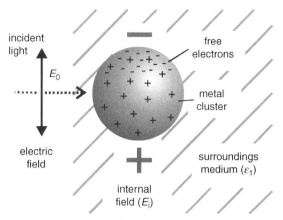

Figure 10.3 Polarization of a metal cluster caused by the electric vector of the incident light wave.

medium is characterized by the dielectric function ε_1. An important parameter is the static polarizability α of the cluster

$$\alpha = 4\pi\varepsilon_0 d^3 \frac{(\varepsilon r_{Cl} + \varepsilon i_{Cl}) - \varepsilon_1}{(\varepsilon r_{Cl} + \varepsilon i_{Cl}) + 2\varepsilon_1} \tag{10.2}$$

with the cluster diameter d. Unfortunately, this solution describes only static conditions. If the cluster is placed in an electromagnetic field, the free electrons will move inside the metal cluster with frequency of the external field. The excitation of the free electrons leads to an internal field. In this case, the dielectric constants must be replaced by their frequency-dependent values. As the magnetic fields do not occur, the complex dielectric function (ε_{Cl}) must be replaced with the frequency-dependent values:

$$\bar{\varepsilon}_{Cl} = \bar{\varepsilon}_{Cl}(\omega) \tag{10.3}$$

where ω is the angular frequency of the incident light wave.

The internal electric field shows a resonance when

$$|\varepsilon r_{Cl}(\omega) + 2\varepsilon_1|^2 + |\varepsilon i_{Cl}(\omega)|^2 = \min \tag{10.4}$$

This means that, in a spherical cluster, the resonance frequency is found by the relationship

$$\varepsilon r_{Cl}(\omega) = -2\varepsilon_1 \tag{10.5}$$

The real part of the dielectric function can be also expressed by the relationship

$$\varepsilon r_{Cl} \approx 1 - \frac{\omega_P^2}{\omega^2} \tag{10.6}$$

where is ω_P the Drude plasma frequency. Consequently, the resonance frequency ω_R of a spherical cluster is given by

$$\omega_R = \frac{\omega_P}{\sqrt{2\varepsilon_1 + 1}} \tag{10.7}$$

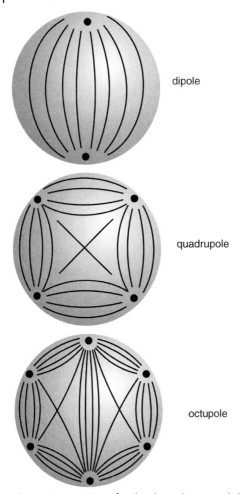

Figure 10.4 Excitation of multipoles within a metal cluster.

A metallic cluster exhibits not only a distinct dipole but also a quadrupole and higher multipole, as depicted in Figure 10.4. The excitation of these dipoles is dependent on the size, the shape and the type of metal, as well as on the wavelength of the incident light. The internal field at the resonance frequency, where the free electrons exhibit a strong collective oscillation, is known as the *surface plasmon*. This term is normally used to describe the excitations at a metal surface, whereas the plasmons in a metal cluster are localized and are thus referred to *localized surface plasmons*. One consequence of the internal field is that there is a strong electromagnetic field in the proximity of the cluster, as shown in Figure 10.5. It should be noted that the term 'localized surface plasmons' also describes the resulting external field of the cluster. As mentioned above, the spectral position of the localized surface

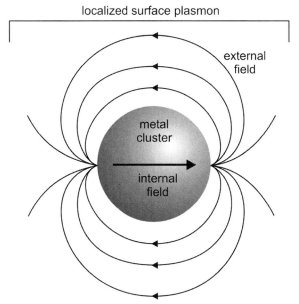

Figure 10.5 The excitation of an internal field causes a strong electromagnetic field around the cluster.

plasmons is dependent on the size, shape and metal type of the cluster, and the dielectric function of the surrounding medium. As an example, Figure 10.6 illustrates the effect of cluster size on the spectral position of the resonance for a gold cluster placed on a glass slide. For larger clusters, higher-order multipoles become important, and this results in a more pronounced shift of the localized surface plasmon resonance (SPR) as the particle size increases. In addition, the position and shape of the resonance are also dependent on dielectric function of the surrounding medium [3–6].

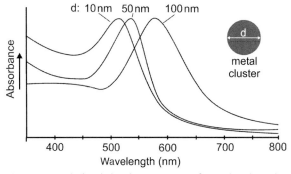

Figure 10.6 Calculated absorbance spectra of a single spherical gold cluster with different sizes.

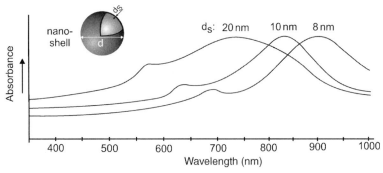

Figure 10.7 Absorbance spectra of nanoshells consisting of a glass core and gold shell.

10.3
Nanoshells

A nanoshell contains different materials in the core and shell. Often, the core material is metallic while the shell forms a thin dielectric layer, and in this case the localized SPR is affected by the shell. When the dielectric constant or the thickness of the shell is changed, however, the resonance conditions are also changed. For example, when the shell material is metallic and the core is a dielectric material, the change in SPR can be dramatic, with the spectral position of the resonance being shifted to longer wavelengths than those in the corresponding solid metal cluster. The same effect occurs when the metal shell thickness is decreased. The calculated absorbance spectra for nanoshells consisting of a glass core and a thin gold shell are shown in Figure 10.7. These nanoshells serve as the main component of many applications in biosensing and semiconducting [2, 7].

10.4
Layer of Clusters

When clusters become closer to each other, the electromagnetic field causes an electromagnetic coupling between the individual units [8]. One simple way to describe this coupling is with a system of two interacting dipolar oscillators, when two principal situations are possible: (i) the orientation of the cluster dipole is in the *same direction*; or (ii) the orientation is in the *opposite direction*. These two configurations are illustrated in Figure 10.8. In the case of perpendicular orientation (Figure 10.8a), the internal dipole vectors have a different polarization, whereas for parallel orientation (Figure 10.8b) the internal dipole vectors exhibit the same polarization. If both dipoles are excited together with the same polarization, then the resulting field is enhanced. In the case of perpendicular orientation when both clusters are excited together, but with different polarization, the resulting field will be weaker – a fact which has a major influence on the spectral position of the SPR. Although the interaction between clusters is also determined by the cluster size and

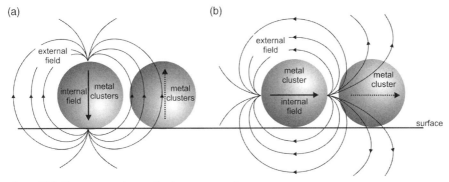

Figure 10.8 Electromagnetic coupling between two clusters. (a) Different polarization; (b) Similar polarization.

distance, as well as by the optical properties of the medium between the clusters, the most critical parameter is polarization of the localized surface plasmons. To illustrate this fact, Figure 10.9 shows the absorbance spectra of two identical gold clusters with the same and different polarizations of the localized surface plasmons. When the

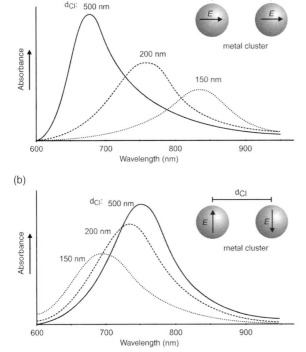

Figure 10.9 Calculated absorbance spectra for two gold clusters at a certain distance. (a) With the same polarization; (b) With different polarization.

internal field of the clusters exhibits the same polarization, the position of the SPR shifts towards longer wavelengths as the inter-cluster distance is increased. In the case of a different polarization a weak blue-shift occurs when the inter-cluster distance is increased. This different optical response – which is also evident for more than two clusters – has major consequences on the sensitivity when a cluster film is used to enhance a weak optical signal. As the red-shift is much stronger than the blue-shift, the cluster that exhibits localized surface plasmons with the same polarization provides a higher sensitivity. A similar behavior can be observed when a cluster film is excited by light with a different polarization of the wave. The substantial difference in electromagnetic fields around clusters caused by an excitation with parallel and perpendicular polarized light is illustrated in Figure 10.10. The direction of the electric vector (E) of the exciting field is the critical parameter. The induced dipoles in the clusters have the same direction as the exciting field. In the case of a perpendicular orientation, E is oriented perpendicular to the surface. The induced dipoles in the clusters must then point in the same direction, and the fields around the clusters develop accordingly. Due to the parallel direction of the dipoles, almost no electric coupling can occur between clusters. However, the

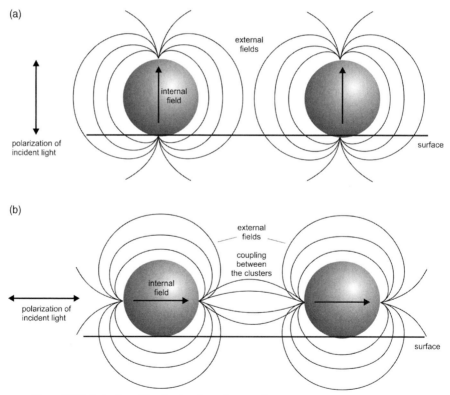

Figure 10.10 Excitation of localized surface plasmons in case of (a) perpendicular orientation and (b) parallel polarization of the incident light wave.

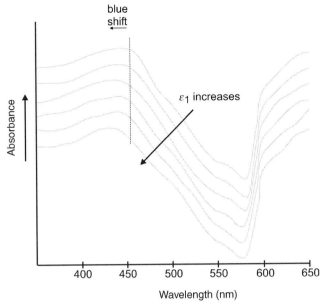

Figure 10.11 Absorbance spectra of a cluster layer for different refractive indices of the surrounding medium (ε_1), measured with unpolarized light.

situation changes dramatically when the orientation of E is parallel to the surface. Now, because of the induced dipoles, the field of the clusters will stretch both towards the substrates and towards the surrounding medium. The direction of the internal dipoles permits a powerful coupling between neighboring clusters, and only in this case will the position of the SPR shift when the refractive index of medium between the clusters changes. The absorbance spectra of cluster layers with different refractive indices of the surrounding medium are shown in Figure 10.11. Since coupling between the clusters occurs only at a parallel polarization of the electric field vector, a spectral shift of the SPR is observed only for parallel polarized light. Upon increasing the refractive index of the medium, however, the field coupling between metal clusters becomes stronger and the SPR shifts towards shorter wavelengths.

Although the optical analysis of cluster arrangements plays an important role, it is impossible to describe the optical properties by extending Equations (10.1) to 10.7 to an arrangement of thousands of clusters. In such a case, an effective medium theory must be applied. Such theories describe the connection and interaction *between* clusters, as well as with their surrounding medium, and provide the 'effective' optical behavior of a cluster arrangement (Figure 10.12).

Several different effective medium theories have been devised, each of which is more or less accurate depending on the cluster material, the size form and the distance between clusters. However, they all assume that the macroscopic material is homogeneous, and generally fail to predict the properties of a composite material.

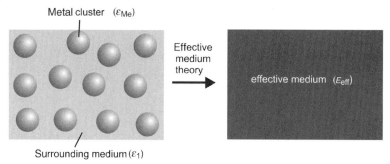

Figure 10.12 The transition from a macroscopically inhomogeneous medium to a homogeneous, optically effective medium.

Details of the three most frequently used effective medium models are summarized in Table 10.1 [9, 10].

The first application of an effective medium theory was made by Maxwell-Garnett to explain the color of a discontinuous metal film. The model was based on clusters which were assumed to be spheres in a host medium; this resulted in an effective dielectric function of the composite material. The *Maxwell-Garnett model* is valid at low volume fractions as it is assumed that the metal clusters are spatially separated. In contrast, the *Bruggeman model* does not distinguish between embedded metal clusters and matrix; rather, the two materials appear in a completely symmetric manner, and consequently the Bruggeman model can easily be extended to more than two components. Free structures of the embedded metal clusters and the matrix and a static treatment of the geometry are also possible using the Bergmann model. The spectral density $g(n,f)$ function (see Table 10.1) carries all of the geometric information and depends only on topology. As a result, the Bergmann model can be used to distinguish between geometric quantities and dielectric properties [11–13].

10.5
Surface-Enhanced Spectroscopy

10.5.1
Surface-Enhanced Raman Scattering

The effect of surface-enhanced Raman scattering (SERS) on small metal clusters has been recognized for more than 30 years. The Raman scattering of a molecule located in close proximity to the surface of a gold or silver cluster is 'enhanced' up to one million-fold. Such an enhancement effect is based on two principal mechanisms:

- The strong electric field of the localized surface plasmons interacts with the electron orbitals of the molecule, which leads to an enhancement of the Raman cross-section by up to four orders of magnitude.

10.5 Surface-Enhanced Spectroscopy

Table 10.1 Most common effective medium model.

Model	Formula	Properties
Maxwell-Garnett	$\dfrac{\varepsilon_{\text{eff}} - \varepsilon_E}{\varepsilon_{\text{eff}} + 2\varepsilon_E} = f_{\text{Me}} \dfrac{\varepsilon_{\text{Me}} - \varepsilon_E}{\varepsilon_{\text{Me}} + 2\varepsilon_E}$ (f_{Me}: volume fraction of the metal clusters)	Suitable for uniform and spherical clusters with a relatively low volume fraction; computations are very efficient; describes poorly composite materials with high volume fraction of the metal clusters.
Bruggeman	$f_{\text{Me}} \dfrac{\varepsilon_{\text{Me}} - \varepsilon_{\text{eff}}}{\varepsilon_{\text{Me}} + 2\varepsilon_{\text{eff}}} + (1 - f_{\text{Me}}) \dfrac{\varepsilon_E - \varepsilon_{\text{eff}}}{\varepsilon_E + 2\varepsilon_{\text{eff}}} = 0$	Self-consistent; provides a quite good correlation between theory and experiment also for higher volume fractions; the result can be a dielectric function of a higher polynomial.
Bergmann	$\varepsilon_{\text{eff}} = \varepsilon_E \left(1 - f_{\text{Me}} \displaystyle\int_0^1 \left(\dfrac{g(n, f_{\text{Me}})(\varepsilon_E - \varepsilon_{\text{Me}})}{\varepsilon_E - n(\varepsilon_E - \varepsilon_{\text{Me}})} \right) dn \right)$ the zeroth and first moment of $g(n, f)$ are $\displaystyle\int_0^1 g(n, f_{\text{Me}}) dn = 1 \quad \int_0^1 n g(n, f_{\text{Me}}) dn = \dfrac{1 - f_{\text{Me}}}{3}$	Best approximation of composite materials; percolations of the cluster are considered; computations are more complex.

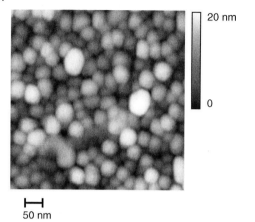

Figure 10.13 Atomic force microscopy image of a silver cluster layer, prepared by evaporation onto a smooth silicon wafer.

- The second effect is related to charge transfer processes; such a chemical interaction may cause an additional enhancement by typically two orders of magnitude.

SERS has been observed for many molecules when adsorbed to gold or silver clusters, or even to nanoscopic rough metal surfaces [14]. The intensity of the Raman scattering is increased as the wavelength is increased. For example, the strongest enhancement for gold clusters may be obtained with a laser excitation wavelength beyond 1000 nm. The SERS spectra obtained are almost completely depolarized. Figure 10.13 shows an atomic force microscopy image of a silver cluster layer used as a substrate for SERS. An example of the SERS effect is represented in Figure 10.14,

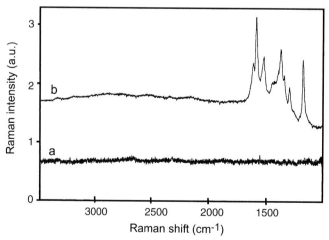

Figure 10.14 Raman spectrum (a) and surface-enhanced Raman spectrum (b) of a very thin film basic fucsine. The Raman spectrum is enhanced 25-fold.

which shows the SER and Raman spectra of a very thin-film basic fucsine on the SERS substrate and a pure silicon wafer.

Clearly, the cluster layer yields an enhancement which is manifested on the most prominent Raman bands. In contrast to the conventional Raman spectrum (Figure 10.14, spectrum a), where no bands can be observed, prominent Raman modes – such as the ring vibration at 1614 cm^{-1}, a deformation vibration of the NH$_2$ groups at 1586 cm^{-1}, and the valence vibration of the C–NH$_2$ groups at 1340 cm^{-1}, appear in the SERS spectrum.

Unfortunately, in recent years the SERS effect has not been used as a routine method in Raman spectroscopy. Although the strong enhancement provides many advantages, such as high sensitivity and the opportunity to characterize thin layers of molecules, SERS does not permit quantitative measurements to be made, despite these very often being required in a routine analysis. The reason for this is that the cluster layers themselves are poorly reproducible. Nonetheless, recent developments have been devoted to overcoming this limitation by using sol–gel clusters embedded in porous glass, or by using photonic crystals with a uniform metal-covered nanostructure.

10.5.2
Surface-Enhanced Fluorescence

Surface-enhanced fluorescence (SEF) is comparable to the SERS effect. The enhancement of the fluorescence signal is also caused by strong interactions between the localized surface plasmons of a metal cluster and the electrons of molecule next to the cluster surface [15]. However, at least two important factors must be considered here:

- First, it is well known that a direct contact between a molecule and a metal leads to a quenching of the fluorescence – this is also known as the 'first layer effect'. Therefore, the optimum enhancement of fluorescence does not derive from molecules that are adsorbed to the metal cluster surface; rather, maximum fluorescence is obtained at a certain distance of few nanometers between the molecule and the cluster surface.

- The second factor includes the size and form of the metal clusters. When the clusters become large, the damping of the fluorescence will be increases, whereas small clusters may not be stable and exhibit only a weak electric field.

The enhancement factors for SEF are comparable to those for SERS. The spectra in Figure 10.14, which were measured at 544 nm excitation, also exhibit a fluorescence which is seen as a very broad signal across the whole spectral range. Although the fluorescence of the first (or more) layer of adsorbed molecules is quenched, the enhancement factor is approximately 50. Experiments with silver clusters covered with a dielectric film of SiO$_2$ a few nanometers thick, produced a much greater fluorescence than did pure silver clusters. As with SERS, SEF has not become a routine method, due not only to the poor reproducibility of the metal cluster but also to the 'first layer effect'.

10.5.3
Surface-Enhanced Infrared Absorption Spectroscopy

Surface-enhanced infrared absorption (SEIRA) spectroscopy also originates from the enhanced electromagnetic field around a metal cluster. The enhancement of infrared absorption is generally on the order of one to two magnitudes [16, 17]. As such enhancement is remarkable for adsorbed molecules on a metal cluster surface, an increase in absorption coefficients and a selection rule for the absorption bands provide additional enhancement and information regarding the orientation of the adsorbed molecule [18, 19]. Figure 10.15 shows a Fourier-transformed infrared (FTIR) spectrum and a SEIRA spectrum of basic fucsine. Both spectra were measured using an attenuated total reflection (ATR) method, with the same measurement parameters. Enhancement of the absorption bands in spectrum b in Figure 10.15 can be clearly seen. Yet, in comparison to the SERS spectrum in Figure 10.14, the strongest band (at $1586\,cm^{-1}$) exhibited only a threefold enhancement.

Due to the surface selection rule, some absorption bands do not appear, or appear only weakly [21]. Shifts of the absorption bands may seem due to the influence of chemisorption of the complex. Although the optimum thickness for SEIRA-active metal cluster films shows some variation from study to study, it is consistently in the range of 5 to 12 nm. The preparation of the SEIRA active surface is straightforward, usually by the evaporation of an ATR crystal surface with gold or silver. A fresh gold surface shows a strong affinity towards many organic substances. In particular, sulfur complexes are chemisorbed immediately and very strongly onto gold cluster surfaces, and this may lead to an unwanted SEIRA spectrum.

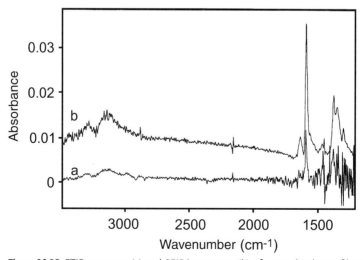

Figure 10.15 FTIR spectrum (a) and SEIRA spectrum (b) of a very thin layer of basic fucsine.

10.6
Biosensing

Metal clusters of gold or silver are stable, inexpensive and simple to prepare, easy to modify, and have optical properties that make them attractive for biosensing and biochips. Today, metal clusters with a functionalized surface are used to detect biomolecules, with an extremely high sensitivity [22, 23]. As mentioned above, the SPR of a metal cluster is also affected by other metal clusters that are in its immediate vicinity. When two or more clusters are brought into proximity, the electric fields will couple, and this will result in a shift of the resonance wavelength. This effect can be easily used in biosensing as an intrinsic enhancement of the detection signal. An example of this is the study of the dynamics of DNA hybridization at the single-molecule level [24]. The principal mechanism of detection is illustrated schematically in Figure 10.16. Here, surface-functionalized metal clusters are used as an anchor for single-stranded DNA (ssDNA). The ssDNA molecules have a biotin molecule attached at one end, and this allows them to bind to the streptavidin-coated 'anchor cluster'. When the surface-bound ssDNA are introduced into a solution with biotin-functionalized clusters, a spectral shift of the SPR occurs, such that gold clusters with a diameter of approximately 40 nm turn from green to orange, while the color of silver metal clusters changes from blue to yellow-green. The limit of detection is in the zeptomolar range. The binding of proteins, interactions between antibodies and antigens and receptor–ligand interactions, can also be detected in this way. The coupling of localized surface plasmons between individual clusters also represents

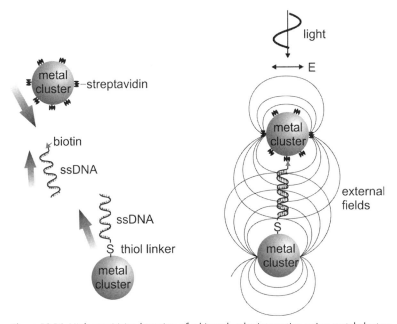

Figure 10.16 High-sensitivity detection of a biomolecular interaction using metal clusters.

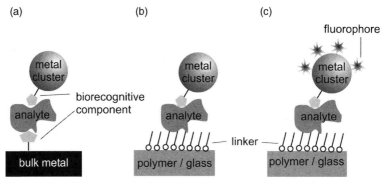

Figure 10.17 Different strategies for metal cluster-enhanced biosensing.

an alternative to the Förster resonance energy transfer (FRET) for monitoring nanometer-scale distances. The plasmon coupling allows measurements to be made over longer time periods and larger distances compared to FRET [20].

Other strategies for biosensing are represented in Figure 10.17. Metal clusters at a defined distance from the surface of a bulk metal (Figure 10.17a) interact with the free electrons of the metal. At a certain distance between the metal cluster and surface, a feedback mechanism enhances the absorption of light, which is then reflected onto the metal surface. The intensity of the absorption is directly proportional to the number of clusters interacting with the metal surface. Metal clusters can be also coupled to a polymer surface (Figure 10.17b); here, the linker molecules not only keep the clusters at a defined distance from the surface but also maintain their spatial distance to other clusters. An optically detectable signal occurs when a certain number of clusters are coupled due to biochemical recognition. Finally, Figure 10.17c shows a similar arrangement where the metal clusters are labeled with fluorophores, and where the fluorescence signal is directly proportional to the bound clusters. The advantage here is an extremely high sensitivity (in the lower femtomolar range) with, under optimal conditions, even single clusters being visible as a color change. The response of the SPR systems is proportional to the product of the adsorbed molecules and the refractive index of the surrounding medium [25]. In addition, metal cluster-based detection provides real-time information on the course of binding over a broad range of binding affinities. Recently, several groups have shown that stable metal cluster layers can be prepared on optical substrates, which makes SPR sensors potentially applicable for continuous-flow detection, as well as for *in vivo* applications in cells and biological fluids [26–29].

More recently, the trend has been towards using nanotechnology to develop sensing surfaces embedded in metal clusters, with such devices improving the sensitivity of a common SPR prism coupler system by a factor of 10. Another novel approach – the use of hydrogels is shown in Figure 10.18. These represent a novel class of polymer that can swell and shrink, depending on their chemical and/or physical environment. As such swelling and shrinkage can be controlled by various molecules, hydrogels may also serve as a host medium for metal clusters. The swelling of a hydrogel containing embedded metal clusters will also cause an increase

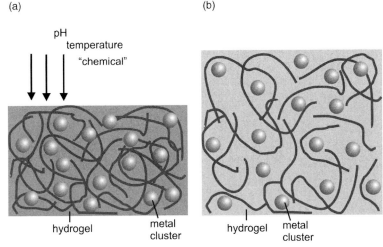

Figure 10.18 Metal clusters embedded in a swelling hydrogel allow sensitive, multiple and reproducible detection of chemical and biochemical parameters.

in the inter-cluster distance, with a spectral shift in color occurring that can easily be measured. An added attraction is that this type of sensor is not only reversible but is also stable over a large number of measurement cycles [30].

References

1 Prasad, P.N. (2004) *Nanophotonics*, John Wiley & Sons Inc, Hokboken, New Jersey.
2 Brongersma, M.L. and Kik, P.G. (eds) (2007) *Surface Plasmon Nanophotonics, Springer Series in Optical Science*, Vol. 131, Springer, Dordrecht.
3 Ohtsu, M. and Hori, H. (1999) *Near-Field Nano-Optics*, Kluwer Academic/Plenum Publishers, New York.
4 Gluodenis, M., Manley, C. and Foss, C.A. Jr (1999) In situ monitoring of the change in extinction of stabilized nanoscopic gold particles in contact with aqueous phenol solutions. *Analytical Chemistry*, **71**, 4554.
5 Steiner, G., Sablinskas, V., Hübner, A., Kuhne, Ch. and Salzer, R. (1999) Surface plasmon resonance imaging of microstructured monolayers. *Journal of Molecular Structure*, **509**, 265.
6 Kreibig, U. and Vollmer, M. (1995) *Optical Properties of Metal Clusters, Springer Series in Material Science*, Vol. 25, Springer, Berlin, Heidelberg, New York.
7 Endo, T., Kerman, K., Nagatani, N., Hiepa, H.M., Kim, D.K., Yonezawa, Y., Nakano, K. and Tamiya, E. (2006) Multiple label-free detection of antigen-antibody reaction using localized surface plasmon resonance-based core-shell structured nanoparticle layer nanochip. *Analytical Chemistry*, **78**, 6465.
8 Thaxton, C.S. and Mirkin, C.A. (2005) Plasmon coupling measures up. *Nature Biotechnology*, **23**, 681.
9 Kuhne, C. (1999) Investigations to an application of surface-enhanced infrared absorption spectroscopy, (in German), *PhD Thesis*, Dresden University of Technology.
10 Ross, D. and Aroca, R. (2002) Effective medium theories in surface enhanced

infrared spectroscopy: The pentacene example. *Journal of Chemical Physics*, **117**, 8095.

11 Sturm, J., Grosse, P. and Theiß, W. (1991) Effective dielectric functions of samples obtained by evaporation of alkali halides. *Zeitschrift für Physik D - Atoms Molecules and Clusters*, **20**, 341.

12 Stroud, D. (1998) The effective medium approximations: Some recent developments. *Superlattices and Microstructures*, **23**, 567.

13 Theiß, W. (1994) *Advances in Solid State Physics*, Vol. 33 (ed. R. Helbig), Vieweg Braunschweig, Wiesbaden.

14 Bordo, V.G. and Ruban, H.-G. (2005) *Optics and Spectroscopy at Surfaces and Interfaces*, Wiley-VCH, Weinheim.

15 Tarcha, P.J., Desaja-Gonzalez, J., Rodriguez-Llorente, S. and Aroca, R. (1999) Surface-enhanced fluorescence on SiO_2-coated silver island films. *Applied Spectroscopy*, **53**, 43.

16 Jensen, T.R., Van Duyne, R.P., Johnson, S.A. and Maroni, V.A. (2000) Surface-enhanced infrared spectroscopy: a comparison of metal island films with discrete and nondiscrete surface plasmons. *Applied Spectroscopy*, **54**, 371.

17 Wanzenböck, H.D., Mizaikoff, B., Weissenbacher, N. and Kellner, R. (1998) Surface enhanced infrared absorption spectroscopy (SEIRA) using external reflection on low-cost substrates. *Fresenius Journal of Analytical Chemistry*, **362**, 15.

18 Nishikawa, Y., Fujiwara, K., Ataka, K.I. and Osawa, M. (1993) Surface-enhanced infrared external reflection spectroscopy at low reflective surfaces and its application to surface analysis of semiconductors, glasses and polymers. *Analytical Chemistry*, **65**, 556.

19 Kellner, R., Mizaikoff, B., Jakusch, M., Wanzenböck, H.D. and Weissenbacher, N. (1997) Surface-enhanced vibrational spectroscopy: a new tool in chemical sensing? *Applied Spectroscopy*, **51**, 495.

20 Sönnichsen, C., Reinhard, B.M., Liphardt, J. and Alivisatos, A.P. (2005) A molecular ruler based on plasmon coupling of single gold and silver nanoparticles. *Nature Biotechnology*, **23**, 741.

21 Griffiths, P.R., de Haseth, J.A. (2007) *Fourier, Transform Infrared Spectrometry*, John Wiley & Sons, Inc, Hoboken, New Jersey.

22 Bauer, G., Pittner, F. and Schalkhammer, Th. (1999) Metal nano-cluster biosensors. *Microchimica Acta*, **131**, 107.

23 Yonzon, C.R., Stuart, D.A., Zhang, X., McFarland, A.D., Haynes, C.L. and Van Duyne, R.P. (2005) Towards advanced chemical and biological nanosensors – An overview. *Talanta*, **67**, 438.

24 Wang, Q., Yang, X. and Wang, K. (2007) Enhanced surface plasmon resonance for detection of DNA hybridization based on layer-by-layer assembly films. *Sensors and Actuators B*, **123**, 227.

25 Steiner, G., Pham, M.T., Kuhne, C. and Salzer, R. (1998) Surface plasmon resonance within ion implanted silver cluster. *Fresenius Journal of Analytical Chemistry*, **362**, 9.

26 Hoa, X.D., Kirk, A.G. and Tabrizian, M. (2007) Towards integrated and sensitive surface plasmon resonance biosensors: A review of recent progress. *Biosensors and Bioelectronics*, **23**, 151.

27 Nath, N. and Chilkoti, A. (2006) Label-free biosensing by surface plasmon resonance of nanoparticles on glass: optimization of nanoparticles size. *Analytical Chemistry*, **76**, 5370.

28 Chau, K.L., Lin, Y.F., Cheng, S.F. and Lin, T.J. (2006) Fiber-optic chemical and biochemical probes based on localized surface plasmon resonance. *Sensors and Actuators B*, **113**, 100.

29 Steiner, G. (2004) Surface plasmon resonance imaging. *Analytical and Bioanalytical Chemistry*, **379**, 328.

30 Hashimoto, N., Hashimoto, T., Teranishi, T., Nasu, H. and Kamiya, K. (2006) Cycle performance of sol-gel optical sensor based on localized surface plasmon resonance of silver particles. *Sensors and Actuators B*, **113**, 382.

11
Impedance Analysis of Cell Junctions
Joachim Wegener

11.1
A Short Introduction to Cell Junctions of Animal Cells

Within the human body there are more than 200 different, highly specialized cell types, each of which has its own individual functions and the corresponding molecular equipment. However, in order to achieve a specific physiological functionality, it often requires the concerted action of a population of cells, whether of the same kind or of a mixed but well-defined population. Within these organized assemblies of cells – known as *tissues* – the cells must interact with each other mechanically by direct cell-to-cell contacts, chemically by secreting chemicals on the one side and responding to these signals on the other, or electrically by means of cell junctions that transmit electrical signals. In an *in vivo* environment the cells also have to interact with their surrounding *extracellular* material for orientation, migration and signaling. This extracellular material – which is known as the extracellular matrix (ECM) – is most often a complex mixture of proteins and carbohydrates embedded in a more or less aqueous environment, depending on the precise location inside the body. The interactions of cells with other cells of the same or a different type, and the interactions of the cells with the ECM, are summarized by the term cell junctions, including both, cell–cell junctions and cell–matrix junctions [1]. The different cell junctions, as found in epithelial cells, are shown schematically in Figure 11.1. Epithelial cells serve as a good example here as they express most of the important cell junctions very prominently, and in a highly organized fashion. It should be noted that other cell types may lack the more specific junctions, such as tight junctions or gap junctions.

From a functional viewpoint, cell junctions can be grouped in three categories which:

- Provide mechanical contacts and stability of the tissue
- Chemically seal extracellular pathways between cells
- Allow direct molecular or ionic exchange between adjacent cells [2].

Nanotechnology. Volume 6: Nanoprobes. Edited by Harald Fuchs
Copyright © 2009 WILEY-VCH Verlag GmbH & Co. KGaA, Weinheim
ISBN: 978-3-527-31733-2

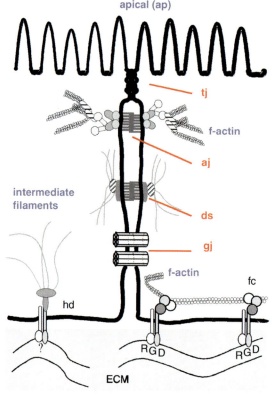

Figure 11.1 Cell–cell and cell–matrix junctions in epithelial cells. The following cell–cell junctions are found in the intercellular cleft between two adjacent epithelial cells: tight junctions (tj), adherens junctions (aj), desmosomes (ds) and gap junctions (gj). The cells are anchored to the extracellular matrix (ECM) by focal contacts (fc) or hemidesmosomes (hd). Cells that provide mechanical stability to cells and tissue are connected to the intracellular filament system of the cytoskeleton, such as the intermediate filaments or the actin cytoskeleton.

However, only specific interactions of the receptor–ligand type mediated by molecular recognition will be addressed in this chapter. Unspecific (electrostatic, electrodynamic, entropic) interactions between two adjacent cells, or between a cell and the surrounding ECM, will not be considered.

11.1.1
Cell Junctions for Mechanical Stability of the Tissue

The mechanical stability of a tissue is provided by cell–cell as well as cell–matrix junctions, which also show a remarkable similarity with respect to their molecular architecture. For both types of junction transmembrane receptors make contact via their extracellular domains to a corresponding protein on the surface of an adjacent cell, or to a binding site within the ECM. On the intracellular site these receptor proteins are connected to the cytoskeleton mediated by highly specialized adaptor proteins. It is this connection of the receptor proteins to the cytoskeleton (an intracellular network of protein filaments) which provides the molecular basis for the mechanical stability of the individual junction that distributes any punctual mechanical load into the entire tissue. With regard to the individual proteins or protein families involved in junction formation, the mechanical junctions can be subdivided as follows.

11.1.1.1 Adherens Junctions

Adherens junctions (aj; see Figure 11.1) are cell–cell junctions that are formed by transmembrane proteins of the *cadherin* family [3]. The name cadherin is derived from the fact that these proteins only bind to their cadherin counterparts on the surface of adjacent cells in the presence of Ca^{2+} (calcium + adhesion). As cadherins on one cell interact with the same cadherins on the opposing cell, the interaction is termed *homophilic*. On the intracellular site, the cadherins are linked to the actin cytoskeleton (also called the *microfilament system*) by a distinct set of linker proteins such as α-, β- or γ-catenin, actinin or paxilin. As shown in Figure 11.1, the adherens junctions with their underlying microfilaments form a very localized structure located close to the upper (apical) pole of the cells, and circumscribe the cell bodies like a belt with bundles of filaments running along on the intracellular site.

11.1.1.2 Desmosomes

Desmosomes (ds in Figure 11.1) have a very similar molecular architecture compared to adherens junctions. The transmembrane proteins are also of the cadherin type, although the adaptor proteins on the intracellular side are different and are connected to the *intermediate filament* instead of the microfilament system [4, 5]. As the intermediate filaments are very different from the microfilaments with respect to their dynamic and mechanical properties, both junctions serve the individual needs of the cells that will not be addressed in detail in this chapter (the reader is referred elsewhere for further details [2, 6]). Both types of junction are synergistically responsible for the mechanical properties of cell–cell adhesion. The desmosomes are located further down the intercellular cleft, just beneath the adherens junctions. However, they do not form a belt-like structure around the cells but rather more punctuate, bullet-like contact sites. The connection between cadherins and intermediate filaments is provided by the adaptor proteins *desmoglein* and *desmocolin*. Although both the adherens junctions and desmosomes tie the intercellular cleft together mechanically, they do not operate as barriers for diffusion along the intercellular cleft, other than confining the cleft width.

11.1.1.3 Focal Contacts

Focal contacts (fc in Figure 11.1) are the most prominent sites of cell–matrix adhesion [7]. They are clusters of individual molecular connections between the cell interior and the ECM. In general, cell–matrix junctions that eventually develop into mature focal contacts are composed of a transmembrane protein that makes contact with the extracellular binding partner. The major family of transmembrane proteins involved in cell–matrix adhesion is the *integrin* family [8]. Integrins are α,β-heterodimeric proteins that extend out of the membrane by approximately 20 nm, and are capable of binding specifically to the ECM proteins. Integrins present in focal contacts are connected to the actin cytoskeleton via linker proteins such as vinculin, paxilin and talin. By means of this molecular construction, an intracellular macromolecular network is connected mechanically to an extracellular macromolecular network, such that the cells and the extracellular environment form a unit that is remarkably resistant to mechanical challenges [8, 9]. Moreover, the binding affinity of integrins in focal adhesion sites can be regulated by the cells either to form or to loosen, and in this way they play a major role in transmembrane signaling in both directions, inside-out and outside-in [2].

11.1.1.4 Hemi-Desmosomes

The hemi-desmosomes (hd in Figure 11.1), as the second class of cell–matrix junctions, differ from the focal contacts in essentially the same way as do desmosomes from adherens junctions. The transmembrane component is provided by a special type of integrin ($\alpha_6\beta_4$) that is exclusively localized in the hemi-desmosomes. Moreover, hemi-desmosomes are connected to the *intermediate filament* system rather than to the microfilaments. As mentioned above, the different properties of these intracellular cytoskeletal networks provide individual functionalities to the two different classes of cell–matrix adhesion sites.

11.1.1.5 Less-Prominent Types of Mechanical Junctions

In addition to the above-mentioned cell junctions there are other, less-prominent molecular assemblies that provide mechanical stability. These include mainly the Ca^{2+}-independent cell adhesion molecules (CAM) as important mediators of cell–cell adhesion, or transmembrane proteoglycans for cell–matrix adhesion. Further details regarding these assemblies are provided elsewhere [2].

11.1.2
Cell Junctions Sealing Extracellular Pathways: Tight Junctions

There is only one type of junction responsible for sealing the extracellular pathway between adjacent cells, and these are known as *tight junctions* (tj in Figure 11.1). To the author's present knowledge, tight junctions are expressed only in epithelial and endothelial cell layers which form the interfacial layers along all inner and outer surfaces of the human body, such as the skin, the gut lining, the bladder or blood vessels. By virtue of their location, it is the predominant physiological task of these tissues to serve as an interface between the two separated compartments, and to control the flux of metabolites or xenobiotics from one compartment to another [10]. Flux control and the exclusion of selected substances is, however, only effective as long as any uncontrolled paracellular diffusion through the intercellular cleft between adjacent cells is limited. Tight junctions provide this seal or *occlusion* of the intercellular cleft and are, thus, also referred to as *occluding junctions* or *zonula occludens*. The tight junctions are located at the apical pole of two apposing epithelial cells (cf. Figure 11.1). Similar to adherens junctions, they are also very focused structures that span no more than 200 nm along the intercellular cleft. Nonetheless, they can be extremely efficient in maintaining chemical or electrochemical gradients in highly specialized tissues such as those which form the blood–brain or the blood–CSF barriers. In other epithelia or endothelia, the barrier function is significantly lower according to the physiological task at the particular location inside the body [11].

Structurally, the tight junctions are not yet fully understood. Many transmembrane or peripheral proteins have been localized almost exclusively to functional tight junctions, including *Occludin*, members of the *Claudin* or *Jam* family, as well as the peripheral proteins ZO-1, ZO-2 and ZO-3, to mention a few. Although a large number of proteins have been identified as constituents of functional tight junctions, various experimental indications exist which suggest that lipids must be involved in junction formation and thus provide some of the junctions' unique properties [12]. (For additional information the reader is referred to [13–17].) Yet, no matter how they are molecularly composed, the formation of tight junctions requires close cell-to-cell-apposition. Tight junctions will only form when a mechanically stable cell–cell adhesion site has been established through adherens junctions and desmosomes before.

11.1.3
Communicating Junctions: Gap Junctions and Synapses

Both, *chemical synapses* and *gap junctions* (gj in Figure 11.1) are cell junctions that are involved in intercellular communication, despite their entirely different structure and molecular architecture. Moreover, it is well known that synapses are only formed at the contact sites between two neurons or a neuron and a muscle cell, whereas gap junctions are expressed by most cell types within the human body, although to very different degrees.

11.1.3.1 Chemical Synapses
In chemical synapses [18] the opposing membranes of two cells approach each other very closely, leaving a water-filled cleft of only 20 nm between them (the synaptic cleft). When a membrane depolarization wave passing along the presynaptic membrane reaches the synapse, transmitter molecules are released from the sender cell, diffuse through the synaptic cleft, and open ligand-gated ion channels in the plasma membrane of the receiver cell (postsynaptic membrane). This incident triggers a depolarization wave in the receiver cell, such that the electrical signal is transmitted. Depending on the length of the axon, the distance between two communicating nerve cells can be up to 1 m [19].

11.1.3.2 Gap Junctions
In contrast to the chemical synapses, gap junctions are simply water-filled channels which locate between two adjacent cells and allow the sharing of molecules with a molar

mass of less than $1000\,\mathrm{g\,mol^{-1}}$ (this is termed metabolic coupling) [20, 21]. From a structural viewpoint gap junctions are composed of two hemi-channels, one provided by each of the opposing cells. Each hemi-channel is known as a *connexon*, which in turn is composed of six *connexins* (each of which is a single span transmembrane protein) arranged in a hexagonal pattern. The central opening in this protein cluster (d = 1.5 nm) provides an aqueous channel between the two connected cells. These channels are not static but can be precisely regulated in their permeability by both cells, the sender and the receiver. As membrane depolarization waves can also be transmitted from one cell to the neighboring cells via gap junctions (e.g. in heart muscle cells), they are also referred to as *electrical synapses*. Notably, the transfer of electrical signals occurs significantly faster via gap junctions than via chemical synapses.

11.2
Established Physical Techniques to Study Cell Junctions

In this section we will provide a rather brief overview of the techniques which have been applied in the past to study cell–cell or cell–matrix junctions from a structural or functional viewpoint. It should be noted, however, that within the chapter we cannot provide a complete survey of all available techniques, nor details of the selected methods. However, further information is available via the references provided in the appropriate passages of the text.

11.2.1
Cell–Matrix Junctions

When studying cell–matrix junctions *in vitro*, the ECM proteins are commonly predeposited on a technical surface such as a Petri dish or a microscope slide, with the cells adhering to this protein-decorated surface. Thus, analyzing *cell–matrix junctions* also means studying *cell–surface junctions* or *cell–substrate junctions* [22]. Today, these three terms are generally used synonymously, depending on the background of the study under discussion.

From the structural viewpoint our current knowledge of cell–matrix or cell–surface junctions is largely based on light microscopy or transmission electron microscopy (TEM) of chemically stabilized (fixed) samples, in combination with the powerful tools of modern molecular biology and immunology. Often, the target molecule suspected of being involved in cell–matrix adhesion is specifically tagged by an antibody – which itself is labeled with a fluorochrome or an electron-dense marker – and then imaged microscopically under various experimental conditions. Studies like this, together with the biochemical analysis of interaction partners, have identified the molecular architecture of cell–matrix junctions as known today and described above.

11.2.1.1 Scanning Probe Techniques
Until now, scanning probe techniques have not contributed a lot to the structural analysis of cell–matrix adhesion sites, as the latter are protected from one side by the

cell body and from the other side by the growth substrate. Nanoprobes simply cannot gain access to these structures in living cells. Most recently, scanning electron microscopy (SEM) has been used in combination with controlled erosion of the sample by a focused ion beam (FIB). Thereby, the organic material of the cell bodies was locally removed in order to obtain a side view of the cell by SEM. These images provided a detailed view on the profile of the cell–substrate interface [23]. Recently, Wrobel *et al.* studied the cell–surface interface using TEM after preparing of thin sections of the sample [24].

11.2.1.2 Nonscanning Microscopic Techniques

Several nonscanning light microscopy techniques have been developed that utilize certain optical effects to provide images from this internal interface between the cell body and the growth substrate. Reflection interference contrast microscopy (RICM) – also referred to as interference reflection microscopy (IRM) – has contributed the most to the existing literature about cell–substrate interactions [25, 26]. This technique allows imaging of the 'footprints' of cells on a substrate rather than only the projections of the cell body. RICM in its basic form is applied to living cells grown on ordinary coverslips, and does not require any staining or fixation. The sample is illuminated from below with an inverted microscope, using monochromatic light. In a first approximation the image is generated from the light reflected either from the glass/medium interface or the lower plasma membrane. Interference of the reflected light then provides an image of the cell–substrate contact area in which the brightness of the pixels code for the optical path difference between the two interfaces. Thus, RICM images map the distance between the lower cell membrane and the glass surface, while time-lapse RICM studies provide a microscopic view of the dynamics of cell–surface junctions with video rate time resolution.

11.2.1.3 Fluorescence Interference Contrast Microscopy

A major improvement with respect to absolute cell–substrate distance measurements was introduced by Braun, Lambacher and Fromherz in 1997 [27–29]. For this technique, which is referred to as fluorescence interference contrast microscopy (FLIC), the cells are grown on silicon substrates that have well-defined steps made from silicon oxide on their surface. The step height ranges between 20 and 200 nm and is, thus, only a fraction of the wavelength of visible light. For the measurement, the adherent cells are stained with a fluorescent dye that integrates into the plasma membranes. As the silicon/silicon oxide interface acts as a mirror, the intensity of the fluorescence light emitted by the dye in the lower membrane is dependent on the distance between the membrane and the silicon surface. The steps on the surface serve as well-defined spacers and make the intensity–distance relationship unique. FLIC microscopy provides the cell–substrate separation distance with an accuracy better than 1 nm.

11.2.1.4 Total Internal Reflection (Aqueous) Fluorescence Microscopy

Total internal reflection fluorescence microscopy (TIRF) and total internal reflection aqueous fluorescence microscopy (TIRAF) are variants of the same microscopic principle, and are based on fluorophore excitation by an evanescent electric field. Here, the cells are grown on a transparent substrate that is illuminated from below with a laser beam at an angle θ relative to the surface normal that is bigger than – or equal to – the critical angle of total reflection, $θ_{crit}$. Under these conditions diffraction phenomena at the interface generate an evanescent wave at the surface. The penetration depth of the associated electric field is rather short, and the field decays within 100 nm of the surface, or slightly beyond. Thus, fluorescence is only excited in those molecules that are close enough to the surface. In TIRF, membrane proteins or other membrane constituents are fluorescently labeled and can be imaged with improved resolution, as fluorescence light from further inside the sample is not excited. For TIRAF measurements a water-soluble fluorescent dye is added to the extracellular fluid. If a cell adheres to the surface it displaces the aqueous phase and thereby the fluorophore from the interface. Thus, cell-covered areas appear dark in TIRAF images. Both techniques have contributed significantly to our understanding about cell–matrix adhesion *in vitro* [30, 31].

11.2.1.5 Quartz Crystal Microbalance

Another emerging tool to study cell–matrix junctions *in vitro* is based on using thickness shear-mode piezo resonators as growth substrates for adherent cells. The interactions of cells with the protein-decorated crystal surface is monitored by reading the resonance frequency and the energy dissipation of the shear oscillation [32]. In principle, this approach has evolved from the so-called quartz crystal microbalance (QCM) technique that is a widely accepted technique for following adsorption reactions at the solid–liquid interface. However, in combination with cells it is not the mass of the cells that determines the signal but rather their anchorage to the oscillating quartz surface, together with the viscoelasticity of the cell bodies. Time-resolved measurements of the resonance frequency can be used to follow the attachment and spreading of cells to the quartz surface that may be precoated with a matrix component of interest [33]. Even though the growth substrate of the cells oscillates mechanically, the technique is considered as being noninvasive as the maximum shear displacement in the center of the resonator is in the order of 1 nm, with a frequency in the MHz regime.

11.2.1.6 Other Techniques

Besides the various techniques mentioned above, several other experimental methods can be used to provide information on cell–matrix adhesion *in vitro*; some details of these are listed very briefly here. Many assays study the forces necessary to remove a cell from a protein-coated surface after it has been allowed to attach and anchor for a predefined time. These approaches can be applied either in an integral manner to a population of cells, or on the single-cell level. The former method is, for instance, realized by exposing the cells to centrifugal forces and counting those that are capable of remaining attached to the matrix proteins on the surface [34]. On the single-cell level, a variant of classical scanning force microscopy – then referred to as cell adhesion force microscopy – is used to measure the forces required to remove a cell from a particular ECM-coated substrate, either by pulling or pushing [35]. Moreover much has been learned from extracellular recordings of neuronal action potentials by means of field effect transistors (FET) regarding the electrochemical properties of the thin cleft between cell and surface as the sealing properties of the junctional area determine the sensitivity of the measurement [36].

11.2.2
Cell–Cell Junctions

The molecular architecture of cell–cell junctions, as shown in Figure 11.1, has been unraveled significantly by using both TEM and SEM after freeze-fracture preparation or other contrasting protocols, as well as with fluorescence microscopy after immunolabeling of the molecular target. When used in conjunction with the molecular biology technique to knock out or knock in a gene of interest, to silence the expression of a given gene or to identify interactions partners, these techniques have provided a comprehensive understanding of the molecular composition of cell–cell junctions. Whilst understanding the molecular arrangements of a given cell junction under stationary conditions is one thing, to follow and identify the changes that occur during regulation or development is another. Thus, a detailed understanding of cell–cell junctions, their interplay and regulation remains an area of intense research worldwide, notably because of their extraordinary biomedical and pharmaceutical relevance. Among the four different types of cell–cell junction (tight junctions, adherens junctions, desmosomes, gap junctions), tight and gap junctions have received most attention with respect to their functional properties, and hence only these will be addressed here. Both of these junctions provide control over the flux of chemicals (metabolites, xenobiotics), either through the interspaces between two adjacent cells or between adjacent cytoplasms.

11.2.2.1 Tight Junctions

Tight junctions, as expressed by endothelial and epithelial cells, form the structural basis for the barrier function of epithelial and endothelial cell layers. A straightforward and popular approach to probe the efficiency of this barrier function – and thus the tightness of the junctions – is a simple *permeation/diffusion assay* [37]. Here, the cells are grown on highly porous filter membranes that support the cell layer mechanically without acting as a significant diffusion barrier themselves. The cell-covered membrane is then placed between two fluid compartments such that any flux of solutes from one compartment to the other must pass the interfacial cell layer. In a typical experiment, a tracer compound is added to one compartment (donor), while samples are taken from the other compartment (acceptor) after well-defined time intervals. From the concentration increase of the tracer in the acceptor compartment, it is possible to calculate the permeation rate P_E that reports on the barrier properties of the cell layer under study. As long as the probe cannot migrate across the cell membranes and is dependent on extracellular diffusion, the experimentally measured flux is predominantly determined by the functional properties of the tight junctions. By using probes of different molecular mass or shape, the size-exclusion properties of the junctions can be inferred from such measurements [38].

The probes can either be radio- or fluorescence-labeled, otherwise their concentration must be determined using chromatography. The readout of this widely established assay is, however, easily compromised by defects in the cell layer (even single cell defects) that may serve as short-cuts for the substrate flux leading to a serious underestimation of the barrier function. Moreover, such a permeation assay is not an *in situ* method but rather requires time for the probe to accumulate in the acceptor compartment [39].

Another method of probing the functional properties of tight junctions relies on measuring the *electrical resistance* of the cell layer, when it is placed on a porous filter membrane between two fluid compartments, as described above. Either compartment is equipped with a pair of Ag/AgCl electrodes, that are used for current injection and voltage reading in either compartment. The resistance value, which is then referred to as the transepithelial or transendothelial electrical resistance (TER), strictly reports on the ionic permeability of the entire cell layer. However, as the extracellular ionic current pathway around the cell bodies through the cell–cell junctions is significantly less resistive in most cases than the ionic current pathways across the dielectric plasma membranes, TER reports on the tightness of the junctions as a first approximation. Unfortunately, the TER approach suffers from the same shortcomings as do readings of the P_E rate (see above), namely that it is very prone to artifacts arising from defects within the cell monolayer. On the other hand, TER readings can provide a snapshot of the barrier function of the junctions, and the readings can be performed in a time-resolved manner. Accordingly, changes in junctional tightness can be monitored more or less in real time with a time resolution that can be reduced to the order of minutes [40].

TER measurements with lateral resolution have been performed by using microelectrodes for local potential measurements while a uniform and homogeneous current density was established across the entire cell layer [41, 42]. By scanning the electrode across the cell layer, it was possible to establish resistance maps that could identify local shortcuts in the barrier function of the cell layers [43]. Moreover, it was possible to quantify the resistance of the *paracellular* current pathway across the tight junctions in contrast to the *transcellular* current pathway across the membranes [42]. In particular, this discrimination between paracellular and transcellular resistance is not possible with the integral TER approach, and may leave data interpretation with some ambiguity as the transcellular current pathway has a high but finite resistance after all. Several modifications of the basal conductance scanning technique have been described and, in scanning ion conductance microscopy (SICM), the lateral resolution has been improved to the submicrometer regime such that the junctional tightness along the periphery of a single cell can now be recorded and analyzed. Korchev and coworkers have even studied the conductance of a single ion channel on a living cell by using SICM [44].

11.2.2.2 Gap Junctions

As for the other junctions, the general structure and composition of gap junctions has been revealed to a large extent by using electron and fluorescence microscopy after labeling individual molecules suspected of being involved. From a functional perspective, junctional performance is routinely monitored by electrical measurements. Either of the two cells that are connected via gap junctions will be impaled by a microelectrode such that current can be injected to flow from one cell the other. When gap junction performance is discussed, the measured resistance is usually expressed as *conductance*. Amazingly, the conductance of a single gap junction channel depends on the individual member of the connexin protein family that forms the transmembrane channels, even though they are highly conserved [45].

Another approach involves the use of micropipettes to inject nonmembrane-permeable fluorescent dyes into the cytoplasm of a cell; the spread of the dye into adjacent cells connected via gap junction channels is then followed. Using this technique allows complete control over the size, surface charge and morphology of the probe, so that the junctional permeability can be extensively analyzed [46]. Along the same lines, but at a much lower experimental cost, it is possible to visualize gap junction conductance using the so-called scrape loading assay. Here, the cell layer is bathed in a buffer that contains the fluorescent gap junction probe. In order to introduce the dye into the cytoplasm of the cells, a needle is moved (scraped) through the cell layer. Those cells lying in the path of the needle are ripped by the moving pipette tip such that their membrane is ruptured. This allows the extracellular dye to enter the cytoplasm through the ruptured membrane, and then diffuse via gap junctions into neighboring cells. The diffusion performance of the dye reports on junctional coupling [47].

11.3
Impedance Spectroscopy

Impedance spectroscopy (IS), which is also referred to as electrical or electrochemical impedance spectroscopy (EIS), represents a versatile approach for probing and characterizing the dielectric and conducting properties of bulk materials, composite samples or interfacial layers. The technique is based on measuring the impedance – that is, the opposition to current flow – of a system while it is excited with low-amplitude alternating current or voltage. The impedance spectrum is obtained by scanning the sample impedance over a broad range of excitation frequencies, typically covering several decades. In continuous-wave IS, the impedance is measured sequentially at each individual frequency. In contrast, with pulse techniques the system is exposed to a superposition of multiple sine waves with different frequencies, at the same time. The impedance spectrum of the system is then extracted from the transient response by means of Fourier transform algorithms. Both data acquisition modes have their unique advantages. Whereas continuous-wave recordings are easier to conduct and have better signal-to-noise ratios, pulsing techniques are much faster, provide a better time resolution and are, thus, more suitable to study dynamic systems that change within seconds.

As long ago as the 1920s, research groups first began to investigate the impedance of tissues and biological fluids, and it was known even then that different tissues exhibit distinct dielectric properties, and that the impedance undergoes changes during pathological conditions or upon changing the cellular environment [48, 49]. Thus, the value of impedance measurements for the analysis of biological samples such as organs, tissues, cell aggregates or even single cells was obvious from an early stage. One of the most important advantages of IS compared to other techniques is its noninvasiveness, as the technique relies entirely on low-amplitude currents and voltages that ensure damage-free examination with a minimum disturbance of the cells or tissues. This noninvasive nature of the method, combined with its high information content, has made it a valuable tool for biomedical research *in vitro*. More recently, *in vivo* applications in clinical settings have also been developed, including electrical impedance tomography (EIT), which is being used increasingly as a noninvasive monitoring tool for heart and/or lung function during surgery [50]. In the following sections, however, attention will be focused on the use of impedance spectroscopy to study cell junctions of animal cells *in vitro*. A broad survey of the use of EIS for biological samples such as tissues and organs is provided elsewhere [50].

11.3.1
Fundamental Relationships in Impedance Analysis

The electrical impedance, Z, is a complex quantity that describes the ability of the system under study to resist the flow of alternating current. In a typical IS experiment a sinusoidal voltage $U(t)$ with angular frequency $\omega = 2\pi f$ is applied to the system and the resulting steady-state current $I(t)$ associated with the voltage is measured. According to Ohm's Law, the impedance is given by the ratio of these two quantities:

$$Z = \frac{U(t)}{I(t)} \tag{11.1}$$

The impedance measurement is typically conducted within a linear voltage–current regime of the sample (i.e. the measured current amplitude is proportional to the amplitude of the applied voltage), so that the resulting current will also be a sine wave with the same frequency ω as the applied voltage signal, but it may be phase-shifted relative to the voltage by the phase angle φ. By introducing a complex notation, Equation 11.1 translates into

$$Z = \frac{U_0}{I_0} \exp(i \cdot \varphi) = |Z| \exp(i \cdot \varphi) \tag{11.2}$$

with U_0 and I_0 representing the amplitudes of voltage and current, respectively, and with $i = \sqrt{-1}$. Thus, at each frequency of interest the impedance is described by two quantities: the magnitude $|Z|$, which is the ratio of the amplitudes of U_0 and I_0, and the phase angle φ between voltage and current.

Instead of presenting the complex impedance Z in polar coordinates, $|Z|$ and φ, it can also be expressed in Cartesian coordinates, with a real (R) and an imaginary (X) component:

$$Z = R + i \cdot X \tag{11.3}$$

with $R = \text{Re}(Z) = |Z| \cdot \cos(\varphi)$

and $X = \text{Im}(Z) = |Z| \cdot \sin(\varphi)$

(11.4)

The real part is called the *resistance R*, and corresponds to the impedance contribution arising from current that is in-phase with the applied voltage. The imaginary part is termed the *reactance X*, and describes the impedance contribution from current which is 90° out-of-phase with the voltage. With respect to biological samples, the resistive portion R of the impedance mirrors either the concentration of ions available for current flow within the sample, or their limited ability to migrate under the influence of the applied electric field. The latter is commonly caused by a geometric confinement of the ionic current pathways by means of insulating cell membranes or cell junctions that occlude or narrow down the aqueous spaces available for current flow. The reactance X arises from the presence of storage elements for electrical charges like, for instance, capacitors or coils in electrical circuits. In biological tissues, the dielectric (or insulating) cell membrane acts as a capacitor, separating two conducting fluids (extracellular and intracellular) by its hydrophobic core, whereas an inductive behavior of biological systems is only described in very rare cases. Thus, as a rule of thumb for animal cells and tissues, it can be said that the measured resistance arises from extracellular or intracellular fluids and their geometric dimensions, whereas the capacitive reactance originates from the cell membranes.

In some cases it is more convenient to use the inverse quantities of Z, R and X, which are referred to as *admittance* $Y = 1/Z$, *conductance* $G = \text{Re}(Y)$ and *susceptance* $B = \text{Im}(Y)$, respectively. In the linear voltage–current regime, the two representations are interchangeable and contain the same information. Accordingly, IS is occasionally entitled admittance spectroscopy.

In IS the impedance is measured over a range of frequencies covering several decades between mHz and GHz, depending on the type of sample and the problem being studied. The frequency regime typically studied to analyze cell junctions ranges from 1 Hz to 1 MHz.

11.3.2
Data Representation and Analysis

When the complex impedance of a system of interest (real and imaginary part) is recorded as a function of frequency, the complete presentation of the data requires a specialized method of data plotting. Most frequently, a so-called Bode diagram is used; here, the impedance magnitude |Z| and the phase shift between current and voltage φ are plotted as a function of frequency on a logarithmic or semi-logarithmic scale, respectively. Figure 11.2a shows such a Bode diagram for an arbitrary electrical network shown in the inset of Figure 11.2b as an example. Alternatively, the imaginary component of the impedance X is plotted versus the real component R in a so-called impedance locus or Wessel diagram (Figure 11.2b). The latter presentation does not provide any further information on the frequency. Normally, an arrow is added to include the direction in which the sampling frequency

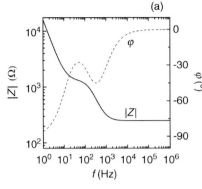 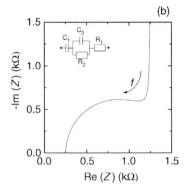

Figure 11.2 Different representations of impedance spectra. (a) Bode diagrams provide impedance magnitude $|Z|$ and phase shift φ between voltage and current as a function of frequency; (b) Impedance loci or Wessel diagrams display the imaginary part of Z as a function of the real part of Z. The arrow indicates the direction of increasing frequency (f). The insert shows the electrical circuit that was used to calculate the impedance data.

increases. The advantage of using impedance loci for data presentation is that the shape of the curve provides information on the electrical structure (perhaps even substructures of the sample), and also whether the behavior of the system deviates from that of ideal electrical networks. For example, a resistive current pathway in parallel to a capacitive one gives rise to a semicircle within an impedance locus that is centered on the *x*-axis. From the center and the radius of the semicircle one can directly extract the resistance and capacitance of this particular structure (cf. Figure 11.2b).

Although the shape of the curve in impedance loci may be useful for direct analysis, the most common method of analyzing experimental impedance spectra is by means of equivalent circuit modeling. Here, the system under study is described by an electrical network (as shown in the insert of Figure 11.2b) that mirrors the (predicted) electrical structure of the system. The system's equivalent circuit is composed of series or parallel connections of impedance elements (resistors, capacitors), as known from electronic circuitry, plus additional impedance elements that have been empirically derived for ionic systems, for example Warburg impedance or constant phase elements (CPEs). These elements have no correspondence in electronic systems. As described above, the impedance spectrum contains much information about the electrical properties of the system, and with experience it is possible to make a 'qualified guess' of a proper model based on the features in the diagrams. For a given equivalent circuit, the frequency-dependent impedance (transfer function) is then derived from the individual components and their interconnection using Ohm's law and Kirchoff's laws. The best estimates for the parameters – that is, the unknown values of the resistors and capacitors within the equivalent circuit – are then iteratively computed by ordinary least-squares algorithms, such as the Levenberg–Marquardt approach. If the impedance and phase spectra of the chosen model fit the

Table 11.1 Individual impedance contributions of ideal and empirical equivalent circuit elements.

Component of equivalent circuit	Parameter	Impedance Z	Phase shift φ
Resistor	R	R	0
Capacitor	C	$1/(i \cdot \omega \cdot C)$	$-\pi/2$
Coil	L	$i \cdot \omega \cdot L$	$+\pi/2$
Constant phase element (CPE)	A, $n (0 \leq \alpha \leq 1)$	$1/(i \cdot \omega)^n \cdot A$	$-n \cdot \pi/2$
Warburg impedance σ	σ	$\sigma \cdot (1-i) \cdot \omega^{-0.5}$	$-\pi/4$

data well, the parameter values are used to describe the electrical properties of the system and its changes throughout an experiment.

In order to find an equivalent circuit model that accurately predicts the impedance of biomaterials, it is often necessary to include nonideal circuit elements – that is, elements for which the parameters are themselves frequency-dependent. Such empirical elements account for ionic phenomena such as adsorption and diffusion that cannot be realized with standard electronic impedance elements. A list of all common circuit elements used to describe biomaterials in terms of their impedance and their phase shift is provided in Table 11.1. The CPE, which represents a nonideal capacitor and is one of these empirical impedance elements, was originally introduced to describe the interface impedance of noble metal electrodes immersed in electrolyte solutions. Although the physical basis of CPE behavior is not fully understood in detail, it is thought to be associated with surface roughness and specific ion adsorption to interfaces. Another empirical element is the Warburg impedance σ, which accounts for the impedance contribution arising from the diffusion limitation of many electrochemical reactions. The parameters that determine the individual impedances of these elements are listed in Table 11.1.

At this point it is important to place a word of caution concerning the equivalent circuit modeling approach. Different equivalent circuit models (which deviate with respect to either the components or the network structure) may produce equally good fits to the experimental data, yet ascribe the sample a very different physical structure. In such a case, independent experiments (microscopy and other spectroscopic approaches) are required to obtain further insight into the electrical structure of the sample and to identify the most appropriate model. It may also be tempting to increase the number of elements in a model to obtain a better agreement between experiment and model. However, the model may then become redundant because the components can no longer be quantified independently. Thus, an overly complex model can easily provide artificially good fits to the impedance data but, at the same time, highly inaccurate values for the individual parameters. Thus, it is sensible to use the equivalent circuit with the minimum number of elements that still describes all details of the impedance spectrum (the nonredundant model) [51].

Another approach towards analyzing impedance data (although less often applied) is based on deriving the current distribution in the system by means of differential equations and boundary values. Solving the differential equations provides the impedance transfer function with the respective model parameters. Fitting of the transfer function to the recorded data then allows extraction of the best estimates for the model parameters. It should be noted that both approaches are essentially only different formalisms.

11.4
Impedance Analysis of Cell Junctions

11.4.1
General Remarks about Experimental Issues

11.4.1.1 Two-Probe versus Four-Probe Measurement
In order to analyze the impedance characteristics of a given electrochemical system, it must be interfaced with electrodes that are required for current injection and voltage sensing at appropriate frequencies. These electrodes are placed at opposite ends of the sample in order to provide the electrical structure of the entire system as a readout. In many cases the measurement can be made with four electrodes, two for current injection and two for voltage sensing. Here, the voltage sensing electrodes are usually placed very close to the sample surface so that only the voltage drop across the sample is measured without contributions from the bathing fluid or other parts of the measurement chamber. The current-injecting electrodes are often placed at the very end of the experimental chamber to avoid any disturbance of the sample when electrochemical reactions occur at the electrode surfaces. When low-amplitude signals are used and no significant faradaic currents (electrochemical reactions at the electrode surface) occur, it is often more convenient to work with two electrodes only. Under these conditions, either electrode is used for both current injection and voltage sensing. The major advantage of such a two-probe measurement is the reduction of electrodes, including the necessary cables and connectors, that must be integrated into the experimental set-up. Moreover, some approaches simply do not allow the use of four electrodes at the required location, as will be detailed below. The disadvantage of two-probe measurements is, however, that the electrical properties of the bathing fluid, the chamber and the electrodes themselves will be included in the experimental data. Thus, it is necessary to find ways to reduce their impact, to perform a meaningful and justified subtraction, or to include them into the model that is used for data analysis.

11.4.1.2 Introducing Electrodes for Impedance Readings into an Animal Cell Culture
In order to apply electrochemical impedance techniques to study cell junctions of animal cells, the initial experimental challenge is to introduce the necessary electrodes without causing major disturbances to normal cell division or differentiation. The measurement requires a homogeneous electric field to be applied across the cell layer or to the individual cells. In recent years, three systems have been established that fulfill these conditions in principle. The most popular set-up makes use of highly

permeable membranes that are manufactured either from polymers such as polycarbonate, or from aluminum oxide, and mechanically support the cell layer at the interface between two fluid compartments. When the electrodes are introduced into these fluid compartments below and above the cell layer, an impedance analysis of the cells on the filter can be carried out. It is important that the electrodes provide a homogeneous electric field across the entire cell layer. As the filter membranes on which the cells are grown often range from several millimeters to centimeters in diameter, it is insufficient to introduce point-like Pt or Ag/AgCl electrodes in either compartment. For meaningful impedance readings, it is necessary to use 2-D electrodes for current injection in order to ensure a homogeneous current penetration through the system. In the past, both, four-probe and two-probe electrode configurations have been used to study cells attached to permeable membranes [40, 52].

A slightly different approach that is, however, only suitable for intact tissues or tissue fragments, uses grids made from Pt or Ag/AgCl to serve simultaneously as the bottom electrode and as a mechanical support for the biological sample. Small pieces of tissue are fixed mechanically to these grids after having been excised from an animal or organ [53]. The electrodes in the upper compartment above the cell layer are easier to realize as they can simply be dipped into the bulk electrolyte above the cell layer. As the pore size of these grids is significantly larger than the size of the individual animal cells (5–20 µm in diameter), these systems cannot be used for cultured cells as these are normally seeded into the culture vessels as single-cell suspensions that will form a continuous cell layer with time.

The third and most recent approach uses thin-film electrodes made from inert noble metals such as gold or inert metal oxides such as indium-tin oxide (ITO). The cells are grown directly on the surface of the electrodes after a layer of adhesive proteins has been preadsorbed; such preadsorption is either performed intentionally before the cells are introduced, or occurs spontaneously from the culture medium. This technique, known as electric cell–substrate impedance sensing (ECIS), was pioneered by Giaever and Keese during the 1980s [54, 55], and today is on the verge of becoming a routine laboratory technology. The measurement principle is illustrated in Figure 11.3 where, as indicated in the insert, the distance between the electrode surface and the cell body is only on the order of 20–200 nm. As the electrical potential within this thin cleft is position-dependent, it is impossible to design a four-probe measurement for these systems. Thus, impedance readings of cells or cell layers that have been grown directly on the surface of thin-film electrodes will always contain contributions from the measuring electrodes and the bathing fluid that must be taken into account. Hence, the experimental set-up and electrode layout must be designed such that these contributions do not mask the impedance of the sample.

The coplanar electrode design shown in Figure 11.3 is composed of two electrodes of very different surface area (factor 1000). The smaller electrode is referred to as the *working electrode*, whereas the larger electrode is called the *counter electrode*. The area ratio between the counter- and working electrodes is typically 500 to 1000. The rationale behind this electrode layout is to make one electrode the bottleneck for the current, so that the total impedance of the system is dominated by the small electrode

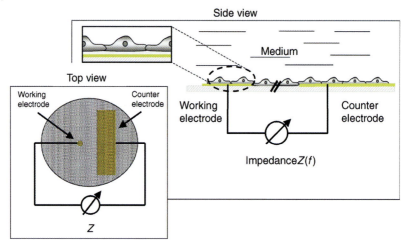

Figure 11.3 Schematic illustrating the principle of impedance analysis of adherent animal cells by means of thin-film electrodes. The cells are grown directly on the surface of the electrodes. The working electrode is made approximately 1000-fold smaller in surface area than the coplanar counter electrode, so that the cell-covered working electrode dominates the readout.

with its small population of cells. Contributions from the counter electrode can be neglected because of its larger area, even though it is also entirely cell-covered. The use of thin-film electrodes, on which the cells are grown, results in several technical advantages:

- Instead of using a dipping electrode that reaches into the bulk electrolyte from above, the second electrode is also a thin-film electrode deposited on the growth surface at sufficient distance from the first electrode (cf. Figure 11.3). A coplanar electrode arrangement avoids the need to open the chamber during the measurement, and also helps to maintain physiological, nonharmful conditions. At first sight, one would suspect that sneak currents might flow underneath the cell bodies in parallel to the surface, from one electrode to the other, without passing the cell layer. However, as the cells are rather close to the surface with only a nanometer-sized, electrolyte-filled cleft between lower cell membrane and electrode surface, this suspected 'sneak pathway' provides a much higher resistance than the current pathway across the cell layer, and is not relevant.

- The area of cell layer under examination largely determines the sensitivity of the measurement. The smaller the electrode (and cell layer), the more sensitive is the readout for the cellular parameters. Whereas, the filter set-up is difficult to miniaturize down to the single cell level, this is technically not problematic for thin-film electrodes. Circular electrodes with diameters of 20 μm (roughly the diameter of an animal cell) can be produced using standard photolithography techniques.

- As the electrodes can be manufactured using standard photolithography procedures, they can be customized for individual assays or experimental needs. Moreover, several electrodes can be arranged to a multiwell format on a common substrate so that several experiments can be performed in parallel. Nowadays, 96-well devices are commercially available.

- Noble metal electrodes in contact with an ionic solution behave, electrically, almost like an ideal capacitor. Small deviations, such as a phase angle smaller than the expected 90°, have been often observed and reported but are not of major importance. When the cells attach and spread on an almost perfectly capacitive electrode, the impedance contribution arising from the thin cleft between the cell membrane and the electrode surface is dependent on the frequency. This frequency dependency of the impedance arising at the site of cell–matrix junctions allows discrimination to be made between this particular impedance and the contribution arising from cell–cell junctions. The latter (most notably the tight junctions) are entirely resistive with respect to their electrical properties and, thus, are frequency-independent. This difference is decisive when assigning cell–cell and cell–matrix junctions their individual impedances. In contrast, when the cells are attached to a porous filter membrane which behaves electrically like a small resistor, the impedance contributions from cell–matrix and cell–cell junctions are both frequency-independent and can no longer be separated. Thus, the use of capacitive thin-film electrodes provides a significant analytical advantage over measurements on porous membranes. With the latter, the impedance contributions of cell–matrix and cell–cell junctions cannot be specified individually but only in combination.

The impedance experiments reported in the following sections were all conducted by using circular thin-film electrodes with diameters between 250 μm and 6 mm. All electrodes were prepared from 100 nm-thick gold or indium-tin oxide films, with the cells being grown directly on the electrode surfaces after an adhesive protein film had been established. The electrodes were held inside a cell culture incubator with a humidified atmosphere, 37 °C and 5% (v/v) CO_2 (the use of CO_2 was necessary to establish a stationary pH inside the bulk electrolyte).

11.4.1.3 Experimental Set-Up

A block diagram of the experimental ECIS set-up, as used for experiments described in the following sections, is shown in Figure 11.4. Here, five working electrodes are arranged side-by-side with one common counter electrode, such that five individual measurements can be performed. The electrode array is placed inside an ordinary cell culture incubator. Outside the incubator, a computer-controlled relay switch allows each of the individual working electrodes to be addressed. The impedance analyzer records the data, while the external function generator is an optional device that can be used to manipulate the cells on the electrode surface by invasive electric fields (electroporation, wounding). The impedance data are acquired in continuous-wave mode using noninvasive sinusoidal voltages of 10 mV_{rms} amplitude. The frequency range for analysis depends on the cell type under study, the surface area of the electrode, and the required time

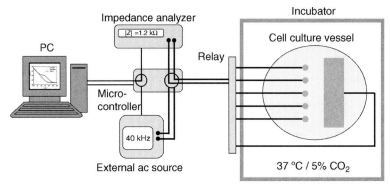

Figure 11.4 Experimental set-up to perform impedance analysis of animal cells grown on the surface of thin-film electrodes such as gold or indium-tin oxide. One or more working electrodes and a common counter electrode are deposited on the bottom of a cell culture dish. The electrode-containing dish is placed inside an ordinary cell-culture incubator to provide physiological conditions. A computer-controlled relay is used to address individual electrodes, the impedance analyzer records the frequency-dependent impedance of the system, and an external ac source may be used for invasive manipulations of the cells on the electrode.

resolution. Full spectral information is typically provided when the impedance is scanned from 1 Hz to 1 MHz.

11.4.2
Time-Resolved Impedance Measurements at Designated Frequencies

The information content of impedance measurements over a range of frequencies during the establishment of a cell monolayer starting from a suspension of single cells is best demonstrated by a 3-D presentation. To account for the complex nature of the impedance Z, the choice can be made to study the impedance magnitude $|Z|$ and the phase angle φ or resistance and reactance or any other representation of the real and imaginary components, as listed above. In the literature about impedance measurements on cultured animal cells by means of thin-film electrodes, one presentation has evolved that provides certain practical advantages [56]. Here, the complex impedance is decomposed in resistance $R = \mathrm{Re}(Z)$ and reactance $X = \mathrm{Im}(Z)$, followed by transformation of the reactance into an equivalent capacitance according to $X = (2\pi \cdot f \cdot C)^{-1}$. In other words, the measured impedance is interpreted as if it had been recorded for a system that behaves like a resistor and a capacitor in series, with the resistance and capacitance of the elements being frequency-dependent. In order to avoid confusion between the parameters of the entire sample and of individual parts of the sample, the total impedance, resistance and capacitance of the entire sample are from now on represented by Z, R and C without subscripts, whereas subordinate components are labeled with appropriate indices.

The time course of the total resistance R and total capacitance C as a function of frequency, when a suspension of epithelial MDCK cells is seeded on the electrodes at

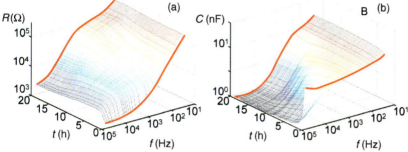

Figure 11.5 Total resistance R (a) and capacitance C (b) of a gold-film electrode ($d = 250\,\mu m$) when a suspension of MDCK cells is seeded on the electrodes at time zero. The seeding density was sufficiently high that no cell division was needed to establish a confluent cell layer. The red lines highlight the resistance and capacitance spectra for a cell-free and a fully cell-covered electrode at the beginning or end of the experiment, respectively.

time zero, is shown in Figure 11.5a and b [57]. The density of the cells was adjusted such that all adhesion sites on the surface were occupied by the settling cells, without any further cell division. At time zero, the resistance and capacitance spectra corresponded to those of an empty electrode (area $= 5 \times 10^{-4}\,cm^2$), whereas the spectra at the very end of the observation time (20 h) corresponded to the spectra of fully established cell layers with all cell junctions being expressed. (Note that the resistance and capacitance are plotted on a logarithmic scale.) When examining the resistance plot, it is easy to recognize that the time course of R is strongly dependent on the monitoring frequency. Whereas, the resistance increases immediately after cell seeding at high frequencies, it follows a biphasic pattern at intermediate frequencies and a retarded increase at the low-frequency end. The data for the total capacitance shows a less involved (but somewhat similar) pattern, with the largest and immediate changes at the high-frequency end and only very minor changes at the low-frequency end. An overlay of the resistance and capacitance spectra of a cell-free (filled symbols) and a cell-covered electrode (open symbols) as extracted from the 3-D profiles, is shown in Figure 11.6a and b, respectively.

The question is how such frequency-dependent resistance and capacitance characteristics arise. The answer to this point is shown in Figure 11.7, where the contributions to the total resistance/capacitance of a cell-free electrode can be described by an equivalent circuit of just two elements (Figure 11.7a): (i) the impedance of the electrode/electrolyte interface (hereafter referred to as the electrode impedance (Z_{el}); and (ii) the constriction resistance (R_{constr}) in series. As discussed above, the impedance of the electrode Z_{el} cannot be modeled accurately by an ideal capacitor, and is therefore represented by the complex impedance Z_{el}. The constriction resistance arises from constricting the electric field from the extended bulk phase down to the size of the electrode. This scales with $1/r$, where r is the radius of the electrode. Although the resistance of the bulk electrolyte is an inherent part of

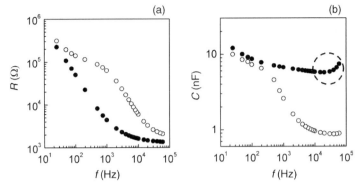

Figure 11.6 Direct overlay of the frequency-dependent resistance (a) and capacitance (b) spectra as recorded for a circular gold-film electrode with a diameter of 250 μm before (filled symbols) and after (open symbols) a monolayer of MDCK cells is established on the electrode surface. The presented spectra have been sliced out of the 3-D presentation shown in Figure 11.5 at times $t = 0\,h$ and $t = 20\,h$. The circle in panel (b) indicates a parasitic increase of the measured capacitance which is due to cables and wiring.

R_{constr} for the electrode sizes studied here the current constriction dominates this parameter. When referring to the spectra shown in Figure 11.6, the electrode impedance Z_{el} determines the total resistance of a cell-free electrode (filled symbols) at low frequencies, whereas R_{constr} dominates at high frequencies. The total capaci-

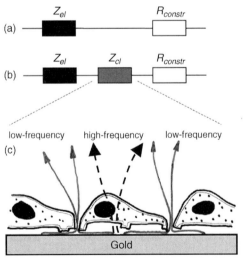

Figure 11.7 Equivalent circuit presentation of a cell-free (a) and a cell-covered (b) gold-film electrode. The cell-free electrode is completely described by a series combination of resistor R_{constr} representing both, the constriction resistance due to the finite size of the electrode and the resistance of the bulk electrolyte, as well as the impedance of the electrode/electrolyte interface, Z_{el}. The impedance contributions of the cell layer, Z_{cl}, arise from the transcellular current pathway through the cells (broken arrows) and the paracellular current pathway around the cell bodies (solid arrows), as sketched in panel (c).

tance of the cell-free electrode arises solely from Z_{el}. It should be noted that the finite slope of the capacitance spectrum indicates that the electrode is not a perfect capacitor.

For a cell-covered electrode the impedance contribution associated with the presence of cell bodies on the electrode surface Z_{cell} must be added in series to the equivalent circuit of the cell-free electrode discussed above (see Figure 11.7b). The impact of Z_{cell} on the total resistance and capacitance is made very clear by comparing the corresponding spectra for a cell-free and a cell-covered electrode, as provided in Figure 11.6. In order to help understand the origin and contributions to Z_{cell}, Figure 11.7c illustrates the frequency-dependent current flow across the cell layer that can be explained in a two-case approach:

- At high frequencies ($f > 10\,kHz$), the current can flow as a displacement current across the membranes, straight through the cell bodies (dashed line) as the plasma membranes behave electrically like capacitors. This is evident from the step-like drop in the spectrum of total capacitance towards higher frequencies (Figure 11.6b). Thus, at these high frequencies C holds information on the capacitance of the plasma membranes C_m, and can be used to monitor its changes during experimental challenges. However, there is another important aspect to this. As the drop in total capacitance at high frequencies is caused by the presence of the plasma membranes on the electrodes, the readings of total capacitance at these high frequencies can be used to determine the electrode coverage. As will be shown below, this relationship between changes in high-frequency capacitance and electrode surface coverage is extremely useful when monitoring cell attachment and spreading to the electrode surface, and in turn the formation of cell–matrix junctions [57].

- When the frequency of the ac signal is lowered well below $10\,kHz$, the plasma membranes become blocked due to their capacitive nature and the current must flow around the cell bodies in order to escape into the bulk electrolyte (straight lines). On bypassing the cell body, the current must pass through the narrow cleft between the lower cell membrane and the electrode surface, and further on through the cell–cell junctions. Thus, the signal holds information on the electrical properties of cell–matrix and cell–cell junctions, and so can be used to monitor these particular structures.

11.4.2.1 *De novo* Formation of Cell–Matrix and Cell–Cell Junctions

From theoretical considerations (as detailed above) and a set of validation experiments, we have learned that capacitance readings at high frequencies (ca. $10\,kHz$ for the electrodes used here) are best for monitoring the *de novo* formation of cell–matrix contacts during attachment and spreading of initially suspended cells, whereas the resistance at frequencies below $1\,kHz$ (usually $400\,Hz$ for the electrodes used here) is best suited for reporting on the formation of cell junctions, in particular barrier forming tight junctions. A typical readout of the resistance R at $400\,Hz$ and the capacitance C at $40\,kHz$ during the establishment of a mature cell layer is summarized in Figure 11.8a and b for a duplicate experiment. Here, a suspension of epithelial

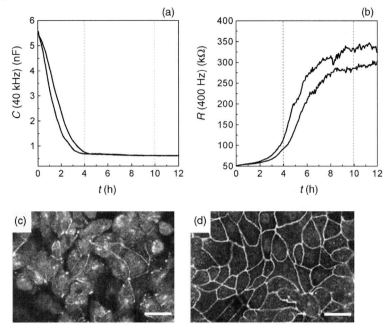

Figure 11.8 Time course of the total capacitance C at a monitoring frequency of 40 kHz (a) and the total resistance R at 400 Hz (b) during attachment, spreading and differentiation of initially suspended MDCK cells. The number of cells seeded on the electrodes was sufficiently high to form a confluent monolayer without cell division. The vertical dashed lines indicate the times at which immunolabeling of duplicate cultures was performed that are shown in panels (c) and (d). The fluorescence micrographs show the MDCK cell monolayer at the indicated times of 4 h (c) or 10 h (d) after seeding on the growth substrate. The cells were stained for the tight junction-associated protein ZO-1. Scale bar = 25 μm.

MDCK cells has been seeded on the electrode at time zero. The initial cell density ($500\,000\,\text{cm}^{-2}$) was sufficiently high so that a complete monolayer of cells could form without any further cell division. (MDCK cells are a well-known epithelial cell line that is used extensively worldwide to study tight junctions.) The capacitance drop presented in Figure 11.8a reports to the attachment and spreading of cells on the electrode surface, which began immediately after the cell suspension had been introduced into the electrode-containing chamber. The cells had been suspended in a serum-containing medium, which led to the spontaneous and immediate adsorption of adhesive proteins from the serum onto the electrode surface. The time course of the total capacitance indicated that the *de novo* formation of cell–matrix junctions was complete, and that the surface was entirely covered with spread cells within 4 h. At approximately the time when cell adhesion was complete, the resistance at 400 Hz began to increase considerably for the next 6 h (Figure 11.8b), indicating the formation of barrier forming tight junctions between adjacent cells. Thus, from a cell biology viewpoint, the formation of tight junctions requires fully established cell–matrix contacts. The minor increase in resistance observed during the initial 4 h of the experiment was caused by the constriction of current flow underneath the cells during attachment and spreading, and was therefore due to the formation of cell–matrix junctions. However, as the drop in capacitance C at 40 kHz correlates linearly with the fractional surface coverage of the electrode, C is to be the more useful parameter to monitor.

In order to back up these impedance data by microscopic studies, two aliquots of the same cell suspension were seeded on ordinary microscope slides and immunostained for the tight junction-associated proteins ZO-1. The fluorescence micrographs in Figure 11.8c and d provide typical snapshots of the distribution of ZO-1 in MDCK cells at 4 h and 10 h after cell inoculation, respectively. Whereas, after 4 h only very few cell–cell contact sites showed a positive staining for ZO-1, the protein was completely allocated to the cell borders after 10 h, consistent with the time course of R at a monitoring frequency of 400 Hz. Taken together, the results of these experiments show that it is possible to monitor the formation of cell–matrix and cell–cell contacts in one and the same cell population by utilizing noninvasive impedance measurements in real time, and that the impedance readings are in line with microscopic studies.

Additional experimental support for the claim that spreading of cells and formation of cell–cell junctions can be monitored individually simply by examining different ac frequencies is provided by an experiment in which, again, MDCK cells were seeded on electrodes that had been precoated with an adhesive protein (fibronectin). In this experiment, one population of cells was suspended in a buffer that contained 1 mM Ca^{2+} as the only divalent cation, while a second population was suspended in buffer containing 1 mM Mg^{2+} as the only divalent cation. The rationale of this approach was that the formation of cell–matrix junctions is dependent on the presence of either Ca^{2+} or Mg^{2+} (or Mn^{2+}), whereas tight junction formation is known to depend on the presence of Ca^{2+}, which cannot be substituted by a different cation. Figure 11.9a shows the time course of the capacitance C at 40 kHz for

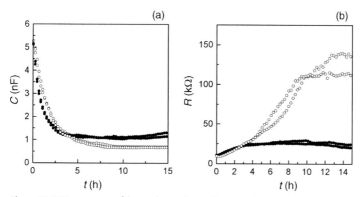

Figure 11.9 Time course of the total capacitance C at a monitoring frequency of 40 kHz (a) and the total resistance R at 400 Hz (b) during attachment, spreading and differentiation of initially suspended MDCK cells in different salt solutions. Cells were suspended in a buffer containing either 1 mM Mg^{2+} (filled symbols) or 1 mM Ca^{2+} (open symbols) as the only divalent cation.

duplicates of both conditions, while Figure 11.9b shows the resistance R at 400 Hz. In the presence of Mg^{2+}, the cells attached and spread slightly faster than in the presence of Ca^{2+}, whereas only in the presence of Ca^{2+} the resistance at 400 Hz increased significantly; indicating that tight junctions were formed only under the latter conditions. Although, from a cell biology viewpoint these experiments did not provide any new insight, they did demonstrate the capability of impedance analysis for monitoring cell junctions noninvasively and in real time. Recently, it was shown that this technology is well suited to unravel the different spreading rates of MDCK cells on various different protein coatings on the electrode surface, with unprecedented time resolution [57].

11.4.2.2 Modulation of Established Cell Junctions

In the experiments described above impedance measurements were used to follow the *de novo* formation of cell junctions when suspended cells were seeded onto the measuring electrodes. Another – perhaps more often conducted – experiment addresses the modulation of already established cell junctions within a confluent monolayer of cells upon exposure to a given stimulus. These stimuli can be biological (bacteria, viruses, cancer cells), physical (electric fields, mechanic shear) or chemical (toxins, drugs) in nature.

As an example of such an experiment, Figure 11.10 traces the time courses of resistance R at 100 Hz and capacitance C at 100 kHz when a confluent layer of epithelial MDCK cells is exposed to a 5 μM solution of Cytochalasin D (CD). CD is a fungal toxin that interferes with the actin cytoskeleton of animal cells. Actin filaments

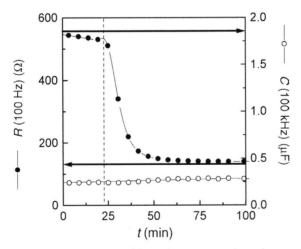

Figure 11.10 Time course of the total resistance R at a monitoring frequency of 100 Hz and total capacitance C at a frequency of 100 kHz before and during a confluent monolayer of MDCK cells was exposed to 5 μM Cytochalasin D (CD). The vertical dashed line indicates the time of CD addition to the cell population; horizontal arrows indicate the corresponding resistance or capacitance value of a cell-free electrode. The arrowheads point to the y-axis to which the arrow refers. Note: this experiment was performed with a working electrode with a surface area of 0.33 cm^2.

(f-actin) are polymers made from globular actin monomers (g-actin), and are highly dynamic structures that grow continuously at one end (polymerization) while shrinking at the other end (de-polymerization), depending on the requirements of the cells in a given physiological situation. Although CD interferes rigorously with actin polymerization, depolymerization is left unaffected so that the actin filaments become disassembled with time. With respect to the impedance readout, exposure to CD leads to a striking drop in the resistance at 100 Hz, whereas the capacitance changes only marginally. In Figure 11.10, the horizontal arrows indicate the values of R and C for the cell-free electrode, with the arrowheads pointing at the axis to which they refer. When translated into a molecular perspective, these time courses reveal that the cell–cell junctions must have completely disassembled during the exposure to CD, whereas the cells remain attached to the surface. Even though cell–matrix junctions might be compromised by the fungal toxin, they can still withstand the inner tension of the cell body and prevent cells from rolling up and losing their matrix anchorage.

The same experiment has been followed using fluorescence microscopy in order to correlate both readouts, comparing a layer of MDCK cells before and 50 min after exposure to 5 μM CD (see Figure 11.11a and b, respectively). For both images, the cells were stained with fluorescently labeled phalloidin. (Phalloidin binds to filamentous but not globular actin, and is therefore the classical stain for studying the actin cytoskeleton when using fluorescence microscopy.) After exposure to CD, the actin filaments visible inside the cell bodies of the control had entirely disappeared, as had the belt of actin filaments that ran around the periphery of the MDCK cells. Only actin aggregates were left in the CD-treated cells which, however, were still anchored to their growth substrate. Fluorescence micrographs of MDCK cells exposed to either 5 μM CD for 50 min or control conditions are shown in Figure 11.11c and d, respectively, although now the tight junction-associated protein ZO-1 is labeled with fluorescent antibodies. Under control conditions the junctional staining completely circumscribed the periphery of the individual cells. In contrast, after CD exposure the junctional staining was punctuated and discontinuous, indicating a loss of junctional tightness. To summarize, these microscopic studies on molecules that contribute to the cellular junctions are perfectly in line with the impedance readings on the time course of junctional disintegration, and their interpretation.

At this point it should be noted that monitoring the capacitance at high frequency only reports on electrode coverage. As long as the *de novo* formation of cell-matrix junctions is considered, coverage of the electrode is indicative of the formation of cell–matrix junctions. However, in an established cell monolayer, cell–matrix junctions may be moderately altered, without causing complete removal of the cell layer from the electrode surface. Under these conditions capacitance readings are insensitive, but improvements can be achieved if the impedance data are recorded over an extended frequency range (instead of only one or two designated frequencies), and when the impedance spectrum of a completely cell-covered electrode is modeled in detail. If the experiment is conducted in this way, even minor changes in the cell–matrix adhesion sites can be identified and followed with time. The model used to analyze the impedance spectra of established cell monolayers is described in the following section.

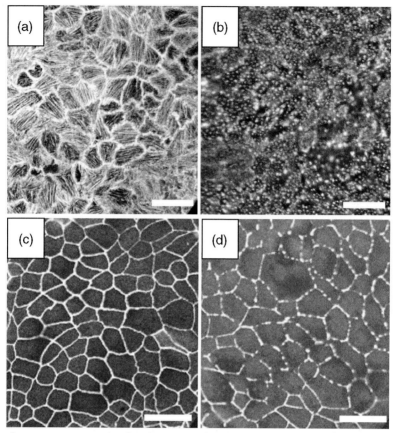

Figure 11.11 Micrographs of confluent MDCK cell layers with fluorescently labeled actin cytoskeleton (a, b) or with an immunofluorescence tag addressing the tight junction-associated protein ZO-1 (c, d). Panels (a) and (c) show control populations; cell layers in panels (b) and (d) were exposed to 5 µM Cytochalasin D for 50 min. Scale bars = 25 µm.

11.4.3
Modeling the Complex Impedance of Cell-Covered Electrodes

In order to interpret the impedance spectra of cell-covered film-electrodes, a physical model has been developed that accounts for the impedance contributions associated with the presence of cells on the electrode surface, as described qualitatively above [58, 59]. In this model (Figure 11.12), the cells are treated as disk-shaped objects with a radius r_c, and are considered to hover at an average distance h above the electrode surface. Hence, two current pathways can be considered:

- At low ac frequencies the current will mainly flow from the electrode through the thin fluid-filled cleft between the cell and the electrode, and leave the cell sheet

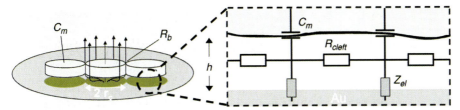

Figure 11.12 Sketch illustrating the physical model used to analyze and predict the complex impedance of cell-covered gold-film electrodes. The resistance R_b represents the resistance of the current pathway between two adjacent cells through the cell–cell junctions; the resistance R_{cleft} represents the resistance of the aqueous cleft underneath the cells at the site of cell–matrix junctions. The capacitance of the plasma membranes is accounted for by C_m. In the model, the cell bodies are treated as cylindrical disks with radius r_c that are separated from the electrode surface by distance h.

through the cell–cell junctions between adjacent cells (straight arrows). The model parameter R_{cleft} accounts for the impedance arising at the site of cell–matrix junctions, whereas R_b represents the resistive properties of the cell–cell junctions.

- At high frequency, the cell membrane, which is modeled as a capacitor C_m, allows a displacement current to pass through the cells (broken arrow). The resistive component of the plasma membrane impedance due to the presence of ion channels is neglected in the calculation as it is significantly higher than the paracellular resistance in most cases.

At intermediate frequencies the current makes use of both pathways and splits up in frequency and position-dependent ratios. Thus, the cell–electrode junction behaves like a 2-D core-coat conductor, with the inner electrolyte core and plasma membrane as well as the electrode surface as outer dielectric coats. Solving the 2-D cable equation with proper boundary conditions provides the following transfer function for the impedance of a cell-covered electrode Z_{cell}:

$$\frac{1}{Z_{cell}} = \frac{1}{Z_{el}} \left(\frac{Z_{el}}{Z_{el}+Z_m} + \frac{\frac{Z_m}{Z_{el}+Z_m}}{\frac{\gamma r_c}{2} \cdot \frac{I_0(\gamma r_c)}{I_1(\gamma r_c)} + R_b \left(\frac{1}{Z_{el}} + \frac{1}{Z_m} \right)} \right) \quad (11.5)$$

$$\text{with} \quad \gamma r_c = r_c \cdot \sqrt{\frac{\rho}{h} \left(\frac{1}{Z_{el}} + \frac{1}{Z_m} \right)} = \alpha \cdot \sqrt{\left(\frac{1}{Z_{el}} + \frac{1}{Z_m} \right)} \quad (11.6)$$

$$\text{and} \quad \alpha^2 = R_{cleft}$$

Here, I_0 and I_1 are modified Bessel functions of the first type of order 0 and 1, R_b is the resistance of cell–cell junctions, ρ is the specific electrolyte resistance inside the cell–electrode junction, and Z_m accounts for the impedance of the transcellular high-frequency pathway across both, the upper and lower membranes $Z_m = 2/(\omega \cdot C_m)$. Finally, the impedance of a cell-free electrode Z_{el} is most accurately modeled by a CPE behavior, thus $Z_{el} = 1/A(i\omega)^n$ with the parameter n ($0 \leq n \leq 1$) as an indicator for the deviation from ideal capacitive behavior, with $n = 1$ [60].

Fitting the above-described model to experimental impedance spectra provides three parameters to describe the changes that occur within the cell layer: R_{cleft} describes alterations at the site of cell–matrix adhesion; R_b describes changes within cell–cell junctions; and C_m reports on changes of the membrane capacitance.

The nonredundant model outlined above has been extended to represent the cellular structure with respect to geometry and electrical structure. More accurately for instance, the above model for disk-like cells has been adapted for elliptical cells, however, the long and small axes of the cellular ellipsoid must be determined independently from microscopic images [61]. Another approach accounts individually for the capacitance of the apical, basal and lateral membrane [62]. However, this latter model introduces two more parameters into the transfer function that cannot be determined independently, thus making it redundant. As can be seen from the microscopic images presented in Figures 11.8 and 11.11, the circular approximation is well justified for MDCK cells.

11.4.4
Spectroscopic Characterization of Cell–Cell and Cell–Matrix Junctions

If a cell layer is studied by continuously recording impedance spectra over a sufficiently extended frequency range during an experimental challenge, the analysis of these spectra with the above-described model will provide the time course of the model parameters R_b, R_{cleft} and C_m, which allows the cell response to be characterized in more detail. For demonstration, the data of MDCK cells challenged with 5 μM CD are reconsidered. The cell response to the CD challenge was monitored by repeatedly recording the impedance spectra, from 1 Hz to 1 MHz. In Figure 11.10, only the time courses of resistance at 100 Hz and capacitance at 100 kHz, respectively, are presented. An analysis of the complete full-width impedance spectra provides the time courses of R_b, R_{cleft} and C_m, but such modeling is only meaningful if the cell layer is still confluent, without significant defects. When the MDCK cells were exposed to 5 μM CD the membrane capacitance did not alter significantly during the time of the study. The changes in R_b and R_{cleft} are shown in Figure 11.13. As already deduced from the resistance readings at 100 Hz, the epithelial barrier function disappeared completely on exposure to CD. R_b fell from almost 100 Ω·cm² to values not significantly different from 0. Although the capacitance readings at 100 kHz correctly indicated that the cells had not detached from the electrode surface, there is nevertheless a drastic change in cell–matrix adhesion. The values for the R_{cleft} fall, from 250 to 25 Ω·cm². Thus, whilst the matrix anchorage is still sufficiently strong to keep the cells attached to the surface, there has been a considerable reorientation of the cell–matrix adhesion sites due to the activity of CD. According to the definition of R_{cleft}, which is

$$R_{cleft} = r_c^2 \frac{\rho}{h} \tag{11.7}$$

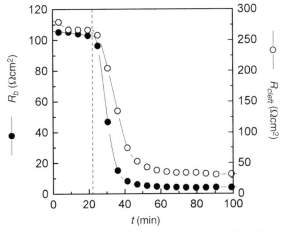

Figure 11.13 Time course of the model parameters R_b and R_{cleft} before and during a confluent monolayer of MDCK cells being exposed to 5 μM Cytochalasin D. The model parameters were extracted from complete impedance spectra recorded over six frequency decades from 1 to 10^6 Hz. The membrane capacitance C_m (data not included) did not change during the course of the experiment.

the observed changes may be due to changes in r_c, ρ or h. As the cell radius cannot be changed very much in a continuous monolayer, the changes must arise from either changes in the specific electrolyte resistance ρ in the cell–electrode junction, or from an increase in the average distance between the lower cell membrane and electrode surface h, or both. This ambiguity cannot be resolved unless one of these two quantities is measured independently. If it is assumed that all changes in R_{cleft} are due to an increase in the average cell–substrate separation distance h (which has been determined as 25 nm under control conditions, using FLIC microscopy), then the observed changes in R_{cleft} translate into a change of h from 25 to 250 nm. However, this seems unlikely, and indicates that both, h and ρ may have changed. This example illustrates that, despite not being an imaging technique, impedance analysis is still sufficiently sensitive to report changes on the nanoscale, even at hidden interfaces such as the cell–electrode junction that are not accessible by scanning probe techniques.

Taken together, the impedance analysis of cell-covered film electrodes provides functional information concerning cell–cell and cell–matrix junctions in a time-resolved and noninvasive, but integral and nonimaging, manner. An approximate calculation using the above-described model, plus realistic estimates for the resolution of each model parameter, reveals that impedance analysis is more sensitive than light microscopy and capable of identifying differences on the nanometer scale. Moreover, the data can be recorded without opening the incubator door and with multiple samples in parallel, both of which are important considerations for screening processes.

Acknowledgments

The author would like to acknowledge the Kurt-Eberhard Bode foundation for generous financial support and Dr. R. Hütterer for careful proofreading.

References

1 Wegener, J. (2002) *Encyclopedia of Life Sciences*, Nature Publishing Group.
2 Alberts, B.A., Bray, D., Lewis, J., Raff, M. Roberts, K. and Watson, J.D. (1995) *Molecular Biology of the Cell*, Wiley Sons Ltd, New York.
3 Hartsock, A. and Nelson, W.J. (2007) *Biochimica et Biophysica Acta*, **1778** (3), 660.
4 Green, K.J. and Simpson, C.L. (2007) *The Journal of Investigative Dermatology*, **127**, 2499.
5 Green, K.J. and Gaudry, C.A. (2000) *Nature Reviews Molecular Cell Biology*, **1**, 208.
6 Insall, R. and Machesky, L. (2001) *Encyclopedia of Life Sciences*, John Wiley & Sons Ltd, Chichester.
7 Lo, S.H. (2006) *Developmental Biology*, **294**, 280.
8 Stupack, D.G. (2007) *Oncology (Williston Park)*, **21**, 6.
9 Chan, K.T., Cortesio, C.L. and Huttenlocher, A. (2007) *Methods in Enzymology*, **426**, 47.
10 Cereijido, M., Shoshani, L. and Contreras, R.G. (2000) *American Journal of Physiology - Gastrointestinal and Liver Physiology*, **279**, G477.
11 Powell, D.W. (1981) *The American Journal of Physiology*, **241**, G275.
12 Grebenkämper, K.G. and Galla, H.-J. (1994) *Chemistry and Physics of Lipids*, **71**, 133.
13 Niessen, C.M. (2007) *The Journal of Investigative Dermatology*, **127**, 2525.
14 Aijaz, S., Balda, M.S. and Matter, K. (2006) *International Review of Cytology*, **248**, 261.
15 Balda, M.S. and Matter, K. (1998) *Journal of Cell Science*, **111**(Pt 5), 541.
16 Matter, K., Aijaz, S., Tsapara, A. and Balda, M.S. (2005) *Current Opinion in Cell Biology*, **17**, 453.
17 Galla, H.J. and Wegener, J. (1996) *Chemistry and Physics of Lipids*, **81**, 339.
18 Garner, C.M. and Nash, J. (2001) Chemical synapses, in *Encyclopedia of Life Sciences*, John Wiley & Sons, Ltd, Chichester, doi: 10.1038/npg. els.0000037.
19 Schweizer, F.E. (2001) Synapses, in *Encyclopedia of Life Sciences*, John Wiley & Sons, Ltd, Chichester, doi: 0.1038/npg. els.0000207.
20 Evans, W.H. and Martin, P.E. (2002) *Molecular Membrane Biology*, **19**, 121.
21 Kumar, N.M. and Gilula, N.B. (1996) *Cell*, **84**, 381.
22 Wegener, J. (2005) *Encyclopedia of Biomedical Engineering* (ed. M. Akay), Wiley & Sons, Hobooken, NJ.
23 Martinez, E., Engel, E., Lopez-Iglesias, C., Mills, C.A., Planell, J.A. and Samitier, J. (2007) *Micron*, **39** (2), 111.
24 Wrobel, G., Holler, M., Ingebrandt, S., Dieluweit, S., Sommerhage, F., Bochem, H.P. and Offenhausser, A. (2007) *Journal of the Royal Society Interface*, **5** (19), 213.
25 Gingell, D. and Todd, I. (1979) *Biophysical Journal*, **26**, 507.
26 Verschueren, H. (1985) *Journal of Cell Science*, **75**, 279.
27 Braun, D. and Fromherz, P. (1997) *Applied Physics A*, **65**, 341.
28 Braun, D. and Fromherz, P. (1998) *Physical Review Letters*, **81**, 5241.
29 Lambacher, A. and Fromherz, P. (1996) *Applied Physics A*, **63**, 207.
30 Truskey, G.A., Burmeister, J.S., Grapa, E. and Reichert, W.M. (1992) *Journal of Cell Science*, **103.2**, 491.

31 Geggier, P. and Fuhr, G. (1999) *Applied Physics A: Materials Science and Processing*, **68**, 505.

32 Wegener, J., Janshoff, A. and Steinem, C. (2001) *Cell Biochemistry and Biophysics*, **34**, 121.

33 Heitmann, V., Reiss, B. and Wegener, J. (2006) *Piezoelectric Sensors* (eds C. Steinem and A. Janshoff), Springer, Berlin.

34 Thoumine, O., Ott, A. and Louvard, D. (1996) *Cell Motility and the Cytoskeleton*, **33**, 276.

35 Sagvolden, G., Giaever, I., Pettersen, E.O. and Feder, J. (1999) *Proceedings of the National Academy of Sciences of the United States of America*, **96**, 471.

36 Braun, D. and Fromherz, P. (2004) *Biophysical Journal*, **87**, 1351.

37 Lohmann, C., Huwel, S. and Galla, H.J. (2002) *Journal of Drug Targeting*, **10**, 263.

38 Zink, S., Rosen, P. and Lemoine, H. (1995) *The American Journal of Physiology*, **269**, C1209.

39 Matter, K. and Balda, M.S. (2003) *Methods*, **30**, 228.

40 Wegener, J., Abrams, D., Willenbrink, W., Galla, H.J. and Janshoff, A. (2004) *Biotechniques*, **37**, 590.

41 Cereijido, M., Gonzalez-Mariscal, L. and Borboa, L. (1983) *The Journal of Experimental Biology*, **106**, 205.

42 Gitter, A.H., Bertog, M., Schulzke, J. and Fromm, M. (1997) *Pflügers Archiv: European Journal of Physiology*, **434**, 830.

43 Giocondi, M.C. and Le Grimellec, C. (1989) *Biochemical and Biophysical Research Communications*, **162**, 1004.

44 Korchev, Y.E., Negulyaev, Y.A., Edwards, C.R., Vodyanoy, I. and Lab, M.J. (2000) *Nature Cell Biology*, **2**, 616.

45 Goldberg, G.S., Valiunas, V. and Brink, P.R. (2004) *Biochimica et Biophysica Acta*, **1662**, 96.

46 Williams, K.K. and Watsky, M.A. (1997) *Current Eye Research*, **16**, 445.

47 el-Fouly, M.H., Trosko, J.E. and Chang, C.C. (1987) *Experimental Cell Research*, **168**, 422.

48 Schwan, H. (1993) *Medical Progress Through Technology*, **19**, 163.

49 Fricke, H. and Morse, S. (1926) *Journal of Cancer Research*, **10**, 340.

50 Grimnes, S. and Martinsen, Ø.G. (2000) *Bioimpedance and Bioelectricity Basics*, Academic Press, Cornwall.

51 Kottra, G. and Fromter, E. (1993) *Pflügers Archiv: European Journal of Physiology*, **425**, 535.

52 Erben, M., Decker, S., Franke, H. and Galla, H.J. (1995) *Journal of Biochemical and Biophysical Methods*, **30** (4), 227.

53 Kottra, G. and Fromter, E. (1984) *Pflügers Archiv: European Journal of Physiology*, **402**, 409.

54 Giaever, I. and Keese, C.R. (1993) *Nature*, **366**, 591.

55 Giaever, I. and Keese, C.R. (1984) *Proceedings of the National Academy of Sciences of the United States of America*, **81**, 3761.

56 Giaever, I. and Keese, C.R. (1986) *IEEE Transactions on Biomedical Engineering*, **BME-33**, 242.

57 Wegener, J., Keese, C.R. and Giaever, I. (2000) *Experimental Cell Research*, **259**, 158.

58 Giaever, I. and Keese, C.R. (1991) *Proceedings of the National Academy of Sciences of the United States of America*, **88**, 7896.

59 Arndt, S., Seebach, J., Psathaki, K., Galla, H.J. and Wegener, J. (2004) *Biosensors and Bioelectronics*, **19**, 583.

60 McAdams, E.T., Lackermeier, A., McLaughlin, J.A., Macken, D. and Jossinet, J. (1995) *Biosensors and Bioelectronics*, **10**, 67.

61 Lo, C.-M. and Ferrier, J. (1998) *Physical Review E*, **57**, 6982.

62 Lo, C.M., Keese, C.R. and Giaever, I. (1995) *Biophysical Journal*, **69**, 2800.

Index

a
acquisition
– frequency   108
– rates   116
– time   108, 116
activation energy   151ff.
– thermal   218
– yield-   155
adhesion   187
adsorbates   5, 293ff.
– atom   293
– lifetime   293, 296
– organic   295f.
adsorption
– nonspecific   189, 191
– selective   187ff.
anisotropy
– energy   33
– in-plane   8
– interface   7f.
– magnetocrystalline   17, 19
– out-of-plane   35
– shape   7f., 35
– surface   7f.
annealing
– low-temperature   244
– stage   239f.
anode layouts   231f.
antiferromagnet
– Cr (001)   18ff.
– Mn   23f.
– spin structures   18ff.
antiferromagnetic   2, 5
– order   40
– surfaces   18
– topological   18f., 21f.
Arrhenius
– fit   153

– kinetics   152f.
– model   152
– plot   153
– relationship   219
 – type   151f.
atom probe (AP)   213ff.
– laser-assisted field evaporation (LATAP)   216
– local electrode (LEAP)   216
– microelectrode   234ff.
– one-dimensional (1-D-AP)   215
– position-sensitive (PoSAP)   215, 250
– pulsed-laser (PLAP)   248ff.
– specimen preparation   236ff.
– three-dimensional (3-D-AP)   215, 233ff.
– wide-angle tomographic (WATAP)   215, 224f.
atom probe tomography (APT), see atom probe (AP)
atomic
– density   227
– scale roughness   227
– terraces   223, 225
– volume   224
atomic force microscopy (AFM)   49ff.
– AM- (amplitude modulation)   61ff.
– feedback system   53, 61, 72f.
– FM- (frequency modulation)   72ff.
– force-vesus-distance curve   58
– in-situ imaging   173, 190
– jump-to-contact   60, 78
– laser beam deflection   52
– modes of operation   58ff.
– NC- (noncontact)   71, 77, 80f.
– scanning process   53f.
– sensitivity   50, 52
– tip-sample forces   53ff.
atomic force tomagraphy (AFT)   213ff.
Auger electron microscopy   280

Nanotechnology. Volume 6: Nanoprobes. Edited by Harald Fuchs
Copyright © 2009 WILEY-VCH Verlag GmbH & Co. KGaA, Weinheim
ISBN: 978-3-527-31733-2

b

band
- edge 10
- gap 10f., 287, 293
- structure 2, 11
Bessel function 353
binding energy 16f., 34f.
Boltzmann
- constant 91, 151
- statistics 116
- transport 275, 277
Born-Oppenheimer approximation 278
Brillouin zone 11, 281
Brownian
- particle 89f., 92f., 103ff.
- sphere 90ff.
Brownian motion 89f., 92, 94, 108ff.
- ballistic 117
- 3-D position-time traces 104
- free diffusive 115, 117
- high-resolution analysis 108ff.
- particle properties 109f.
- time scale 99f.
- trapping potential 109f.

c

Cahn-Hilliard theory 244f.
cantilever 49ff.
- bimorph 182
- current-voltage response curve (I-V curve) 124
- deflection 50, 52f., 62
- /heater-surface distance 131
- leg 129
- microfabrication techniques 123f., 129, 184
- Q-factor 72
- resonance frequencies 50
- silicon 51, 72, 74, 81, 124, 129
- silicon nitride 123
- spring constant 50f., 59f.
- stiffness 141
- surface distance 129ff.
- thickness 51
- torsion 51f.
CCD camera 215, 231, 262ff.
cell
- cell contact 204, 207
- epithelial MDCK 344ff.
- living 198f., 204
- living MDCKII 207
- living Schwann 204
- phantom cell 262, 269
- structures 197
- tomograms 266f.
- Wigner-Seitz 228
cell junctions 325ff.
- analysing techniques 330ff.
- animal cell culture 340ff.
- cell adhesion molecules (CAM) 328f., 354
- cell- 325f., 333, 343, 347, 353ff.
- cell-matrix junctions 325f., 330, 343, 347, 354ff.
- electrode junction 353
- *de novo* formation 347f.f.
- extracellular matrix (ECM) 325ff.
- gap junctions 329f., 334
- impedance analysis 340ff.
- mechanical stability of tissue 326ff.
- modulation 350ff.
- synapses 329f.
- tight junctions 328f., 333
chemisorption 35
chiral magnetic order 39ff.
chirality 21f., 36
clusters 307ff.
- absorbance spectra 311ff.
- biosensing 321ff.
- diameter 309
- electromagnetic coupling 313
- excitation of localized surface plasmon 313f.
- excitation of multipoles 310f.
- gold 311ff.
- layer 312, 315
- metal 307ff.
- polarization 309, 312ff.
- silver 318
- single 308
- size 311
- spherical 309
- surface-enhanced spectroscopy 316ff.
coating
- antiferromagnetic 8
- dip- 182f.
- ink 182ff.
- local vapor 183
- magnetically 7
- polymer 127
- solution 182f.
- spin- 190
- thin film 7f.
combinatorial chemistry 189f.
composition
- 1-D composition profile 239ff.
- gradient 239
- map 246
- profile 244, 253

– stoichiometric 241
concentration
– fields 247
– profiles 241, 243, 247
conduction
– ballistic 144
– ion 198
conductance 3, 8
– differential 14, 19
– spectroscopy of the differential 14
– spin-averaged 4
conductivity
– sample 127
– thermal 126ff.
– thermal conductivity map 126
coupling
– antiferromagnetic 2, 18f., 35
– dipolar 26
– exchange 19, 33f.
– frictional 278
– island spin 34
– spin-orbit 40
cross correlation function (CCF) 266
cryoelectron tomography, see electron tomography
crystalline periodicity 218
Curie temperature 16, 36
current
– ion 198, 200, 203, 206f.
– saturation 202f.
current squeezing effect 198, 200

d

Debye-Weller factors 226
defects 226, 246
deformation
– elastic 55
– force-induced 198
degradation
– GMR, see giant magnetoresistance 243
– low-temperature 243
density functional theory (DFT) 34, 296
density of states (DOS) 2, 276
deposition 173
– direct 189f.
– molecular 175
– selective 187, 190
depth
– penetration 247
– profile 237, 254
– scale 223f., 226
desorption 250, 286
– induced by electronic transitions (DIET) 275

– induced by multiple electronic transitions (DIMET) 275
– rate 220
detection
– angle 222f.
– bandwidth 112
– efficiency 231
– interferometric particle position 90, 100f., 103ff.
– position detection configuration 116f., 230, 233
– propability 224f., 231
detector
– delay line 232f.
– impact position 228f.
– optical poSAP 231f.
– optical TAP 231f.
– position-sensitive ion 230, 233
– single-ion 215, 230
– time- and position-sensitive 220
– time-to-flight (ToF) 280
– wedge and strip 232ff.
dielectric
– constant 131
– function 308f., 311
– layer 312
3-D force spectroscopy 79f.
differential scanning calorimetry (DSC) 148
diffusion 89f., 97
– anomalous 173, 175
– coefficient 89, 115, 175, 245, 247
– constant 95, 115
– critical diffusion depth 241f.
– grain boundary 243, 245
– interdiffusion 239ff.
– lateral 175
– mode 242
– optical tracking interferometer (OTI) 114ff.
– reactive 239
– theory 174
– volume 239
domain
– boundary 24f.
– dendritic-like 17
– flux closure arrangement 29
– Landau-type configurations 29
– structure 5, 9, 17f., 27
– thickness 29
– vortex structures 29ff.
domain walls 12f., 19f., 26, 35ff.
– Bloch-type character 36
– chirality 21f., 36
– width 19ff.
doping 253f.

– distribution 252, 254
double layer (DL) 9, 13
– domain walls 36, 38
– Fe 27, 35f.
– Fe/W(110) 26f.
– nanowires 27
– stripes 13, 26, 36
Drude plasma frequency 309
dynamic force microscopy (DFM) 60, 62, 64, 66, 70, 72f.
dynamic force spectroscopy (DFS) 71, 78
dynamic mechanical analysis (DMA) 148
dynamic mechanical thermal analysis (DMTA) 121, 144

e

eigenfrequency 64, 74
electric vector 308f.
elecrochemical impedance spectroscopy (EIS) 335f.
electrode extraction 235
electrolyte 199, 201
electron
– ballistic transport 278, 286
– cloud 231
– defect scattering 279
– density 3, 275
  – electron (e-e) collisions 278
  – electron (e-e) scattering 274ff.
– energy 273f.
– excitation energies 275f.
– free electron gas (FEG) 275f.
– hole lifetime 285f.
  – hole scattering 276
– hopping 39
– inelastic lifetime 275f.
– momentum 273f.
– phonon(e-ph) coupling 278f., 286f.
– phonon (e-ph) interaction 278
– photoexcited 10
– relaxation 275f., 285ff.
– scattering 5, 285
– secondary electron multiplication 217, 230
– spin-polarized 35
– transfer time 296
– tunneling 217f.
electron energy loss spectroscopy (EELS) 238f.
electron tomography 259ff.
– automated cryoelectron tomography 262f.
– basic principles 260ff.
– cryptomograms 265ff.
– 3-D density distribution 261f.
– molecular interpretation 265f.
– projection images 260ff.
– resolution 263f.
– signal-to-noise ratio (SNR) 263f.
– specific labeling 266ff.
– visualization of tomograms 263f.
electronic friction 275
Eliashberg function 278f.
energy
– barrier 218
– dissipation 65, 70f.
– kinetic 223, 234
– sublimation 219
energy-dispersive X-ray (EDX) 239
energy-dispersive X-ray spectroscopy (EDS) 238
environmental scanning electron microscopy (ESEM) 178
error-function shape 244
etching
– by sputtering 237
– chemical 189
– mask 237
evaporation 216
– anisotropic 233
– barrier 251
– field 213f., 217f., 220, 225f., 250
– probability 223, 225, 228f.
– pulsed-laser 249ff.
– rate 219f., 224
– sequence 227
– simulated 229
– threshold 219f., 226f.
– trigger field 249f.
exciton 297ff.

f

fatigue test 250
Fermi
– energy 3
– level 2, 10, 18, 35, 275f.
– liquid theory (FLT) 275ff.
Fermi-Dirac distribution 275
ferrimagnet 18
ferromagnets 5, 11
– rare-earth metals 15ff.
– transition metals 17f.
ferromagnetic
– electrodes 2f.
– layers 3
– Mn 24
– nanoparticles 2
– tip 5
field emission gun 263
field evaporation 213f., 217f.

field ion micrograph 217f., 222
field ion microscopy (FIM) 213ff.
– field-of-view 245
first reaction product 239
fluctuation
– dissipation theorem 91
– laser intensity 106
– partical's thermal 105f.
fluorescence
– image 181, 191, 208
– labeled 186, 190
– multi-channel 183
fluorescence interference contrast microscopy (FLIC) 331
fluid
– density 109, 111
– fluctuations 91
– incompressible 90f., 93, 95, 105
– inertia 92f., 96, 111f., 114f.
– molecules 90f.
– Newtonian 90, 114
– velocity 95
– velocity field 93
– viscosity 91f., 109, 111f.
– viscous 95
force
– adhesion 56, 60, 187
– attractive 55f., 59, 68, 70
– capillary 55
– chemical binding 57
– constant 94, 111
  – distance curves 80
– elastic 56
– electrostatic 55, 57
– external 94
– frictional 57, 62, 89, 91, 93
– ionic repulsion 55
– Lennard-Jones 78f.
– magnetic 57
– mechanical 198
– of inertia 222
– Pauli repulsion 55
– pull-off 142f.
– repulsive 55f., 59, 68, 78
– thermal 90f.
– tip-sample 53ff.
– van der Waals 54ff.
focused ion beam (FIB) 237f., 285
– dual-beam 254
– milling 253
Fourier
– methods 225f.
– series 75
– transformation 94, 105f.

frequency
– acquisition 108
– angular 309
– corner 108, 110
– driving 73
– long-term 108
– low-frequency noise (drift) 106, 108
– modulation 70, 127
– phonon 278
– pulse 235
– resonance 66, 73f., 206, 276, 284
– shift 72ff.
friction force microscopy (FFM) 58
full width at half maximum (FWHM) 235, 289f.
functional
– molecules 176f.
– protein nanoarrays 191

g
gain factor 75
giant magnetoresistance (GMR) 2, 242
– amplitude 242
– breakdown 245
– degradation 243, 245
– sensor layers 242
Gibbs free energy 242
glas transition
– region 161f.
– temperature 146, 153, 162
glassy state 145
grain boundaries (GB) 243, 245ff.
– arrangement 246
– three-level cascade process 247
– triple junction 247
– triple line 245ff.
– width 247

h
Hamaker constant 55
hardness testing, *see* indentation
heat capacity 277
heat loss
– conductive 129
– path 126f., 129ff.
heat transfer 127, 142f., 146
– mechanisms 129
– silicon tip 132ff.
heatable probes 121ff.
– silicon heater 128f.
– temperature calibration 124
– Wollaston wire probe 122
heating efficiency 140f.
Heisenberg-Hamiltonian 39

helicity 10f.
high-resolution magnetic imaging 5, 7
high-throughput 186, 267
high-voltage pulsing 248f., 251
HOMO (highest occupied molecular orbital) 296f.
Hooke's law 49
humidity 175, 177f.
– phospholipid ink 192
hydrodynamic
– back-flow 112
– fluctuations 89f.
– interactions 90
– memory effect 92, 108, 114

i

image potential state (IS) 273f.
imaging
– compressing factor 222, 224f.
– gas 217
impedance spectroscopy (IS) 335ff.
– analysis of cell junctions 340ff.
– bode diagram 338
– cell junctions 325ff.
– constant phase elements (CPEs) 338f.
– electric cell-substrate impedance sensing (ECIS) 341, 343
– Levenberg-Marquardt approach 338
– modeling of complex impedance 352ff.
– time-resolved 344ff.
– Warburg impedance 338f.
– Wessel diagram 337f.
indentation 144ff.
– depth 146, 152
– image 146
– inking multiple tips 182f
– kinetics 151f.
– permanent 150
– rubbery 151f.
– shape 150
– temperature-load plots 146
– time 146f.
  – writing temperature 146f.
– writing threshold 145f., 148ff.
infrared spectroscopy 186
ink transport 171, 173ff.
– covalent reaction 176f.
– driving forces 176
– external driving forces 179
– fountain pens 184
– humidity 178f.
– ink wells 183, 192
– meniscus formation 178
– microfluidic ink-delivery systems 183

– nanopipettes 184f.
– noncovalent driving force 177
– theoretical models 173ff.
– tip coating 182ff.
insulating barrier 2, 4, 11
interaction
– collective molecular 189
– coulombic 39
– Dzyaloshinskii-Moriya 39
– energy map 80
– exchange 19, 33f.
– force 49, 54f.
– Heisenberg exchange 39
– hydrodynamic 90
– intermolecular 173
– long-range 56, 78f.
– mechanical 197f., 200, 204
– pattern 259, 267
– short-range 56, 78f.
– spin-orbit 10, 39
– tip-sample 54, 69, 72f., 76, 78f.
– tip-surface 61, 139
interface
– air-water 175
– Al/Co 239f.
– broadening 245
– chemistry 244
– curvature 243, 245f.
– phonon scattering 139
– roughness 237, 243
– silicon-polymer 139
– silicon-silicon 139
– temperature 140
– thermodynamics 244f.
– tip-sample 129
– tip-substrate 140
– tip-surface 127ff.
– width 244
interference reflection microscopy (IRM) 331
interferometer
– Mach-Zehnder 281, 291
– Michelson 281
intrinsic
– domain wall width 21
– material parameters 19
– sample properties 8f.
ion
– implantation 253
– sputtering 236
ionic
– curve 218f.
– impact 216f., 223
– impact position 221, 223

– potential 218f.
– trajectories 220f., 223, 227, 234
ionization
– energy 218
– field 216
– localization 217
– rate 217
irradiation damage 237

j
junction
– cell 325ff.
– Fe-Al2O3-Fe 4
– GaAs-insulator-ferromagnetic tunnel 11
– planar 2
– tunnel 8, 11, 15

k
kinetic energy distribution 297

l
Langevin equation 90f., 93ff.
laser
– pulsed 214, 248ff.
– pumping 297
– ultra-short laser pulses 273
lateral force microscopy (LFM) 173, 189
lattice 226f.
– constant 229f.
– low indexed lattice plane 227, 229
– plane resolution 224f.
layer
– Al/Cu/Al trilayer 236f.
– interlayer 18
– passivation 188f.
– planar 237
– single molecular 172
LCD (liquid crystal display) 299
lithography 144, 161ff.
– active arrays of parallel PPN 181f.
– applications 163ff.
– characterization of DPN pattern 185ff.
– dip-pen nanolithography (DPN) 161ff.
– electrochemical dip-pen lithography (E-DPN) 179f., 182
– electron-beam (EBL) 163, 171, 187, 236f., 283
– fabrication process 176
– maskless (ML2) 163
– parallel dip-pen lithography 181, 185f., 189
– passive arrays of parallel PPN 181
– scanning probe (SPL) 161, 163, 171
– 'scratch'-type 162

– thermal dip-pen lithography (tDPN) 165, 176, 179f., 182
local density of states (LDOS) 13
lock-in amplifier 6, 62, 127, 203
low-energy electron diffraction (LEED) 280
luminescence recombination 12
LUMO (lowest occupied molecular orbital) 296f.

m
magnetic
– dipole interaction 8
– map 19f., 32f., 37
– moment 8
– nanoscale wires 26ff.
– superstructure 24
– unit cell 23f.
magnetic random access memory (MRAM) 35
magnetism
– inter-atomic 39
– intra-atomic noncollinear 38f.
magnetization
– direction 5, 13, 17, 24
– isalnd 33f.
– orientation 5, 15f.
– parallel 3
– saturation 5f., 37
– state 33f.
magnetization axis
– in-plane 5f., 10, 8, 23f., 26ff.
– out-of-plane 5f., 10, 8f., 27f., 41
magneto-optic-coefficient 11
magnetoresistance 243
– local tunneling 14f.
– tunneling 2, 5
magnetostriction 6
magnification 223f.
– image 220
– local magnification effects 227
Markovian
– non- 116
– process 92
Maxwell's equation 308
mean square displacement (MSD) 94, 96ff.
membranes
– artificial 197f., 204ff.
– black lipid (BLM) 204f.
– cellular 198
– free-standing 198
– monolayer 204
– pore-suspending artificial 198, 204f., 207
– proteins 198, 203
meniscus

– formation 178
– water 175, 184
microelectromechanical systems (MEMS) 155, 160
microfabrication techniques 123f., 129, 171
microscopy light path 101
microvilli 198, 203
mode
– amplitude modulation (AM), *see* tapping mode
– constant amplitude 72
– constant excitation 72
– contact 58f., 62
– dynamic 60ff.
– frequency modulation (FM) 60, 72f.
– layer-by-layer 225f.
– non-contact 70f.
– potential 75
– static 58ff.
model
– apex 228
– ball 217f.
– Bergmann 317
– Bruggeman 316f.
– charge exchange 219
– Debye 278
– DMT (Derjaguin, Muller and Toporov) 56f., 79
– geometric model of a truncated cone 220ff.
– Hertz theory 54, 79
– Hertz-puls-offset 54, 56
– image hump 219
– JKR (Johnson, Kendall and Roberts) 56, 142
– Jullière 2
– Lennard-Jones potential 56, 78f.
– Maxwell-Garnett 316f.
– spring-mass- 63
modulation transfer function (MTF) 264
molecular
– dynamics simulations 176
– interaction patterns 267
– recognition 188
monolayer (ML) 7, 18, 28f., 285
– domain walls 36, 38
– Fe/Mo(110) 28
– Fe/W(110) 26f.
– island 174
– Mn 23, 40f.
multichannel plate (MCP) 216, 230f
– functional principles 230f.
multilayer
– periodicity 245
– structures 192, 224f., 237

n

nanoindentation, *see* thermomechanical nanoindentation
nanolithography, *see* lithography
nanopipette 198ff.
– bent 199
– current-distance behavior 201f.
– fabrication 200
– geometry 200f.
– optical reflectivity 206
– resistance 201f.
– transversally oscillated 199
nanoplasmonics, *see* clusters
nanoshells 312
nanostructures 187
nanotribology 161
nanowear, *see* wear
Navier-Stokes equation 92f.
Newton's
– force balance equation 89
– second law 91
nonadiabatic process 278
nonconductive materials 248ff.
noncontact atomic force microscopy (NC-AFM), *see* atomic force microscopy (AFM)
noncontact image 198, 200, 204
nuclear pore complex (NPC) 268
nucleation 239, 241f.
– barrier 242
– polymorphic nucleation mechanism 241
– process 241
– stages 239

o

Oerstedt field effect 29
Ohm's law 336
optical path
– detection light path 101
– infrared (IR) 101f.
– visible 102
optical tracking interferometer (OTI) 90, 92, 94, 96, 99ff.
– bandwidth 114
– calibration 108
– diffusion 114ff.
– experimental set-up 100ff.
– position signal processing 105ff.
– resolution 105f.
– sample preparation 103
– single particle tracking 114
– thermal noise statistics 116
optical trapping potential 106, 108, 114

optical weak trap   90, 110
optically pumped   10f., 12
oscillation amplitude   65f., 71, 199

p

pair correlation functions   225
particle
– colloidal   90, 96
– density   113
– fluctuations   91
– inertia   112
– mass   113
– nonfree   97
– properties   109ff.
– radius   113
pattern
– dimension   189f.
– DPN   185ff.
– functional chemical   187
– recognition   268f.
patterning   144, 173, 189
– binary ink mixtures   177
– biofunctional molecules   177, 185
– characterization of DPN   185ff.
– DNA   190
– nanoscale   144
phase separation   190
phonon   132ff.
– back-scattering   137, 140
– density of state   278f.
– electron-phonon coupling   252
– frequency   278
– mean free path   133, 136, 139
– scattering   133, 136f., 139, 144
– spectrum   280
– thermal equilibrium   135
– tunneling   144
high-resolution photoelectron
    spectroscopy   287
photoemission   11, 273ff.
– autocorrelation map   283f.
– electron microscopy (PEEM)   274, 281f.
– inverse photoemission spectroscopy   273
– lateral photoemission distribution   290
– lifetime map   290
– map retardation effects   284
– multiphoton   276
– phase-averaged two-photon (2PPE)   282
– phase-resolved two-photon (2PPE)   282
– plasmonic processes, *see* plasmons
– polarization-shaped femtosecond
    pulses   299f.
– spectroscopy   273ff.
– spin-polarized   35

– time-resolved two-photon (TR-2PPE)   273, 282ff.
– two-photon (2PPE)   273ff.
– two-temperature model   277f.
– volume excitation   289ff.
photoionization   250
Planck's constant   131, 286
plasma oscillation   276
plasmon   276f.
– excitation   284, 290
    – induced coupling   284
    – localized surface plasmons (LSP)   276f., 284, 291f., 310, 313f.
– surface plasmon polariton (SPP)   291ff.
– surface plasmon resonance (SPR)   311f.
pointing instabilities   106
Poisson's equation   228
polymer   145f.
– crosslink density   148f., 158
– deformation   145, 151f., 154, 164
– degradation   161f.
– Diels-Adler   163f.
– endurance   158
– hardness   148f.
– indentation   145ff.
– mobility   159
– monomer friction   150f.
– partially crystalline   150
– rubbery deformation   151f.
– thermal degradation   161
– viscoelastic deformation   145, 150
– volatilization   164f.
potential
– curve   218f.
– energy barrier   80
– energy profile   80
– harmonic   90, 99, 108, 115, 117
– harmonic cantilever   73
– ionic   219f.
– segregation   247
– tip-sample interaction   73f.
power spectral density (PSD)   94ff.
probe tip
– amorphous CoFeNiSiB   5
– antiferromagnetic   8ff.
– apix   7f., 13
– aspect ratio   236
– bulk   5, 8
– conical   136
– Cr-coated   9
– diameter   7
– endurance   159
– Fe-coated   23f.
– ferromagnetic   4f., 13

– GaAs   10f.
– geometry   177f.
– Gd-coated   24, 27
– GdFe   9
– heat transport   141ff.
– heater, *see* heatable probes
– inking   172ff.
– length   6
– magnetic thin film   7f.
– microfabrication techniques   184
– Mn-coated W   24f.
– modes of operation   13ff.
– nanofountain probe (NFP)   184
– Ni   12
– nonmagnetic   12f., 23
– optically pumped   10ff.
– polycrystalline tungsten   7
– radius   224
   – sample contact   132
   – sample distance   5, 198, 200, 202f., 206f.
– sensitivity   5ff.
– silicon heater   128f., 132ff.
– silicon nitride   181
– soft magnetic   6
– solid-solid tip-sample contact   132
– spin-polarized electron tunneling   5
– substrate   175
– surface contact   132, 138, 142
– surface distance   202
– temperature-sensitive, *see* heatable probes
– thin-film   5, 7
– tungsten   7, 23f., 222
– volcano-like   184
– voltage   216f.

q
quadrant photodiode (QPD)   102ff.
quantum mechanical tunneling   2
quartz crystal microbalance (QCM)   332

r
radiation damage   260, 262
reconstruction   224f., 227, 229
– algebraic reconstruction techniques (ARTs)   261
– 3-D atomic   214f., 239f., 254
– simulated   227
– simultaneous iterative reconstruction techniques (SIARTs)   261
– spatial   226
– tomographic   223, 267
– volume   220, 226, 241, 243
reflection coefficient   137

reflection interference contrast microscopy (RICM)   331
reflectron geometry   234
refractive indices   108, 112
relaxation
– alpha-   159
– elastic stress   154
– excited electrons   285ff.
– monomer   150
– processes   80
– time   95, 110, 275
remagnetization process   6
resistance
– electrical   122f., 124f.
– interface   137, 139f.
– inverse   4
– Sharvin   135
– tip   140
resolution
– atomic   18, 23, 35, 39, 79, 81, 215
– depth   226
– electron tomography   263f.
– lateral   31, 123, 127f., 140f., 171
– lattice plane   224f.
– mass   234, 252
– molecular   63, 259
– spatial   90, 105f., 117, 123, 127f., 215, 225, 250
– temperature   127
– temporal   105, 116
– three-dimensional (3-D) spatial   213, 254
– time   127, 250, 335

s
scanning confocal microscopy (SCM)   208
scanning electron microscopy (SEM)   7, 50f., 124, 280
scanning force microscopy (SFM)   121f.
scanning ion conductance microscopy (SICM)   185, 197f., 334
– basic set-up   199
– modulated-scan technique   202ff.
– shear force distance control   206
– /SNOM set-up   208
scanning near-field optical microscopy (SNOM)   184, 186, 199, 208
scanning probe microscopy (SPM)   121f.
– data rate   159f.
– heated-probe   131
– thermal   122
scanning thermal microscopy (SthM)   123, 126ff.
– data-storage applications   155ff.
– probe   123, 128

scanning tunneling microscopy (STM) 4, 21, 23, 71, 280
– topographic image 26ff.
screw dislocation 20ff.
secondary ion mass spectroscopy (SIMS) 254
semiconductor 10f., 141, 166, 177, 214, 237, 250
– band-gap 10
– surface 176
sensor
– angular 242
– deflection 52
– displacement 72
– force 190
– interferometric 62
– position 242
shape memory 158
shear
– force 199
– modulus 51
shear force microscopy 199, 207
signal-to-noise ratio (SNR) 260ff.
– electron tomography 263f.
silicon 144
– anisotropic etching 51
– density 51
– wafer 71
spatiotemporal femtosecond dynamics 290
spin
– adatom 33f.
– antiferromagnetic model configuration of spin moment 25
– atomic 31f.
– conserving 2
– density wave (SDW) 40f.
– dependent scattering 2
– down electrons 3
– frustration 20ff.
– injection efficiency 12
– orbit splitting 10
– orbit scattering 39
– orientation 2, 12f., 18
– polarization 3, 5, 10f., 16, 23, 33f.
– relaxation lifetime 12
– sensitivity 8
– spiral rotating 41
spin-polarized
– electron tunneling 4f., 10f., 13
– scanning tunneling microscopy (SP-STM) 1ff.
– scanning tunneling spectroscopy (SP-STS) 15f.
spreading
– monolayer 174

– phenomena 173
– process 204
Stefan-Boltzmann equation 131
Stokes
– Boussinesq friction force 95, 98
– drag 92
– friction 89, 93f., 96
– law 92
stress-strain curve 150, 154
substrate
– chemisorption 176f.
– covalent reaction 177
– physisorption 175f.
– planar wafer 237
– porous silicon 205
– roughness 177f.
– tungsten 236
superconductor 15, 276
surface
– boundary 103
– 3-D 79
– ferromagnetic 35
– iso-concentartion 225, 253f.
– magnetic 5, 31ff.
– metall 33
– nickel-coated 180
– oxidation 180
– passivation 187
– state 15ff.
surface-enhanced fluorescence (SEF) 319
surface-enhanced infrared absorption spectroscopy (SEIRA) 320
surface-enhanced Raman scattering (SERS) 276, 316ff.
system
– adsorbate/substrate 293
– heterogeneous 227f.

t
tapping mode 61ff.
– bistable tapping regime 69
– tapping regime 69
template 187
– DPN 188, 190
– library 269f.
– matching 269
templating 187f.
thermal
– conduction 130, 132
– coupling 129
– energy 217
– excitation 219
– images 127
– radiation 129f., 132

– stability 242
thermal resistance 126, 129ff.
– diffusive 144
– interface 139, 142
– spreading 138f., 144
– tip 141ff.
– water meniscus 132
thermomechanical
– erasing 156f.
– nanoindentation 145ff.
– properties 145
– reading 156f.
– writing 155f.
thin film
– ferromagnetic 2, 8, 11
– interfaces 239
– magnetic 7f.
– metallic 239
– Mn_3Ni_2 24
– organic 190
– polymer 138f.
– silicon 139
– thickness 138
time-to-digital converter (TDC) 232f.
time-to-flight (ToF) mass spectroscopy 215, 220, 233, 280
total internal reflection fluorescence (aqueous) microscopy (TIRAF) 331f.
transendothelial electrical resistance (TER) 334
transepithelial electrical resistance (TER)-spectroscopy 197
transmission probability 10
transmission electron microscopy (TEM) 213, 237ff.
– high-resolution (HR-TEM) 239, 242
trapping
– efficiency 101
– force contant 90
– potential 94, 106, 108ff.
– stiffness 110ff.

tunneling
– barrier 11
– conductance 4, 11
– current 2f., 6, 13, 16, 29
– spectra 16f.
tunneling magnetoresistance (TMR) 2, 4f.

u

ultra-high-vacuum (UHV) 5, 60, 70, 74f., 163, 179
– chamber 7, 216f.
– gas-aggregation 285

v

velocity autocorrelation function (VAF) 94, 96f., 105ff.
– anti-correlations 110, 112
very-large-scale integration (VLSI) 123, 160
vibration amplitude 206f.
Vogel temperature 152

w

wave packet propagation 293f.
wear
– nanowear testing 161
– rate 158
– ripple pattern 162
– tracks 162
WLF (Williams-Landel-Ferry) 151ff.
– equation 151
– kinetics 151ff.

x

extreme ultraviolet (XUV) 281
X-ray photoelectron spectroscopy (XPS) 186

y

yield
– shear yielding 150
– stress 150, 154
Youngs' modulus 51